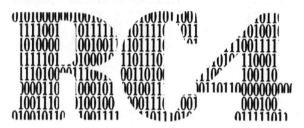

RC4

Stream Cipher and Its Variants

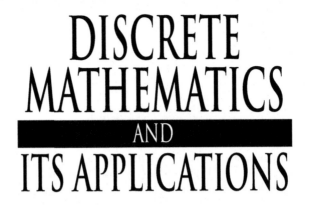

DISCRETE MATHEMATICS AND ITS APPLICATIONS

Series Editor

Kenneth H. Rosen, Ph.D.

Titles (continued)

Richard A. Mollin, RSA and Public-Key Cryptography

Carlos J. Moreno and Samuel S. Wagstaff, Jr., Sums of Squares of Integers

Dingyi Pei, Authentication Codes and Combinatorial Designs

Kenneth H. Rosen, Handbook of Discrete and Combinatorial Mathematics

Douglas R. Shier and K.T. Wallenius, Applied Mathematical Modeling: A Multidisciplinary Approach

Alexander Stanoyevitch, Introduction to Cryptography with Mathematical Foundations and Computer Implementations

Jörn Steuding, Diophantine Analysis

Douglas R. Stinson, Cryptography: Theory and Practice, Third Edition

Roberto Togneri and Christopher J. deSilva, Fundamentals of Information Theory and Coding Design

W. D. Wallis, Introduction to Combinatorial Designs, Second Edition

W. D. Wallis and J. C. George, Introduction to Combinatorics

Lawrence C. Washington, Elliptic Curves: Number Theory and Cryptography, Second Edition

DISCRETE MATHEMATICS AND ITS APPLICATIONS

Series Editor KENNETH H. ROSEN

Stream Cipher and Its Variants

Goutam Paul

Jadavpur University
Kolkata, India

Subhamoy Maitra

Indian Statistical Institute
Kolkata, India

CRC Press
Taylor & Francis Group
Boca Raton London New York

CRC Press is an imprint of the
Taylor & Francis Group, an **informa** business

A CHAPMAN & HALL BOOK

CRC Press
Taylor & Francis Group
6000 Broken Sound Parkway NW, Suite 300
Boca Raton, FL 33487-2742

First issued in paperback 2019

ISBN-13: 978-1-4398-3135-9 (hbk)
ISBN-13: 978-0-367-38216-2 (pbk)

Library of Congress Cataloging-in-Publication Data

Paul, Goutam.
   RC4 stream cipher and its variants / Goutam Paul, Subhamoy Maitra.
      p. cm. -- (Discrete mathematics, its applications)
   "A CRC title."
   Includes bibliographical references and index.
   ISBN 978-1-4398-3135-9 (hardcover : alk. paper)
   1. Stream ciphers. 2. Internet--Security measures. 3. Data encryption (Computer science) 4. Computer security. I. Maitra, Subhamoy, 1970- II. Title.

QA76.9.A25P385 2012
005.8--dc23
                                       2011039582

Visit the Taylor & Francis Web site at
http://www.taylorandfrancis.com

and the CRC Press Web site at
http://www.crcpress.com

Dedicated to

Siddheshwar Rathi

who brought our attention to RC4. Rathi was a budding researcher at the Cryptology Research Group of the Indian Statistical Institute, Kolkata. Due to his untimely demise on October 28, 2006, he did not see this book published.

Contents

Foreword

I am delighted to introduce the first complete book on RC4, the most popular and widely deployed software stream cipher.

Whereas a number of excellent books exist on general cryptology as well as on block cipher design and analysis, there is still a dearth of good textbooks in the stream cipher domain except for a few on LFSR-based hardware stream ciphers. In this situation, I warmly welcome the introduction of a dedicated book on the RC4 stream cipher. Being the simplest of all stream ciphers, RC4 is a wonderful takeoff platform for new researchers who want to embark on the design and analysis of software stream ciphers.

Goutam Paul earned his Ph.D. under the supervision of Subhamoy Maitra who, in turn, earned his under mine. Being the father and grandfather of them in the research geneology, I have known both of them closely for a long time. In the past, eminent cryptologists such as Adi Shamir, Eli Biham, Bart Preneel, Serge Vaudenay, Vincent Rijmen and many other outstanding researchers have published important results on RC4 cryptanalysis. Still, after twenty years of research, many questions on RC4 remain unanswered. In recent years, Goutam and Subhamoy (with several other co-researchers) have published some of the important results related to RC4 and its variants. They continue to work and I expect more results will be included in future editions of this book.

This book contains not only the authors' own research materials, it has nicely assimilated the results of others to make it a complete treatise on RC4 and its variants. I believe this book will be widely accepted as an important research monograph, by both the students and experienced researchers in the field of cryptology.

Bimal Roy
Founder, Cryptology Research Society of India,
and *Director*, Indian Statistical Institute

Preface

The main theme of this book is the analysis and design issues of the stream cipher RC4 and its variants in the shuffle-exchange paradigm of stream cipher.

Why a Book on RC4?

The RC4 stream cipher was designed by Ron Rivest for RSA Data Security in 1987. It was a propriety algorithm until 1994 when it was allegedly revealed on the Internet. RC4 is often referred to as "ARCFOUR" or "ARC4" (meaning Alleged RC4, as RSA Data Security has never officially released the algorithm). Presently, RC4 stands out as one of the most popular among the state-of-the-art software stream ciphers with wide industrial applications. As part of network protocols such as Secure Sockets Layer (SSL), Transport Layer Security (TLS), Wired Equivalent Privacy (WEP), Wi-Fi Protected Access (WPA) etc., it is used for encrypting Internet traffic. The cipher is also used in Microsoft Windows, Lotus Notes, Apple Open Collaboration Environment (AOCE), and Oracle Secure SQL.

Two decades have passed since the inception of RC4. Although a variety of other stream ciphers have been discovered subsequent to RC4, it is still the *most widely deployed algorithm* due to its simplicity, ease of implementation, speed and efficiency.

Though there are many modern stream ciphers (like the eSTREAM candidates), so far not much analysis has been done on any of them. The available materials on these ciphers, according to us, could hardly be expanded to a complete book. On the other hand, RC4 has undergone more than twenty years of investigation, producing lots of interesting results in the literature. Further, RC4 is perhaps the *simplest of all the state-of-the-art stream ciphers.* It involves a one-dimensional integer array and two integer indices and requires less than ten lines of code in C for its implementation. Though the simple structure of RC4 lures immediate cryptanalytic attempts, it is a fact that even after so many years of analysis, RC4 is *not yet completely broken* and can safely be used with certain precautions. This leads us to believe that RC4 can serve as a very good model for the study of software stream ciphers in general.

Ours is the *first book on RC4*. Apart from this, we are aware of only three Master's theses [110, 118, 177] and a doctoral thesis [130] on RC4. Though there is a book on WLAN security by Tews and Klein [179], it contains only a few specific attacks on RC4 relevant to WLAN; it has intersection with only Chapter 7 of our book and it is not a full treatise on RC4.

Organization of the Book

The first chapter provides a brief introduction to the vast field of cryptology.

In Chapter 2, a review of stream ciphers, both hardware and software, is presented, along with a description of RC4.

Chapter 3 deals with a theoretical analysis of RC4 KSA, including different types of biases of the permutation bytes toward the secret key bytes as well as toward absolute values. The next chapter (Chapter 4) focuses on how to reconstruct the secret key from known state information.

Chapter 5 contains analysis of the RC4 PRGA in detail including a sketch of state recovery attacks.

The subsequent three chapters deal with three popular attacks on RC4 stream cipher. Chapter 6 concentrates on distinguishing attacks, Chapter 7 describes the attacks on the use of RC4 in Wired Equivalent Privacy (WEP) protocol (part of IEEE 802.11) and Chapter 8 considers fault attacks on RC4.

In Chapter 9, several variants of RC4 are presented with their advantages and disadvantages. Chapter 10 describes stream cipher HC-128, which can be considered as the next level of evolution after RC4 in the software stream cipher paradigm.

In the last chapter (Chapter 11), after a brief summary, some points on the safe use of RC4 are mentioned.

Exercises in a monograph style reference text are not relevant and so we have avoided cooking them up. However, for avid readers and researchers, we have added a list of open research problems at the end of each chapter (except for the Chapters 1, 2 and 11).

Prerequisites

This book presumes that the reader has adequate knowledge in the following disciplines at the undergraduate level:

1. Combinatorics,

2. Data Structures,

3. Algorithms,

4. Probability and Statistics.

One may refer to [148], [74], [30], [70] and [72] for elementary introductions to the above topics. The reader may also go through the books [119, 174] for detailed exposition on cryptology.

Target Audience

Though the book is primarily a monograph, it can be easily used as a reference text for a graduate as well as an advanced undergraduate course on cryptology. This work would cater to the three-fold needs of students, teachers and advanced researchers who want to study stream ciphers in general with RC4 as a case study, and also to those who want to specialize in RC4-like stream ciphers based on arrays and modular addition.

Acknowledgments

Most of the materials covered in this book are based on our own research publications on RC4 and its variants. We heartily thank Sourav Sen Gupta, Santanu Sarkar, Shashwat Raizada, Subhabrata Sen, Rudradev Sengupta, Riddhipratim Basu, Tanmoy Talukdar, Shirshendu Ganguly, Rohit Srivastava, and Siddheshwar Rathi, who contributed as our co-authors in many of those papers. Extra thanks to Sourav for his input toward aesthetic improvements of the presentation in many places and to Santanu for detailed proofreading.

We express our sincere gratitude to all our colleagues at the Cryptology Research Group (CRG) of the Indian Statistical Institute, Kolkata, for boosting us with constant encouragement. In particular, we would like to thank Prof. Palash Sarkar for some fruitful discussions.

Apart from our own research work, we also explain the state-of-the-art materials related to RC4 and its variants. We sincerely acknowledge the contributions of the authors of these publications in preparing the book.

The first author would like to thank his parents Nanda Dulal Paul and Jharna Rani Paul, and his wife Soumi for sacrificing their share of his time that they longed to spend with the author during the preparation of the book. Their interminable enthusiasm played a key role in accomplishing this task.

Special thanks from the second author go to his parents Kalimay Maitra and Manasi Maitra, brother Rajarshi and wife Arpita for their keen and un-failing support. He would like to express his appreciation of his children Sonajhoori and Shirsa who ungrudgingly gave up the many hours he spent away from their playful company.

Goutam Paul and Subhamoy Maitra

List of Symbols

$x \oplus y$	Bit-wise XOR of two (equilength) binary strings x and y
$x \| y$	Concatenation of two bitstrings x and y
$f \circ g$	Composition of two functions f and g
$\gg$	Right shift operator
$\ll$	Left shift operator
$\ggg$	Right rotation operator
$\lll$	Left rotation operator
$ROTL32(w, n)$	Left rotate the 32-bit word w by n bits
$ROTR32(w, n)$	Right rotate the 32-bit word w by n bits
b_R^n	Byte b after right shifting by n bits
b_L^n	Byte b after left shifting by n bits
$\|x\|$	Length in bits of the integer x
$bin_n(x)$	n-bit binary representation of the number x
$w^{(i)}$	i-th least significant byte of the word w
$[w]^i$	i-th least significant bit of the word w
$+=$	Increment operator
$\overset{p}{=}$	Equality holds with probability p
$\overset{w.h.p.}{=}$	Equality holds with high probability (close to 1)
$\mathbb{Z}_N$	The set $\{0, 1, \ldots, N-1\}$ of integers modulo N
$[a, b]$	The set $\{a, a+1, \ldots, b\}$ of integers from a to b
$K[a \ldots b]$	The array K from index a to index b
$K[a \uplus b]$	The sum $K[a] + K[a+1] + \ldots + K[b]$ for the array K
$\binom{n}{m}$	The binomial coefficient ${}^nC_m$
$O\left(f(n)\right)$	Big Oh order notation
$P(A)$ or $Prob(A)$	The probability of an event A
$E(X)$	The expectation of a random variable X
$V(X)$	The variance of a random variable X
$\sigma(X)$	The standard deviation of a random variable X
$\mathcal{B}er(p)$	The Bernoulli distribution with success probability p
$\mathcal{B}(n, p)$	The Binomial distribution with the number of trials n and a success probability p in each trial
$\mathcal{N}(\mu, \sigma^2)$	The Normal distribution with mean μ and variance σ^2
$\Phi(.)$	The standard normal cumulative distribution function

List of Figures

List of Tables

List of Algorithms

Chapter 1

Introduction to Cryptology

Society is becoming more information-driven every day. Naturally, the necessity of keeping valuable information secret from unauthorized access is growing as well. Here comes the role of *cryptology*, the art and science of hiding information.

Each one of us, knowingly or unknowingly, uses cryptology in our daily lives. ATM machines, credit card transactions, prepaid recharge of mobile phones, online banking, the digital gate-pass in offices for access control, copyright protection etc., are only a few examples where cryptology finds its applications. Modern military systems also make enormous use of cryptology to secure tactical and strategic data links, such as in fighter plane control systems, missile guidance, nuclear control etc.

1.1 Etymology and Some Historical Facts

Cryptography is derived from Greek κρύπτω (*krýpto*) which means "hidden" and γράφω (*gráfo*) which means "to write." Thus, the term "cryptography" means the art and science of making ciphers. On the other hand, the term "cryptanalysis" means the art and science of breaking ciphers. The subject *cryptology* comprises cryptography and cryptanalysis.

The oldest evidence of the subject is found in Egyptian *hieroglyphs* dating back to 2000 B.C. The Jews around 500 B.C used a simple *substitution cipher* called *Atbash* for the Hebrew alphabet. Around 400 B.C., the Spartans, the military superpower of the then Greece, used a device called *Scytale* to perform a *transposition cipher* for communication between two military commanders. Roman emperor Julius Caesar (100–44 B.C.) used a simple substitution with the normal alphabet to send secret messages to his soldiers in the battlefield. Even the famous Indian text *Kama Sutra*, written some time between 4th and 6th century A.D., lists cryptography as an art men and women should know and practice. During World War II, German military used a cryptosystem called *Enigma* that was broken by Polish Mathematician Marian Rejewski. Around that time Americans also used a system called M-209 and the Germans used another system called *Lorenz*. Due to rapid advances in the field of

1

digital computers and networking, the last three decades have experienced a tremendous growth in the application of cryptography in diverse domains. The books [80, 170] provide interesting historical facts about cryptography.

1.2 Primary Goals of Cryptography

The objective of modern cryptography is mainly four-fold [119, Chapter 1]. Below, we initiate the reader into each of these goals through examples.

1. *Confidentiality or Privacy:* Suppose, one makes an online payment through credit card. Then the credit card information should be read only by the intended merchant, and not by any eavesdropper. Confidentiality means keeping the content hidden from all but those authorized to have it. Not only transmitted messages, but stored data may also need to be protected against unauthorized access (e.g. by hackers).

2. *Integrity:* Suppose, A gives his laptop to B for a day and asks B not to alter contents of folder $\mathcal{F}$. If B adds or deletes or modifies anything inside $\mathcal{F}$, A should be able to detect it. Integrity means detecting unauthorized alteration of data.

3. *Authentication:* Suppose two police officers are entering into a communication via telephone. Before sending any important message, they must identify each other. Authentication simply means verifying identity. Usually, two broad classes of authentication are considered. Validating the identity of a person, a machine or an object, is called *entity authentication* or *identification*. Corroborating the source of information is called *data origin authentication* or *message authentication*.

4. *Non-repudiation:* Suppose A sells a property to B and both sign an agreement. It may happen that A later claims that the signature on the agreement is not his. A method involving a "trusted" third party (TTP) may be needed to resolve such a dispute. Non-repudiation means preventing an entity from denying previous commitments or actions.

Cryptography provides many other services that became quite relevant in recent times. Some of these are related to the four principal goals listed above. Let us discuss a few important ones.

- *Digital Signature:* Suppose A signs a check in favor of B. When B submits the check to the bank, the bank authority matches A's signature on the check with that in their records before withdrawing the money from A's account. Similarly, a digital signature [141] scheme is a method of signing a message stored in electronic form. Such a signature should be verifiable like a hand-written signature. Also, special care needs to be taken to prevent forgery. Digital signatures can be used for authentication and non-repudiation.

- *Digital Certificate*: If a candidate wants to convince an interviewer about some awards that he/she has achieved, he/she can do so by presenting a certificate from a "trusted" third party (often, the award-issuing authority). The Digital Certificate [47] borrows the same idea in the digital domain. For example, a digital certificate can bind the identity of an entity to its public key. The entities may use a trusted third party (TTP) to certify the public key of each entity. The TTP has a private signing algorithm S_T and a verification algorithm V_T assumed to be known by all entities. The TTP verifies the identity of each entity, and signs a message consisting of an identifier and the entity's authentic public key.

- *Secret Sharing*: Suppose there is a locker and three managers are in charge of it. It is required that that no single manager should be able to open the locker, but any two out of the three managers together can. This is an example of secret sharing [160]. In military and Government systems, such mechanisms are often required. A simple method of secret sharing among n persons so that any k $(< n)$ together can access the secret would be to choose a $(k-1)$-degree polynomial $P_{k-1}(x)$, and distribute to each of the n persons one distinct point $(x, P_{k-1}(x))$ on the plot of $P_{k-1}(x)$ versus x. Since the polynomial is completely characterized by any set of k distinct points, the scheme works. Getting less than k distinct points does not allow to compute the polynomial and hence the secret cannot be obtained.

- *Multicast Security*: Suppose a pay-TV movie is to be broadcast only to a privileged set S of authentic subscribers. Such one-to-many secure communication in a network of users is the main theme of multicast security [79]. The set S is, in general, not known before the scheme is initialized.

- *Multiparty Computation*: Suppose a group of people want to know who is the richest among them, but no one wants to reveal to the others how much wealth he/she possesses. In a general multiparty computation [39] scenario, n participants $p_1, p_2, \ldots, p_n$ each have private data $d_1, d_2, \ldots, d_n$ respectively, and they want to compute a global function $f(d_1, d_2, \ldots, d_n)$ without revealing the individual d_r's to each other.

- *Zero-Knowledge Proof*: A person may want to convince somebody else that he or she possesses certain knowledge, without having to reveal even a single bit of information about that knowledge. Such a requirement can be fulfilled by what is called a zero-knowledge proof [56, 57]. Zero-knowledge proofs provide a powerful tool for the design of cryptographic protocols [55, Chapter 4].

1.3   Basic Terminology

In the classical model of cryptography, two persons, conventionally named Alice and Bob, communicate over a public channel, i.e., a channel accessible to anybody. Alice acts as the sender and Bob plays the role of the receiver. The information that Alice wants to send to Bob can be text or numerical data in any natural or human-readable language and is called *plaintext*. Alice uses a parameter, called a *key*, to transform this plaintext into encrypted form which is called *ciphertext* and sends it over the public channel. If an adversary, conventionally called Oscar, can see the entire ciphertext in the channel by eavesdropping, but is not able to determine the corresponding plaintext, then this transmission is called *secure*. However, Bob knows the key and using it he can recover the plaintext from the corresponding ciphertext easily. The transformation of the plaintext to ciphertext is called *encryption* and the reverse process is called *decryption*. A system that transforms the plaintext to ciphertext and vice versa is called a *cryptosystem*. On the other hand, *cryptanalysis* aims at analyzing a cryptosystem to exploit its weaknesses so as to gain some knowledge about the plaintext from the ciphertext without knowing the key or retrieving the secret key from some known information. The subject *cryptology* comprises of both *cryptography* and *cryptanalysis*.

We can formalize the above concepts as follows. Let M be a finite set of possible plaintexts, called the *message-space* and C be a finite set of possible ciphertexts, called the *ciphertext-space* and K be a finite set of possible keys, called the *key-space*. Then, *encryption* is a function

$$E_{k_e}(m) : M \times K \to C$$

that takes as input a plaintext $m \in M$ and produces a ciphertext $c \in C$, according to the key $k_e \in K$. On the other hand, *decryption* is a function

$$D_{k_d}(m) : C \times K \to M$$

that takes as input a ciphertext $c \in C$ and produces a plaintext $m \in M$, according to the key $k_d \in K$. k_e and k_d are called the *encryption key* and the *decryption key* respectively. A *cipher* or a *cryptosystem* is a pair (E, D) of two functions, such that for any plaintext $m \in M$ and any encryption key $k_e \in K$, there exists a decryption key $k_d \in K$ such that $D_{k_d}(E_{k_e}(m)) = m$.

1.4 Cryptographic Attack Models

From the viewpoint of cryptanalysis, there are four basic models [174, Page 26] of attacks. In these basic models, it is assumed that the attacker has access to the details of the encryption and decryption algorithms.

1. *Known Ciphertext Attack* or *Ciphertext-only attack*: The attacker knows the ciphertext of several messages encrypted with the same key or several keys and his goal is to recover the plaintext of as many messages as possible or to deduce the key (or keys).

2. *Known Plaintext Attack*: The attacker knows {ciphertext, plaintext} pair for several messages and his goal is to deduce the key to decrypt further messages.

3. *Chosen Plaintext Attack*: The attacker can choose the plaintext that gets encrypted. This type of situation is possible, for example, when the attacker has access to the encryption device for a limited time.

4. *Chosen Ciphertext Attack*: A decryption oracle is available to the attacker and the attacker gets the plaintexts corresponding to a series of ciphertexts of his choice. Based on this information, the attacker may decrypt new ciphertexts or determine the secret key.

For stream ciphers, the above four attack models essentially boil down to two distinct attacks, namely, the *key recovery attack* (retrieving the secret key from the keystream) and the distinguishing attack (identifying a non-random event in the keystream). We will discuss each of these in more detail in later chapters.

The above attacks are all *passive* attacks, where the adversary only monitors the communication channel. A passive attacker only threatens confidentiality of data. There is another type of attack, called an *active* attack, where the adversary attempts to alter or add or delete the transmissions on the channel. An active attacker threatens data integrity and authentication as well as confidentiality.

1.5 Cryptographic Security

The concept of cryptographic security is attributed to Kerckhoffs. Kerckhoffs' Principle [82] states that the security of a cipher should rely on the secrecy of the key only. It is assumed that the attacker knows every detail of the cryptographic algorithm except the key. With this fundamental premise, there exist several notions of cryptographic security [174, Page 45].

1. *Unconditional Security* or *Perfect Secrecy*: A cryptosystem is called unconditionally secure if it cannot be broken, even with infinite computational resources.

2. *Computational Security*: A cryptosystem is called computationally secure if the best known algorithm for breaking it requires at least N operations, where N is some specified, very large number.

3. *Provable Security*: A cryptosystem is called provably secure if it is as difficult to break as solving some well-known and supposedly difficult (such as NP-hard [30]) problem. It is important to note that a cryptosystem based on a hard problem does not guarantee security. For example, the worst case complexity for solving a problem may be exponential, but the average case complexity or the complexity for some specific instances of the problem may be polynomial.

1.6   Private and Public Key Cryptosystems

There are two types of cryptosystems: (1) private or symmetric and (2) public or asymmetric.

1. *Private Key Cryptosystem* or *Symmetric Key Cryptosystem*. Here we have a single secret key $k = k_e = k_d$, shared between the sender (Alice) and the receiver (Bob). Alice and Bob agree upon the key prior to communication (either through face-to-face discussion when they were together, or through some secure channel).

   Private or symmetric key cryptosystems are further divided into two classes:

   (a) *Block Cipher*. It is a memory-less (or state-less) permutation algorithm that breaks a plaintext m into successive blocks $m_1, m_2, \ldots$, and encrypts each block with the same key k. Thus,

   $$E_k(m) = E_k(m_1)E_k(m_2)\ldots.$$

   The decryption function is the inverse permutation algorithm.

   (b) *Stream Cipher*. It is a finite state machine with internal memory that breaks a plaintext m into successive characters or bits $m_1, m_2, \ldots$, and encrypts each m_r with a different key k_r from a keystream $k = k_1, k_2, \ldots$. Thus,

   $$E_k(m) = E_{k_1}(m_1)E_{k_2}(m_2)\ldots.$$

The decryption function is the same finite state machine that generates the same keystream, with which the plaintext is extracted from the ciphertext.

A block cipher typically consists of a complex Substitution-Permutation Network (SPN) [81] or a Fiestel structure [99] and involves complicated mixing of message and key bits. On the other hand, a stream cipher is typically defined on the binary alphabet $\{0, 1\}$ and a plaintext is bitwise XOR-ed with the keystream to generate the ciphertext, i.e.,

$$E_{k_r}(m_r) = m_r \oplus k_r.$$

During the decryption operation, the ciphertext is bitwise XOR-ed with the same keystream to get back the plaintext.

2. *Public Key Cryptosystem* or *Asymmetric Key Cryptosystem*. There are two major disadvantages of a private key cryptosystem. First, the common key has to be agreed between Alice and Bob before the communication begins. Second, n communicating parties require $\binom{n}{2}$ keys to communicate with each other. The latter leads to high key storage requirement. The public key cryptosystem takes care of both these issues. Bob (or any receiver) fixes two distinct keys, one for encryption and one for decryption. He publishes the encryption key k_e, called the *public key*, that can be used by anybody who likes to send a message to Bob. He keeps the decryption key e_d, called the *private key* secret and uses it to decipher any message which has been encrypted by the public key. Thus, for n communicating parties, only $2n$ keys are required, a (private key, public key) pair for each person.

In 1973, National Bureau of Standards (now the National Institute of Standards and Technology, or NIST) of the United States published a set of requirements for a cryptosystem. This led to the development of the famous block cipher, called *Data Encryption Standard* (DES) [45]. In 1999, it was replaced by an improved block cipher called the *Advanced Encryption Standard* (AES) [34,46]. An overview of stream ciphers is discussed in the next chapter.

The idea of a public key cryptosystem was first proposed by Diffie and Hellman [37, 38] in 1976. Since then many public key cryptosystems [120] have been invented, the most important and famous one being the *RSA Cryptosystem* [144] due to Rivest, Shamir and Adleman in 1977.

How the secret key can be established between the sender and the receiver who are geographically far apart is a pertinent problem in the private key cryptosystem. There are two broad categories of key establishment protocols [23,25], namely, *key distribution* and *key agreement*. In the first category, a trusted third party (TTP) acts as a central server that generates the requested key and distributes it among the individual communicating parties. In the second category, there is no need for a central server to generate the keys; the parties agree upon the key by collectively executing a decentralized protocol.

Public key cryptography plays an important role in key agreement protocols. Another use of public key systems is in *digital signatures*. Such signature schemes often use what we call a *hash function*

$$h : \{0,1\}^* \to \{0,1\}^n,$$

that takes as input a message m of arbitrary length and produces a *hash* or *message digest* $h(m)$ of a fixed size n. In a public key cryptosystem, *digital certificates* are used to authenticate the public keys. A *Public Key Infrastructure* or PKI [1] is a secure system that is used for managing and controlling these certificates. *Identity-Based Encryption* or IBE [54, 161] is an alternative to PKI, that removes the need for certificates by generating the public key of a user by applying a public hash function on the user's identity string (for example, an email ID).

Unlike private key cryptosystems, public key cryptosystems do not require prior communication of the secret key between the sender and the receiver. However, the performance of public key cryptosystems is in general slower than that of private key cryptosystems. In 1985, Koblitz and Miller proposed a new domain of public key cryptosystem called *Elliptic Curve Cryptography* or ECC [90, 121] based on the arithmetic on elliptic curves. ECC is usually much faster than other traditional public key systems (see [64] for details). The security of all public key systems relies on the difficulty of solving certain problems for which no efficient polynomial time algorithm is known. For example, the security of RSA depends on the hardness of factoring a large integer that is formed by multiplying two large primes (called the *integer factorization* problem), the security of the *ElGamal* scheme depends on the difficulty of computing the logarithm of an integer modulo a prime (called the *discrete log* problem), and that of ECC relies on a variant of the discrete log problem in the domain of elliptic curves.

So far we have discussed issues based on the classical (i.e., Turing) model of computation. In 1982, a new era in computer science began when Richard Feynman asserted that a quantum system can be used to do computations [49]. In 1985, this idea was formalized by David Deutsch [36], that initiated a new field of study called *Quantum Computation*. In 1984, Charles Bennett and Gilles Brassard demonstrated the use of *Quantum Cryptography* for the first time, by using quantum mechanical effects to perform key distribution [11]. The books [24, 129] contain comprehensive treatises on quantum computation and quantum cryptography.

In 1994, Peter Shor published his revolutionary paper [166, 167] that proved that integer factorization and discrete log problems can be solved efficiently (i.e., in polynomial time) using the quantum computation model. So far, the quantum computer has mostly remained a theoretical concept and very little development has been made toward the actual implementation. If, however, scientists are successful in manufacturing a quantum computer, then the traditional public key cryptosystems like RSA and ECC would collapse. Thus, a prime target of modern cryptography is to investigate how secure communications can be realized to cope with adversaries with access to quantum computers. This has led to the advent of a new domain of cryptography called *Post-Quantum Cryptography* [12].

Chapter 2

Stream Ciphers and RC4

2.1 Introduction to Stream Ciphers

In general, stream ciphers have faster throughput and are easier to implement compared to block ciphers. However, a specific instance of a block cipher may be more efficient than a specific instance of a stream cipher in terms of speed and implementation.

Let $m_1, m_2, \ldots$ be the message bits, $k_1, k_2, \ldots$ the keystream bits and $c_1, c_2, \ldots$ the corresponding ciphertext bits. For a typical stream cipher, the encryption is performed as

$$c_r = m_r \oplus k_r$$

and the decryption is performed as

$$m_r = c_r \oplus k_r.$$

This bitwise encryption-decryption is also called the *Vernam cipher* after the name of its inventor. When a different keystream is used with each different plaintext message, the Vernam cipher is called a *one-time pad*. One time pads have the property of *perfect secrecy* [164], i.e., the conditional probability of a message given the ciphertext is the same as the probability of the message; or, in other words, the ciphertext reveals no information about the plaintext.

Any finite state machine can generate a pseudo-random sequence of a finite period only. A plaintext message of length larger than this period would cause violation of the perfect secrecy condition. Hence, ideal one-time pads cannot be realized in practice.

A practical implementation of stream cipher involves "seeding" a finite state machine with a finite length key and then deriving the current keystream bit (or byte or word) as a function of the current internal state. The generated keystream bit sequence should satisfy the standard randomness tests [88]. Linear Feedback Shift Registers (LFSR) are frequently used as keystream generators of stream ciphers. A detailed discussion on LFSR is available in [62]. There exists significant literature on the analysis and design of LFSR-based stream ciphers that employ Boolean functions (see [32, 142] and the references therein for details) as non-linear elements. Apart from hardware-friendly LFSR-based stream ciphers, there have been many works on software-based stream ciphers as well. Detailed surveys on different kinds of stream ciphers are available in [149, 150].

There are two types of stream ciphers. The simplest type, in which the keystream is constructed from the key, independent of the plaintext string, is called *synchronous* stream cipher. Whereas, if each keystream element (bit or byte or word) depends on the previous plaintext or ciphertext elements as well as the key, it is called *self-synchronous* stream cipher. In a synchronous stream cipher, both the sender and the receiver must be synchronized for proper decryption. In other words, they must use the same key and operate at the same state given that key. Synchronization may be lost if ciphertext elements are inserted or deleted during the transmission, causing the decryption to fail. For re-synchronization, one needs additional techniques such as re-initialization, placing special markers at regular intervals in the ciphertext, or trying all possible keystream offsets (if the plaintext contains enough redundancy). In self-synchronous stream ciphers, the decryption depends only on a fixed number of preceding ciphertext characters. Thus, they are capable of re-establishing proper decryption automatically after loss of synchronization, at the cost of only a fixed number of irrecoverable plaintext characters. RC4 is a synchronous stream cipher, whereas Mosquito [33] is an example of a self-synchronous stream cipher.

In a stream cipher, the same key always produces the same keystream. Hence, repeated use of the same key is just as bad as reusing a one-time pad. One approach to handle this problem is to renew the secret key from time to time. But this involves key exchange overhead. An alternative remedy is the use of *initialization vectors*. An Initialization Vector (IV) is a random value that changes with each session of the cipher. The IV is combined with the secret key to form the effective key for the corresponding session of the cipher, called a *session key*. Different session keys make the output of the stream cipher different in each session, even if the same key is used.

During 2000–2003, a European research project called New European Schemes for Signatures, Integrity and Encryption (NESSIE) [127] was established to identify secure cryptographic primitives. Six stream ciphers were submitted to the NESSIE project, but none was selected for the portfolio. In Asiacrypt 2004, Shamir raised the pertinent question [162]: "Stream Ciphers: Dead or Alive?" In the same year, a 4-year European research initiative was launched under the name European Network of Excellence for Cryptology (ECRYPT) and a project called eSTREAM was started with an aim to identifying "new stream ciphers that might become suitable for widespread adoption." The submissions to eSTREAM fall into either or both of two profiles:

1. Profile 1: "Stream ciphers for software applications with high throughput requirements."

2. Profile 2: "Stream ciphers for hardware applications with restricted resources such as limited storage, gate count, or power consumption."

As of September 2008, seven stream ciphers have been selected for the final portfolio. Among them, HC-128, Rabbit, Salsa20/12 and SOSEMANUK belong to the software profile and the other three, namely, Grain v1, MICKEY v2 and Trivium belong to Profile 2. A comprehensive survey on all of these ciphers is available in [145]. Thus the current eSTREAM portfolio is as follows.

Profile 1 (SW)	Profile 2 (HW)
HC-128	Grain v1
Rabbit	MICKEY v2
Salsa20/12	Trivium
SOSEMANUK	

It is expected that research on the eSTREAM submissions in general, and the portfolio ciphers in particular, will continue. It is also possible that changes to the eSTREAM portfolio might be needed in the future.

2.2 Attack Models for Stream Cipher Cryptanalysis

Below we briefly describe the common attack models with respect to which the security analysis of stream ciphers are performed.

2.2.1 Brute Force Key Search

For a key of size l bits, a brute force search consists of trying all 2^l keys to check which one leads to the observed data. It is the most basic attack against any symmetric key cryptosystem.

The following techniques can speed-up the brute force search.

1. *Time Memory Trade-Off* (TMTO): Hellman showed [68] that by precomputing some values and storing them in memory one can perform a brute force attack on block ciphers faster. Later, in [7], such TMTO attacks were shown to be applicable for stream ciphers as well.

2. *Sampling Resistance*: In [20], this property of stream cipher related to the TMTO attack was introduced. It measures how easy it is to find keys that generate keystreams with a certain property of the keystream output.

2.2.2 Weak Keys

The secret key completely determines the output sequence of a stream cipher. If the keys are "weak," in the sense that they leak secret key information in the keystream, then the attacker may get relevant information about the key by analyzing the keystream.

Key Weaknesses originate from different scenarios.

1. *Key-Output Correlation*: It is desirable that a single bit flip in the secret key flips each bit of the keystream with probability $\frac{1}{2}$. Otherwise, secret key bits may be recovered in a complexity less than the brute force search.

2. *Related Key*: Given a known relationship between two keys should not produce a known relationship in the keystream. Otherwise, such relationship can be used to mount what is called a related key attack.

3. *IV Weakness*: If the IVs are not used with proper care, they may leak secret key information in the keystream.

2.2.3   Distinguishers

A distinguisher is an algorithm that takes as input n bits of data, which are either taken from the keystream of a stream cipher or are completely random data, and decides which one of these two categories the input belongs to. If this decision is correct with a probability more than the random guess, then we have a distinguisher of complexity n. More details on these types of attacks (called *distinguishing attacks*) can be found in Chapters 6 and [187].

2.3   Hardware Stream Ciphers

Initially, stream ciphers were targeted toward hardware only. In hardware domain, mostly LFSRs are used as linear elements and combining functions (may be with some amount of memory) are used as nonlinear elements.

Let us first discuss bit-oriented LFSRs. An example is shown in Figure 2.1.

The recurrence relation connecting the LFSR output bit and the internal state bits is given by $s_{t+6} = s_{t+4} \oplus s_{t+1} \oplus s_t$. The corresponding connection polynomial over $GF(2)$ is given by $x^6 + x^4 + x^1 + 1$, which is a primitive polynomial of degree 6. It is known that a primitive polynomial of degree d provides maximum length cycle, i.e., of length $2^d - 1$. Such sequences are well known as *m-sequences* [96].

By itself, an LFSR is not cryptographically secure, but it is a useful building block for pseudo-randomness. In the domain of communications, pseudo-random sequences are known as p-n sequences. LFSRs find easy and efficient implementation in hardware, using registers (Flip Flops) and simple logic gates. Research in LFSR-based stream ciphers has incurred deep mathemati-

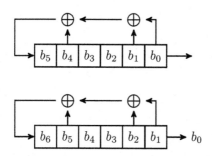

FIGURE 2.1: LFSR: one step evolution

cal development for a long time, leading to elegant results in the area of linear complexity [62].

A *nonlinear combiner model* takes n LFSRs of different lengths (may be pairwise co-prime). They are initialized with non-zero seeds. In each clock, n-many outputs from the LFSRs are taken and fed as n-inputs to an n-variable Boolean function. Some memory element may optionally be added.

For nonlinear filter-generator model, n-many outputs from different locations of the LFSR are fed as n-inputs to an n-variable Boolean function. Here too, some memory element may optionally be added. The Boolean function and memory together form a Finite State Machine.

We now mention a few recent state-of-the-art hardware stream ciphers. E_0 is a stream cipher used in the Bluetooth protocol for link encryption. It uses 4 LFSRs of lengths 25, 31, 33 and 39 bits and one FSM to combine their outputs. Several vulnerabilities [53,60,69,98] have already been discovered for this cipher. A5/1 is a stream cipher used to provide over-the-air voice privacy in the GSM cellular telephone standard. A5/1 consists of three short binary LFSRs of length 19, 22, 23 denoted by R1, R2, R3 respectively. The LFSRs are clocked in an irregular fashion. The works [13,20,41,59] constitute important attacks against this cipher. Grain [66] is a stream cipher primitive that is designed to be accommodated in low end hardware. It contains three main building blocks, namely a linear feedback shift register (LFSR), a nonlinear feedback shift register (NFSR), both of 80-bit size, and a filter function. Given certain attacks on Grain [115], a new cipher in a similar direction, called Grain-128 is proposed in [67]. In the hardware implementation of the Grain family of stream ciphers, the gate count is very low and it is expected that these ciphers can be accommodated in very low end hardwares, even in RFID tags. SNOW 3G is chosen as the stream cipher for the 3GPP encryption algorithms [43]. It is an LFSR based stream cipher with 32-bit words with a 128-bit key. Except the fault analysis [35], no other attacks on this cipher are known. The most recent addition to the family of hardware stream ciphers is ZUC [44], designed by the Data Assurance and Communication Security Research Center of the Chinese Academy of Sciences. This cipher has been proposed for inclusion in the 4G mobile standard called LTE (Long Term Evolution). ZUC uses a 16-word LFSR where each word is of 31 bits. The key size is 128 bits and the keystream is generated in 32-bit words. A finite state machine works as a non-linear core for this cipher.

A few non-LFSR-based hardware stream ciphers have also been designed. For examples, the work [128] presents a family of stream ciphers called Vest, based on non-linear parallel feedback shift registers (NLFSR) and non-linear Residue Number System (RNS) based counters.

2.4 Software Stream Ciphers

A software stream cipher typically consists of two modules. The first module takes as input the secret key (and optionally an IV) and expands it into a secret internal state. This is called key scheduling. The second module generates the keystream from the internal state. At every state transition, a pseudo-random keystream bit or byte or word is output depending on the design of the cipher.

Here we discuss RC4, the most widely used software stream cipher. Variants of RC4 are discussed in Chapter 9. The eSTREAM finalist HC-128 can be considered as the next generation of software stream cipher evolution after RC4 and RC4-like ciphers. It is not easy to maintain an array of 2^{32} locations to implement a 32-bit instance of RC4. More security margin for 32-bit cipher obviously requires more time/memory. HC-128 is a 32-bit cipher that satisfies these requirements. Use of arrays and modular addition in HC-128 is in similar line as RC4, but many more additional operations are incorporated. We devote Chapter 10 to the discussion of HC-128 in detail.

2.4.1 RC4 Stream Cipher

The RC4 stream cipher was designed by Ron Rivest for RSA Data Security in 1987. It is believed to be a propriety algorithm. In 1994, it was allegedly revealed on the internet [6]. Currently, RC4 stands out to be one of the most popular among state-of-the-art software stream ciphers with varied industrial applications. It is used for encrypting the internet traffic in network protocols such as Secure Sockets Layer (SSL), Transport Layer Security (TLS), Wired Equivalent Privacy (WEP), Wi-Fi Protected Access (WPA) etc. The cipher is also used in Microsoft Windows, Lotus Notes, Apple Open Collaboration Environment (AOCE), and Oracle Secure SQL.

RC4 data structure consists of an S-Box

$$S = (S[0], \ldots, S[N-1])$$

of length $N = 2^n$, where each entry is an n-bit integer. Typically, $n = 8$ and

$N = 256$. S is initialized as the identity permutation, i.e.,

$$S[i] = i \quad \text{for } 0 \leq i \leq N - 1.$$

A secret key κ of size l bytes (typically, $5 \leq l \leq 16$) is used to scramble this permutation. An array

$$K = (K[0], \ldots, K[N-1])$$

is used to hold the secret key, where each entry is again an n-bit integer. The key is repeated in the array K at key length boundaries. For example, if the key size is 40 bits, then $K[0], \ldots, K[4]$ are filled by the key and then this pattern is repeated to fill up the entire array K. Formally, we can write

$$K[y] = \kappa[y \bmod l], \quad \text{for } 0 \leq y \leq N - 1.$$

The RC4 cipher has two components, namely, the Key Scheduling Algorithm (KSA) and the Pseudo-Random Generation Algorithm (PRGA). The KSA uses the key K to shuffle the elements of S and the PRGA uses this scrambled permutation to generate pseudo-random keystream bytes.

Two indices, i and j, are used in RC4. i is a deterministic index that is incremented by 1 (modulo N) in each step and j serves as a pseudo-random index that is updated depending on the secret key K and the state S.

The KSA initializes both i and j to 0, and S to be the identity permutation. It then steps i across S looping N times, and updates j by adding the i-th entries of S and K. Each iteration ends with a swap of the two bytes in S pointed by the current values of i and j.

Input: Secret key array $K[0 \ldots N-1]$.
Output: Scrambled permutation array $S[0 \ldots N-1]$.

Initialization:
for $i = 0, \ldots, N-1$ **do**
   | $S[i] = i$;
   | $j = 0$;
end

Scrambling:
for $i = 0, \ldots, N-1$ **do**
   | $j = (j + S[i] + K[i])$;
   | $\text{Swap}(S[i], S[j])$;
end

Algorithm 2.4.1: RC4 KSA

The PRGA also initializes both i and j to 0. It then loops over four operations in sequence: it increments i as a counter, updates j by adding $S[i]$, swaps the two entries of S pointed by the current values of i and j, and

Input: Key-dependent scrambled permutation array $S[0 \dots N-1]$.
Output: Pseudo-random keystream bytes z.

Initialization:
$i = j = 0$;

Output Keystream Generation Loop:
$i = i + 1$;
$j = j + S[i]$;
Swap($S[i]$, $S[j]$);
$t = S[i] + S[j]$;
Output $z = S[t]$;

Algorithm 2.4.2: RC4 PRGA

outputs the value of S at index $S[i] + S[j]$ as the value of z. Figure 2.2 gives a pictorial representation of one keystream byte generation.

The n-bit keystream output z is XOR-ed with the next n bits of the message to generate the next n bits of the ciphertext at the sender end. Again, z is bitwise XOR-ed with the ciphertext byte to get back the message at the receiver end.

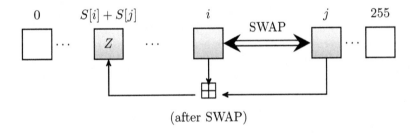

(after SWAP)

FIGURE 2.2: Output byte generation in RC4 PRGA.

Any addition, used in the RC4 description or in addition of key and permutation bytes in this book, is in general addition modulo N unless specified otherwise.

A simple implementation of RC4 in C is given in Appendix A.

Two decades have elapsed since the inception of RC4. A variety of other stream ciphers have been discovered after RC4, yet it is the most popular and frequently used stream cipher algorithm due to its simplicity, ease of implementation, byte-oriented structure, speed and efficiency. On the contrary, the eSTREAM candidates have much complicated structures in general and they work on word (32 bits) oriented manner. Even after many years of cryptanalysis, the strengths and weaknesses of RC4 are of great interest to the community.

Notations Related to RC4 used in this Book

We denote the permutation, the deterministic index and the pseudo-random index be denoted by S, i, j for the KSA and by S^G, i^G, j^G for the PRGA respectively. t is used to denote the index in S^G from where the keystream byte z is chosen.

We use subscript r to these variables to denote their updated values in round r, where $1 \leq r \leq N$ for the KSA and $r \geq 1$ for the PRGA. Thus, for the KSA,

$$i_r = r - 1.$$

For the PRGA,

$$i_r^G = r \bmod N$$

and

$$z_r = S_r^G[t_r].$$

According to this notation, S_N is the permutation after the completion of the KSA.

By S_0 and j_0, we mean the initial permutation and the initial value of the index j respectively before the KSA begins and by S_0^G and j_0^G, we mean the same before the PRGA begins. Note that S_0^G and S_N represent the same permutation.

Sometimes, for clarity of exposition, we omit the subscripts and/or the superscripts in S, i, j, t, z. This will be clear from the context.

We use the notation S^{-1} for the inverse of the permutation S, i.e., if $S[y] = v$, then $S^{-1}[v] = y$.

The identity

$$f_y = \frac{y(y+1)}{2} + \sum_{x=0}^{y} K[x]$$

will be used frequently in this book.

Other notations will be introduced as and when necessary.

2.5   On Randomness and Pseudo-Randomness

When should something be called *random* has been a long-standing philosophical debate. Without going into historical anecdotes, we briefly discuss the three primary approaches to randomness prevalent in the scientific paradigm.

The first approach is rooted in Shannon's information theory [163], that equates randomness with lack of information. In this model, perfect randomness is associated with uniform distribution, i.e., a quantity is called random if all of its values are equally likely. The more non-uniform a distribution is, the more it is deviated from randomness.

The second approach, propounded by Solomonoff [171], Kolmogorov [91] and Chaitin [27], views randomness as lack of pattern and structure. In this approach, the shorter the description of an object (objects are understood to be binary strings), the more it is random. Here description means an input string (i.e., a program) which when fed to a fixed Universal Turing Machine generates the object. If the length of the shortest such program is greater than or equal to the length of the object, then the object is called perfectly random. In other words, randomness is equivalent to incompressibility.

The third approach is computational. In this view, a distribution that cannot be efficiently distinguished from the uniform distribution is considered random. Often the term "pseudo-random" is used to express this particular notion of randomness. Contrary to the previous two approaches, here randomness is not an inherent property of objects, or distributions, rather it is relative to the computational resources of the observer. This model allows for an efficient and deterministic generation of (pseudo-)random bitstrings from shorter seeds.

RC4, and as a matter of fact, any stream cipher, generates a pseudo-random keystream from a (relatively) short secret key based on the third view of randomness above.

Chapter 3

Analysis of Key Scheduling

In this chapter, we present theoretical analysis of the Key Scheduling Algorithm (KSA) of RC4. We start with an exposition on several biases of permutation bytes toward secret key bytes in Section 3.1. In 1995, Roos [146] observed that the initial permutation bytes, obtained just after the KSA, are biased to some combinations of the secret key bytes. After presenting an explicit formula for the probabilities with which the permutation bytes at any stage of the KSA are biased to the secret key, we analyze a generalization of the RC4 KSA corresponding to a certain class of update functions. This reveals an inherent weakness of the shuffle-exchange kind of key scheduling. Interestingly, the nested permutation entries, such as $S[S[y]], S[S[S[y]]]$, and so on, are also biased to the secret key for the initial values of y. We end the first section with a discussion on this issue.

In Section 3.2, we shift focus on non-randomness of permutation S. Under reasonable assumptions, each permutation byte after the KSA is significantly biased toward many values in the range $\mathbb{Z}_N$. This weakness was first noted by Mantin [107] and later revisited in [122, 135]. We present a proof of these biases and demonstrate that permutation after the KSA can be distinguished from random permutation without any assumption on the secret key. The last part of Section 3.2 is devoted to the *anomaly pairs*, i.e., the (index,value) pairs where the theoretical computations and the empirical data regarding the biases differ significantly.

In RC4, an interesting phenomenon is that there exists a non-uniformity in the expected number of times each value of the permutation is touched by the indices i, j. The theoretical analysis of this fact first appeared in [101]. We present this in Section 3.3 of this chapter.

This chapter ends with a discussion on key collisions of RC4 [111].

3.1 Bias of Permutation toward Secret Key

In this section we analyze the KSA to prove different types of biases of the permutation bytes toward combinations of secret key bytes. We begin with the theoretical derivation of the probability expression for Roos' observation [146]. Next, we move on to describe other biases as well.

3.1.1   Roos' Biases

Roos [146] and Wagner [185] reported the initial empirical works on the weakness of the RC4 KSA, in which they identified several classes of weak keys. The following important technical result appears in [146, Section 2, Result B].

Proposition 3.1.1. *[Roos, 1995] The most likely value of the y-th element of the permutation after the KSA for the first few values of y is given by* $S_N[y] = f_y$.

The experimental values of the probabilities $P(S_N[y] = f_y)$ reported in [146] is provided in Table 3.1 below.

y	$P(S_N[y] = f_y)$
0-7	0.370 0.368 0.362 0.358 0.349 0.340 0.330 0.322
8-15	0.309 0.298 0.285 0.275 0.260 0.245 0.229 0.216
16-23	0.203 0.189 0.173 0.161 0.147 0.135 0.124 0.112
24-31	0.101 0.090 0.082 0.074 0.064 0.057 0.051 0.044
32-39	0.039 0.035 0.030 0.026 0.023 0.020 0.017 0.014
40-47	0.013 0.012 0.010 0.009 0.008 0.007 0.006 0.006

TABLE 3.1: The probabilities experimentally observed by Roos.

When y is small enough (e.g., $y <$ the key length l), then Roos [146, Section 2] argues that there is a high probability that none of the two elements being swapped has been involved in any previous exchanges, and so it can safely be assumed that $S_y[y] = y$ before the swap happens in round $y + 1$. Thus, the most likely value of $S[1]$ is $K[0] + K[1] + 0 + 1$ and that of $S[2]$ is $K[0] + K[1] + 0 + 1 + 2$. Since the probability that a particular single element of S would not be indexed by j throughout KSA is $(\frac{N-1}{N})^N$, Roos' arguments can be formalized as below.

Corollary 3.1.2. *After the KSA, the most likely values of the second and the third elements of the permutation are given as follows.*

1. $P(S_N[1] = K[0] + K[1] + 1) \approx (\frac{N-1}{N})^N$.

2. $P(S_N[2] = K[0] + K[1] + K[2] + 3) \approx (\frac{N-1}{N})^N$.

Proof sketch of the above also appeared in [118, Section 3.4.1], but in [133, Section 2], the expressions for the probabilities $P(S_r[y] = f_y)$ for all values of the index y in $[0, N-1]$ and for all rounds r, $1 \le r \le N$, were theoretically derived for the first time. Related analysis in this area was performed independently in [180, Section 4] and [184, Section 3] around the same time of [133]. However, these efforts are mostly tuned to exploit the WEP scenario, where the secret key is used with IV.

Below we are going to present the general formula (Theorem 3.1.5) that estimates the probabilities with which the permutation bytes after each round of the RC4 KSA are related to certain combinations of the secret key bytes, namely f_y. The result gives a theoretical proof explicitly showing how these probabilities change as functions of y. Further, the result holds for any arbitrary initial permutation (need not be an identity permutation).

Though j is updated using a deterministic formula, it is a linear function of the secret key bytes. If the secret key generator produces the secret keys uniformly at random, which is a reasonable assumption, then the distribution of j would also be uniform.

The proof of Theorem 3.1.5 depends on the following results.

Lemma 3.1.3. *Assume that the index j takes its value from $\mathbb{Z}_N$ independently and uniformly at random at each round of the KSA. Then, for $0 \le y \le N-1$,*

$$P\left(j_{y+1} = \sum_{x=0}^{y} S_0[x] + \sum_{x=0}^{y} K[x]\right) \approx \left(\frac{N-1}{N}\right)^{1+\frac{y(y+1)}{2}} + \frac{1}{N}.$$

Proof: For $y \ge 0$, let E_y denote the event that

$$j_{y+1} = \sum_{x=0}^{y} S_0[x] + \sum_{x=0}^{y} K[x]$$

and let A_y denote the event that

$$S_x[x] = S_0[x] \quad \text{for all } x \in [0, y].$$

Let $\bar{A}_y$ stand for the complementary event of A_y. We have

$$P(E_y) = P(E_y|A_y) \cdot P(A_y) + P(E_y|\bar{A}_y) \cdot P(\bar{A}_y).$$

We also have $P(E_y|A_y) = 1$ and $P(E_y|\bar{A}_y) \approx \frac{1}{N}$ due to random association. We show by induction that

$$P(A_y) = \left(\frac{N-1}{N}\right)^{\frac{y(y+1)}{2}}.$$

The base case $P(A_0) = 1$ is trivial. For $y \ge 1$, we have

$$
\begin{aligned}
P(A_y) &= P\left(A_{y-1} \wedge (S_y[y] = S_0[y])\right) \\
&\approx P(A_{y-1}) \cdot P(S_y[y] = S_0[y]) \\
&= \left(\frac{N-1}{N}\right)^{\frac{(y-1)y}{2}} \cdot P(S_y[y] = S_0[y]).
\end{aligned}
$$

Conditioned on the event that all the values $j_1, j_2, \ldots, j_y$ are different from y, which happens with probability $(\frac{N-1}{N})^y$, $S_y[y]$ remains the same as $S_0[y]$. Note that if at least one of the values $j_1, j_2, \ldots, j_y$ hits the index y, the probability that $S_y[y]$ remains the same as $S_0[y]$ is too small. Hence

$$P(S_y[y] = S_0[y]) \approx \left(\frac{N-1}{N}\right)^y.$$

This completes the proof.   ∎

Lemma 3.1.4. *Assume that the index j takes its value from $\mathbb{Z}_N$ independently and uniformly at random at each round of the KSA. Then, for $0 \le y \le r-1$, $1 \le r \le N$,*

$$P(S_r[y] = S_0[j_{y+1}]) \approx \left(\frac{N-y}{N}\right) \cdot \left(\frac{N-1}{N}\right)^{r-1}.$$

Proof: During round $y+1$, the value of $S_y[j_{y+1}]$ is swapped into $S_{y+1}[y]$. Now, the index j_{y+1} is not involved in any swap during the previous y many rounds, if it is not touched by the indices $\{0, 1, \ldots, y-1\}$, as well as if it is not touched by the indices $\{j_1, j_2, \ldots, j_y\}$. The probability of the former is $(\frac{N-y}{N})$ and the probability of the latter is $(\frac{N-1}{N})^y$. Hence,

$$P(S_{y+1}[y] = S_0[j_{y+1}]) \approx \left(\frac{N-y}{N}\right) \cdot \left(\frac{N-1}{N}\right)^y.$$

After round $y+1$, index y is not touched by any of the subsequent $r-1-y$ many j values with probability $(\frac{N-1}{N})^{r-1-y}$. Hence,

$$
\begin{aligned}
P(S_r[y] = S_0[j_{y+1}]) &\approx \left(\frac{N-y}{N}\right) \cdot \left(\frac{N-1}{N}\right)^y \cdot \left(\frac{N-1}{N}\right)^{r-1-y} \\
&= \left(\frac{N-y}{N}\right) \cdot \left(\frac{N-1}{N}\right)^{r-1}.
\end{aligned}
$$

∎

Theorem 3.1.5. *Assume that the index j takes its value from $\mathbb{Z}_N$ independently and uniformly at random at each round of the KSA. Then, for $0 \le y \le r-1$, $1 \le r \le N$,*

$$P(S_r[y] = f_y) \approx \left(\frac{N-y}{N}\right) \cdot \left(\frac{N-1}{N}\right)^{\left[\frac{y(y+1)}{2}+r\right]} + \frac{1}{N},$$

where

$$f_y = S_0\left[\sum_{x=0}^{y} S_0[x] + \sum_{x=0}^{y} K[x]\right].$$

Proof: Let A_y denote the event as defined in the proof of Lemma 3.1.3. Now, $S_r[y]$ can be equal to f_y in two ways. One way is that the event A_y and the event $S_r[y] = S_0[j_{y+1}]$ occur together (recall that $P(E_y|A_y) = 1)$). Combining the results of Lemma 3.1.3 and Lemma 3.1.4, we get the contribution of this part to be approximately

$$P(A_y) \cdot P(S_r[y] = S_0[j_{y+1}]) = \left(\frac{N-1}{N}\right)^{\frac{y(y+1)}{2}} \cdot \left(\frac{N-y}{N}\right) \cdot \left(\frac{N-1}{N}\right)^{r-1}$$

$$= \left(\frac{N-y}{N}\right) \cdot \left(\frac{N-1}{N}\right)^{[\frac{y(y+1)}{2}+(r-1)]}.$$

Another way is that neither of the above events happen and still $S_r[y]$ equals $S_0\left[\sum_{x=0}^{y} S_0[x] + \sum_{x=0}^{y} K[x]\right]$ due to random association. The contribution of this second part is approximately $\left(1 - (\frac{N-y}{N}) \cdot (\frac{N-1}{N})^{[\frac{y(y+1)}{2}+(r-1)]}\right) \cdot \frac{1}{N}$. Adding these two contributions, we get the total probability to be approximately

$$\left(\frac{N-y}{N}\right) \cdot \left(\frac{N-1}{N}\right)^{[\frac{y(y+1)}{2}+(r-1)]}$$

$$+ \left(1 - \left(\frac{N-y}{N}\right) \cdot \left(\frac{N-1}{N}\right)^{[\frac{y(y+1)}{2}+(r-1)]}\right) \cdot \frac{1}{N}$$

$$= \left(1 - \frac{1}{N}\right) \cdot \left(\frac{N-y}{N}\right) \cdot \left(\frac{N-1}{N}\right)^{[\frac{y(y+1)}{2}+(r-1)]} + \frac{1}{N}$$

$$= \left(\frac{N-y}{N}\right) \cdot \left(\frac{N-1}{N}\right)^{[\frac{y(y+1)}{2}+r]} + \frac{1}{N}.$$

$\blacksquare$

Corollary 3.1.6. *The bias of the final permutation after the KSA toward the secret key is given by*

$$P(S_N[y] = f_y) \approx \left(\frac{N-y}{N}\right) \cdot \left(\frac{N-1}{N}\right)^{[\frac{y(y+1)}{2}+N]} + \frac{1}{N}, \text{ for } 0 \le y \le N-1.$$

Proof: Substitute $r = N$ in the statement of Theorem 3.1.5. $\blacksquare$

Note that if the initial permutation S_0 is identity, then we have $f_y = \frac{y(y+1)}{2} + \sum_{x=0}^{y} K[x]$. Table 3.2 shows the theoretical values of the probability $P(S_N[y] = f_y)$ in this case which is consistent with the experimental values provided in Table 3.1.

Index 48 onwards, both the theoretical as well as the experimental values tend to $\frac{1}{N} = 0.00390625 \approx 0.0039$, for $N = 256$. This is the probability of the equality between two randomly chosen values from a set of N elements.

y	$P(S_N[y] = \frac{y(y+1)}{2} + \sum\limits_{x=0}^{y} K[x])$
0-7	0.371 0.368 0.364 0.358 0.351 0.343 0.334 0.324
8-15	0.313 0.301 0.288 0.275 0.262 0.248 0.234 0.220
16-23	0.206 0.192 0.179 0.165 0.153 0.140 0.129 0.117
24-31	0.107 0.097 0.087 0.079 0.071 0.063 0.056 0.050
32-39	0.045 0.039 0.035 0.031 0.027 0.024 0.021 0.019
40-47	0.016 0.015 0.013 0.011 0.010 0.009 0.008 0.008

TABLE 3.2: The probabilities following Corollary 3.1.6.

3.1.2   Intrinsic Weakness of Shuffle-Exchange Type KSA

In the KSA of RC4, the deterministic index i is incremented by one and the pseudo-random index j is updated by the rule

$$j = j + S[i] + K[i].$$

In our notation, we write, for $0 \le y \le N - 1$,

$$j_{y+1} = j_y + S_y[y] + K[y].$$

Here, the increment of j is a function of the permutation and the secret key. One may expect that the correlation between the secret key and the permutation can be removed by modifying the update rule for j. However, it turns out that for a certain class of rules of this type, where j across different rounds is distributed uniformly at random, there always exists significant bias of the permutation at any stage of the KSA toward some combination of the secret key bytes with non-negligible probability. Though the proof technique is similar to that in Section 3.1.1, it may be noted that the analysis in these proofs focus on the weakness of the particular "form" of RC4 KSA.

The update of j in the KSA can be modeled as an arbitrary function u of

1. the current values of i, j,

2. the i-th and j-th permutation bytes from the previous round, and

3. the i-th and j-th key bytes.

Using our notations, we may write

$$j_{y+1} = u\big(y, j_y, S_y[y], S_y[j_y], K[y], K[j_y]\big). \tag{3.1}$$

For future reference, let us call the KSA with this generalized update rule the GKSA.

Lemma 3.1.7. *Assume that the index j takes its value from $\mathbb{Z}_N$ independently and uniformly at random at each round of the GKSA. Then, one can always construct functions $h_y(S_0, K)$, which depends only on y, the secret key bytes and the initial permutation, such that for $0 \le y \le N - 1$,*

$$P\big(j_{y+1} = h_y(S_0, K)\big) = \left(\frac{N-1}{N}\right)\pi_y + \frac{1}{N},$$

where the probabilities π_y depends only on y and N.

Proof: Consider Equation (3.1). Due to the swap in round y, we have $S_y[j_y] = S_{y-1}[y-1]$; so

$$j_{y+1} = u\big(y, j_y, S_y[y], S_{y-1}[y-1], K[y], K[j_y]\big).$$

Let $h_y(S_0, K)$ have the following recursive formulation.

$$h_0(S_0, K) = u\big(0, 0, S_0[0], S_0[0], K[0], K[0]\big)$$

and for $1 \le y \le N - 1$,

$$h_y(S_0, K) = u\big(y, h_{y-1}(S_0, K), S_0[y], S_0[y-1], K[y], K[h_{y-1}(S_0, K)]\big).$$

For $y \ge 0$, let E_y denote the event that

$$j_{y+1} = h_y(S_0, K).$$

For $y \ge 1$, let A_y denote the event that

$$u\left(x, j_x, S_x[x], S_{x-1}[x-1], K[x], K[j_x]\right)$$
$$= u\left(x, j_x, S_0[x], S_0[x-1], K[x], K[j_x]\right)$$

for all $x \in [1, y]$. Let A_0 denote the trivial event $S_0[0] = S_0[0]$ or

$$u\big(0, 0, S_0[0], S_0[0], K[0], K[0]\big) = u\big(0, 0, S_0[0], S_0[0], K[0], K[0]\big),$$

which occurs with probability $\pi_0 = P(A_0) = 1$. Let $\bar{A}_y$ stand for the complementary event of A_y. We have

$$P(E_y) = P(E_y|A_y) \cdot P(A_y) + P(E_y|\bar{A}_y) \cdot P(\bar{A}_y).$$

We also have $P(E_y|A_y) = 1$ and $P(E_y|\bar{A}_y)$ approximately equals $\frac{1}{N}$ due to random association. By induction on y, we will show how to construct the the probabilities π_y recursively such that $P(A_y) = \pi_y$. For $y \ge 1$, suppose $P(A_{y-1}) = \pi_{y-1}$. Now the event A_y occurs, if the following two hold:

1. the event A_{y-1} (with probability π_{y-1}) occurs, and

2. one or both of the events $S_y[y] = S_0[y]$ (with probability $(\frac{N-1}{N})^y$) and $S_{y-1}[y-1] = S_0[y-1]$ (with probability $(\frac{N-1}{N})^{y-1}$) occurs, depending on whether the form of u contains one or both of these terms respectively.

Thus, $\pi_y = P(A_y)$ can be computed as a function of y, N, and π_{y-1}, depending on the occurrence or non-occurrence of various terms in u. This completes the proof of the induction as well as that of the Lemma. ■

Theorem 3.1.8. *Assume that the index j takes its value from $\mathbb{Z}_N$ independently and uniformly at random at each round of the GKSA. Then, one can always construct functions $f_y(S_0, K)$, which depend only on y, the secret key bytes and the initial permutation, such that for $0 \leq y \leq r - 1$, $1 \leq r \leq N$,*

$$P(S_r[y] = f_y(S_0, K)) \approx \left(\frac{N-y}{N}\right) \cdot \left(\frac{N-1}{N}\right)^r \cdot \pi_y + \frac{1}{N}.$$

Proof: Let us first show that

$$f_y(S_0, K) = S_0[h_y(S_0, K)],$$

where the function h_y's are given by Lemma 3.1.7. Let A_y denote the event as defined in the proof of Lemma 3.1.7. Now, $S_r[y]$ can equal $S_0[h_y(S_0, K)]$ in two ways.

Case I: One possibility is that the event A_y and the event $S_r[y] = S_0[j_{y+1}]$ occur together (recall that $P(E_y|A_y) = 1)$). Combining Lemma 3.1.7 and Lemma 3.1.4, we find that the probability of this case is approximately $\left(\frac{N-y}{N}\right) \cdot \left(\frac{N-1}{N}\right)^{r-1} \cdot \pi_y$.

Case II: Another possibility is that neither of the above events happen and still $S_r[y] = S_0[h_y(S_0, K)]$ due to random association. The contribution of this part is approximately $\left(1 - \left(\frac{N-y}{N}\right) \cdot \left(\frac{N-1}{N}\right)^{r-1} \cdot \pi_y\right) \cdot \frac{1}{N}$.

Adding the above two contributions, we get the total probability to be approximately $\left(\frac{N-y}{N}\right) \cdot \left(\frac{N-1}{N}\right)^r \cdot \pi_y + \frac{1}{N}$. ■

Next we demonstrate some special cases of the update rule u as examples of how to construct the functions f_y's and the probabilities π_y's for small values of y using Lemma 3.1.7. In all the following cases, we assume S_0 to be an identity permutation and hence $f_y(S_0, K)$ is the same as $h_y(S_0, K)$.

Example 3.1.9. *Consider the KSA of RC4, where*

$$u(y, j_y, S_y[y], S_y[j_y], K[y], K[j_y]) = j_y + S_y[y] + K[y].$$

We have

$$
\begin{aligned}
h_0(S_0, K) &= u(0, 0, S_0[0], S_0[0], K[0], K[0]) \\
&= 0 + 0 + K[0] \\
&= K[0].
\end{aligned}
$$

Moreover, $\pi_0 = P(j_1 = h_0(S_0, K)) = 1$. For $y \geq 1$,

$$
\begin{aligned}
h_y(S_0, K) &= u(y, h_{y-1}(S_0, K), S_0[y], S_0[y-1], K[y], K[h_{y-1}(S_0, K)]) \\
&= h_{y-1}(S_0, K) + S_0[y] + K[y] \\
&= h_{y-1}(S_0, K) + y + K[y].
\end{aligned}
$$

Solving the recurrence, we get

$$
h_y(S_0, K) = \frac{y(y+1)}{2} + \sum_{x=0}^{y} K[x].
$$

From the analysis in the proof of Lemma 3.1.7, we see that in the recurrence of h_y, $S_y[y]$ has been replaced by $S_0[y]$ and j_y has been replaced by $h_{y-1}(S_0, K)$. Hence, we would have

$$
\begin{aligned}
\pi_y &= P(S_y[y] = S_0[y]) \cdot P(j_y = h_{y-1}(S_0, K)) \\
&= \left(\frac{N-1}{N}\right)^y \cdot \pi_{y-1}.
\end{aligned}
$$

Solving this recurrence, we get

$$
\pi_y = \prod_{x=0}^{y} \left(\frac{N-1}{N}\right)^x = \left(\frac{N-1}{N}\right)^{\frac{y(y+1)}{2}}.
$$

Example 3.1.10. *Consider the update rule*

$$
u(y, j_y, S_y[y], S_y[j_y], K[y], K[j_y]) = j_y + S_y[j_y] + K[j_y].
$$

Here,

$$
\begin{aligned}
h_0(S_0, K) &= u(0, 0, S_0[0], S_0[0], K[0], K[0]) \\
&= 0 + 0 + K[0] \\
&= K[0]
\end{aligned}
$$

and $\pi_0 = P(j_1 = h_0(S_0, K)) = 1$. For $y \geq 1$,

$$
\begin{aligned}
h_y(S_0, K) &= u(y, h_{y-1}(S_0, K), S_0[y], S_0[y-1], K[y], K[h_{y-1}(S_0, K)]) \\
&= h_{y-1}(S_0, K) + S_0[y-1] + K[h_{y-1}(S_0, K)] \\
&= h_{y-1}(S_0, K) + (y-1) + K[h_{y-1}(S_0, K)].
\end{aligned}
$$

From the analysis in the proof of Lemma 3.1.7, we see that in the recurrence of

h_y, $S_{y-1}[y-1]$ and j_y are respectively replaced by $S_0[y-1]$ and $h_{y-1}(S_0, K)$. Thus, we would have

$$\pi_y = \left(\frac{N-1}{N}\right)^{y-1} \cdot \pi_{y-1}.$$

Solving this recurrence, one can get

$$\pi_y = \prod_{x=1}^{y} \left(\frac{N-1}{N}\right)^{x-1}$$

$$= \left(\frac{N-1}{N}\right)^{\frac{y(y-1)}{2}}.$$

Example 3.1.11. *For one more example, suppose*

$$u(y, j_y, S_y[y], S_y[j_y], K[y], K[j_y]) = j_y + y \cdot S_y[j_y] + K[j_y].$$

As before,

$$
\begin{aligned}
h_0(S_0, K) &= u(0, 0, S_0[0], S_0[0], K[0], K[0]) \\
&= 0 + 0 \cdot S[0] + K[0] \\
&= 0 + 0 + K[0] \\
&= K[0]
\end{aligned}
$$

and $\pi_0 = P(j_1 = h_0(S_0, K)) = 1$. *For* $y \geq 1$,

$$
\begin{aligned}
h_y(S_0, K) &= u(y, h_{y-1}(S_0, K), S_0[y], S_0[y-1], K[y], K[h_{y-1}(S_0, K)]) \\
&= h_{y-1}(S_0, K)]) + y \cdot S_0[y-1] + K[h_{y-1}(S_0, K)] \\
&= h_{y-1}(S_0, K)]) + y \cdot (y-1) + K[h_{y-1}(S_0, K)].
\end{aligned}
$$

As in the previous example, here also the recurrence relation for the probabilities is

$$\pi_y = \left(\frac{N-1}{N}\right)^{y-1} \cdot \pi_{y-1},$$

whose solution is

$$\pi_y = \prod_{x=1}^{y} \left(\frac{N-1}{N}\right)^{x-1}$$

$$= \left(\frac{N-1}{N}\right)^{\frac{y(y-1)}{2}}.$$

The above discussion shows that the design of RC4 KSA cannot achieve further security by changing the update rule by any rule from a large class.

3.1.3   Biases of Nested Permutation Entries to Secret Key

We have, so far, explained the biases of the elements $S[y]$ toward the secret key. The next natural question to ask is whether there exists any bias
if we consider accessing the elements of S by more than one level of indexing. The work [103] established that the entries $S[S[y]]$ are indeed correlated to the secret key for different values of y. As the KSA proceeds, the
probabilities $P(S[y] = f_y)$ decrease monotonically, whereas the probabilities
$P(S[S[y]] = f_y)$ first increases monotonically until the middle of the KSA and
then decreases monotonically until the end of the KSA.

Let us first discuss how $P(S_r[S_r[1]] = f_1)$ varies with round $r, 1 \leq r \leq N$,
during the KSA of RC4. Refer to Figure 3.1 that demonstrates the nature of
the curve with an experimentation using 10 million randomly chosen secret
keys. The probability $P(S_r[S_r[1]] = f_1)$ increases till around $r = \frac{N}{2}$ where
it gets the maximum value around 0.185 and then it decreases to 0.136 at
$r = N$. On the other hand, as we saw in Theorem 3.1.5, $P(S_r[1] = f_1)$
decreases continuously as r increases during the KSA.

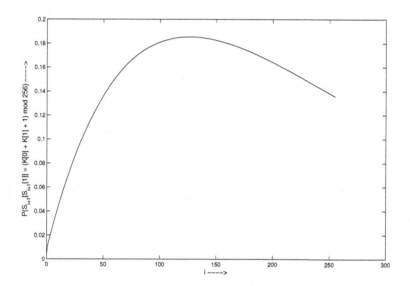

FIGURE 3.1: $P(S_{i+1}[S_{i+1}[1]] = f_1)$ versus i ($r = i + 1$) during RC4 KSA.

We first present theoretical analysis for the base case for $r = 2$, i.e., after
round 2 of the RC4 KSA.

Lemma 3.1.12. $P(S_2[S_2[1]] = K[0] + K[1] + 1) = \frac{3}{N} - \frac{4}{N^2} + \frac{2}{N^3}.$
Further, $P(S_2[S_2[1]] = K[0] + K[1] + 1 \wedge S_2[1] \leq 1) \approx \frac{2}{N}.$

Proof: The proof is based on three cases.

Case I: Let $K[0] \neq 0, K[1] = N - 1$. The probability of this event is $\frac{N-1}{N^2}$. Now

$$
\begin{aligned}
S_2[1] &= S_1[K[0] + K[1] + 1] \\
&= S_1[K[0]] \\
&= S_0[0] \\
&= 0.
\end{aligned}
$$

So,

$$
\begin{aligned}
S_2[S_2[1]] &= S_2[0] \\
&= S_1[0] \\
&= K[0] \\
&= K[0] + K[1] + 1.
\end{aligned}
$$

Note that $S_2[0] = S_1[0]$, as $K[0] + K[1] + 1 \neq 0$.

Moreover, in this case, $S_2[1] \leq 1$.

Case II: Let $K[0] + K[1] = 0, K[0] \neq 1$, i.e., $K[1] \neq N - 1$. The probability of this event is $\frac{N-1}{N^2}$. Now

$$
\begin{aligned}
S_2[1] &= S_1[K[0] + K[1] + 1] \\
&= S_1[1] \\
&= S_0[1] \\
&= 1.
\end{aligned}
$$

Note that $S_1[1] = S_0[1]$, as $K[0] \neq 1$. So,

$$
\begin{aligned}
S_2[S_2[1]] &= S_2[1] \\
&= 1 \\
&= K[0] + K[1] + 1.
\end{aligned}
$$

Also, in this case, $S_2[1] \leq 1$ holds.

Case III: $S_2[S_2[1]]$ could be $K[0] + K[1] + 1$ by random association except for the two previous cases.

Out of that, $S_2[1] \leq 1$ will happen in $\frac{2}{N}$ proportion of cases.

Thus

$$
\begin{aligned}
P(S_2[S_2[1]] = K[0] + K[1] + 1) &= \frac{2(N-1)}{N^2} + \left(1 - \frac{2(N-1)}{N^2}\right)\frac{1}{N} \\
&= \frac{3}{N} - \frac{4}{N^2} + \frac{2}{N^3}.
\end{aligned}
$$

Further,

$$
\begin{aligned}
P(S_2[S_2[1]] &= K[0] + K[1] + 1 \wedge S_2[1] \leq 1) \\
&= \frac{2(N-1)}{N^2} + \frac{2}{N}\left(1 - \frac{2(N-1)}{N^2}\right)\frac{1}{N} \\
&= \frac{2}{N} - \frac{4(N-1)}{N^4} \\
&\approx \frac{2}{N}.
\end{aligned}
$$

∎

Lemma 3.1.12 shows that after the second round ($i = 1, r = 2$), the event ($S_2[S_2[1]] = K[0] + K[1] + 1$) is not a random association.

Next, we present the inductive case. Let

$$
p_r = P(S_r[S_r[1]] = K[0] + K[1] + 1 \wedge S_r[1] \leq r - 1), \text{ for } r \geq 2.
$$

Lemma 3.1.13. *For $r \geq 3$, we have*

$$
p_r = \left(\frac{N-2}{N}\right)p_{r-1} + \frac{1}{N}\cdot\left(\frac{N-2}{N}\right)\cdot\left(\frac{N-1}{N}\right)^{2(r-2)}.
$$

Proof: After the $(r-1)$-th round is over,

$$
p_{r-1} = P(S_{r-1}[S_{r-1}[1]] = K[0] + K[1] + 1 \wedge S_{r-1}[1] \leq r - 2).
$$

The event

$$
\big((S_r[S_r[1]] = K[0] + K[1] + 1) \wedge (S_r[1] \leq r - 1)\big)
$$

can occur in two mutually exclusive and exhaustive ways:

Case I: $\big((S_r[S_r[1]] = K[0] + K[1] + 1) \wedge (S_r[1] \leq r - 2)\big)$, and

Case II: $\big((S_r[S_r[1]] = K[0] + K[1] + 1) \wedge (S_r[1] = r - 1)\big)$.

Let us calculate the contribution of each part separately.

In the r-th round, $i = r - 1 \notin \{0, \ldots, r - 2\}$. Thus, Case I occurs if we already had

Case I.a $\big((S_{r-1}[S_{r-1}[1]] = K[0] + K[1] + 1) \wedge (S_{r-1}[1] \leq r - 2)\big)$ and

Case I.b $j_r \notin \{1, r - 1\}$.

Hence, the contribution of this part is $p_{r-1}(\frac{N-2}{N})$.

Case II occurs if after the $(r-1)$-th round, we have the following.

$$
S_{r-1}[r-1] = r - 1, S_{r-1}[1] = K[0] + K[1] + 1 \text{ and } j_r = 1,
$$

causing a swap between the elements of S at the indices 1 and $r - 1$.

Case II.a We have $S_{r-1}[r-1] = r-1$, if the location $r-1$ is not touched during the rounds $i = 0, \ldots, r-2$. The probability of this is at least $(\frac{N-1}{N})^{r-1}$.

Case II.b The event $S_{r-1}[1] = K[0] + K[1] + 1$ can happen in the following way. In the first round (when $i = 0$), $j_1 \notin \{1, K[0] + K[1] + 1\}$ so that $S_1[1] = 1$ and $S_1[K[0] + K[1] + 1] = K[0] + K[1] + 1$ with probability $(\frac{N-2}{N})$. After this, in the second round (when $i = 1$), we will have $j_2 = j_1 + S_1[1] + K[1] = K[0] + K[1] + 1$, and so after the swap, $S_2[1] = K[0] + K[1] + 1$. Now, $K[0] + K[1] + 1$ remains in index 1 from the end of round 2 till the end of round $(r-1)$ (when $i = r-2$) with probability $(\frac{N-1}{N})^{r-3}$. Thus,

$$P(S_{r-1}[1] = K[0] + K[1] + 1) = \left(\frac{N-2}{N}\right) \cdot \left(\frac{N-1}{N}\right)^{r-3}.$$

Case II.c In the r-th round (when $i = r-1$), j_r becomes 1 with probability $\frac{1}{N}$.

Hence,

$$P\big((S_r[S_r[1]] = K[0] + K[1] + 1) \wedge (S_r[1] = r-1)\big)$$

$$= \left(\frac{N-1}{N}\right)^{r-1} \cdot \left(\frac{N-2}{N}\right) \cdot \left(\frac{N-1}{N}\right)^{r-3} \cdot \frac{1}{N}$$

$$= \frac{1}{N} \cdot \left(\frac{N-2}{N}\right) \cdot \left(\frac{N-1}{N}\right)^{2(r-2)}.$$

Adding the contributions of Cases I and II, we get

$$p_r = \left(\frac{N-2}{N}\right) p_{r-1} + \frac{1}{N} \cdot \left(\frac{N-2}{N}\right) \cdot \left(\frac{N-1}{N}\right)^{2(r-2)}.$$

■

The recurrence in Lemma 3.1.13 along with the base case in Lemma 3.1.12 completely specify the probabilities p_r for all $r \in [2, \ldots, N]$.

Theorem 3.1.14.

$$P(S_N[S_N[1]] = K[0] + K[1] + 1) \approx \left(\frac{N-1}{N}\right)^{2N}.$$

Proof: Approximating $\frac{N-2}{N}$ by $(\frac{N-1}{N})^2$, the recurrence in Lemma 3.1.13 can be rewritten as

$$p_r = a p_{r-1} + a^{r-1} b,$$

where $a = (\frac{N-1}{N})^2$ and $b = \frac{1}{N}$. The solution of this recurrence is given by

$$p_r = a^{r-2} p_2 + (r-2) a^{r-1} b, \qquad r \geq 2.$$

Substituting the values of p_2 (from Lemma 3.1.12), a and b, we get

$$p_r = \frac{2}{N}\left(\frac{N-1}{N}\right)^{2(r-2)} + \left(\frac{r-2}{N}\right)\left(\frac{N-1}{N}\right)^{2(r-1)}.$$

Putting $r = N$ and noting the fact that

$$P\big((S_N[S_N[1]] = K[0] + K[1] + 1) \wedge (S_N[1] \leq N-1)\big)$$
$$= P(S_N[S_N[1]] = K[0] + K[1] + 1),$$

we get

$$P(S_N[S_N[1]] = K[0] + K[1] + 1)$$
$$= \frac{2}{N}\left(\frac{N-1}{N}\right)^{2(N-2)} + \left(\frac{N-2}{N}\right)\left(\frac{N-1}{N}\right)^{2(N-1)}.$$

The second term (≈ 0.1348 for $N = 256$) dominates over the negligibly small first term (≈ 0.0011 for $N = 256$). So replacing $\frac{N-2}{N}$ by $(\frac{N-1}{N})^2$ in the second term, one can write

$$P(S_N[S_N[1]] = K[0] + K[1] + 1) \approx \left(\frac{N-1}{N}\right)^{2N}$$

∎

So far, we have explained the non-random association between $S_N[S_N[1]]$ and f_1. An obvious generalization of this is the association between $S_N[S_N[y]]$ and f_y, and moving further, the association between $S_N[S_N[S_N[y]]]$ and f_y, for $0 \leq y \leq N-1$ and so on. This gives new forms of biases compared to the previously studied biases of $S_N[y]$ toward f_y. Experimental observations show that these associations are not random (i.e., much more than $\frac{1}{N}$) for initial values of y (see Figure 3.2).

Next we present the theoretical analysis of the biases of $S_r[S_r[y]]$ toward f_y for small values of y. The results involved in the process are tedious, and one needs to approximate certain quantities to get the closed form formula.

Lemma 3.1.15. *For* $0 \leq y \leq 31$,

$$P\big((S_{y+1}[S_{y+1}[y]] = f_y) \wedge (S_{y+1}[y] \leq y)\big)$$
$$\approx \left(\frac{1}{N} \cdot \left(\frac{N-1}{N}\right)^{\frac{y(y+1)}{2}}\right) \cdot \left(y\left(\frac{N-2}{N}\right)^{y-1} + \left(\frac{N-1}{N}\right)^{y}\right).$$

Proof: $S_{y+1}[y] \leq y$ means that it can take $y+1$ many values $0, 1, \ldots, y$. Consider two cases:

Case I: $S_{y+1}[y] < y$ and

Case II: $S_{y+1}[y] = y$.

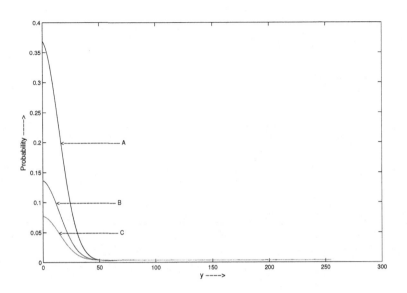

FIGURE 3.2:   A: $P(S_N[y] = f_y)$,   B: $P(S_N[S_N[y]] = f_y)$,   C: $P(S_N[S_N[S_N[y]]] = f_y)$ versus y $(0 \le y \le 255)$.

Suppose $S_{y+1}[y] = x$, $0 \le x \le y - 1$. Then $S_{y+1}[x]$ can be equal to f_y, if all of the following four events occur.

Case I.a From round 1 (when $i = 0$) to x (when $i = x - 1$), j does not touch the indices x and f_y. Thus, after round x, $S_x[x] = x$ and $S_x[f_y] = f_y$. This happens with probability $(\frac{N-2}{N})^x$.

Case I.b In round $x + 1$ (when $i = x$), j_{x+1} becomes equal to f_y, and after the swap, $S_{x+1}[x] = f_y$ and $S_{x+1}[f_y] = x$. The probability of this event is $P(j_{x+1} = f_y) = \frac{1}{N}$.

Case I.c From round $x + 2$ (when $i = x + 1$) to y (when $i = y - 1$), again j does not touch the indices x and f_y. Thus, after round y, $S_y[x] = f_y$ and $S_y[f_y] = x$. This occurs with probability $(\frac{N-2}{N})^{y-x-1}$.

Case I.d In round $y + 1$ (when $i = y$), j_{y+1} becomes equal to f_y, and after the swap,

$$
\begin{aligned}
S_{y+1}[y] &= S_y[f_y] \\
&= x
\end{aligned}
$$

and

$$
\begin{aligned}
S_{y+1}[S_{y+1}[y]] &= S_{y+1}[x] \\
&= S_y[x] \\
&= f_y.
\end{aligned}
$$

Following Lemma 3.1.3, this happens with probability

$$
P(E_y) = \left(\frac{N-1}{N}\right)^{1+\frac{y(y+1)}{2}} + \frac{1}{N}.
$$

For small values of y, this is approximately equal to

$$
P(A_y) = \left(\frac{N-1}{N}\right)^{\frac{y(y+1)}{2}},
$$

which gives a simpler expression (see the proof of Lemma 3.1.3 for the definitions of the events E_y and A_y). We consider $0 \le y \le 31$ for good approximation.

Considering the events I.a, I.b, I.c and I.d to be independent, we have

$$
\begin{aligned}
&P\big((S_{y+1}[S_{y+1}[y]] = f_y) \wedge (S_{y+1}[y] = x)\big) \\
&= \left(\frac{N-2}{N}\right)^x \cdot \frac{1}{N} \cdot \left(\frac{N-2}{N}\right)^{y-x-1} \cdot \left(\frac{N-1}{N}\right)^{\frac{y(y+1)}{2}} \\
&= \left(\frac{1}{N}\right) \cdot \left(\frac{N-2}{N}\right)^{y-1} \cdot \left(\frac{N-1}{N}\right)^{\frac{y(y+1)}{2}}.
\end{aligned}
$$

Summing for all x in $[0, \ldots, y-1]$, we have

$$
\begin{aligned}
&P\big((S_{y+1}[S_{y+1}[y]] = f_y) \wedge (S_{y+1}[y] \le y-1)\big) \\
&= \left(\frac{y}{N}\right) \cdot \left(\frac{N-2}{N}\right)^{y-1} \cdot \left(\frac{N-1}{N}\right)^{\frac{y(y+1)}{2}}.
\end{aligned}
$$

Next, suppose $S_{y+1}[y] = y$. Now $S_{y+1}[S_{y+1}[y]]$ can be equal to f_y, if all of the following three events occur.

Case II.a f_y has to be equal to y. This happens with probability $\frac{1}{N}$.

Case II.b Index y is not touched by j in any of the first y rounds. This happens with probability $\left(\frac{N-1}{N}\right)^y$.

Case II.c In the $(y+1)$-th round, $j_{y+1} = f_y$ so that there is no swap. This happens approximately with probability $\left(\frac{N-1}{N}\right)^{\frac{y(y+1)}{2}}$ (as explained in Case I.d above).

Hence,

$$P\big((S_{y+1}[S_{y+1}[y]] = f_y) \wedge (S_{y+1}[y] = y)\big)$$
$$= \left(\frac{1}{N}\right) \cdot \left(\frac{N-1}{N}\right)^{y} \cdot \left(\frac{N-1}{N}\right)^{\frac{y(y+1)}{2}}.$$

Adding the contributions of Case I ($0 \leq S_{y+1}[y] \leq y-1$) and Case II ($S_{y+1}[y] = y$), we get

$$P\big((S_{y+1}[S_{y+1}[y]] = f_y) \wedge (S_{y+1}[y] \leq y)\big)$$
$$= \left(\frac{1}{N} \cdot \left(\frac{N-1}{N}\right)^{\frac{y(y+1)}{2}}\right) \cdot \left(y\left(\frac{N-2}{N}\right)^{y-1} + \left(\frac{N-1}{N}\right)^{y}\right).$$

■

For $0 \leq y \leq N-1$, $1 \leq r \leq N$, let

$$q_r(y) = P\left((S_r[S_r[y]] = f_y) \wedge (S_r[y] \leq r-1)\right).$$

Lemma 3.1.16. *For $0 \leq y \leq 31$, $y + 2 \leq r \leq N$, we have*

$$q_r(y) = \left(\frac{N-2}{N}\right) q_{r-1}(y) + \frac{1}{N} \cdot \left(\frac{N-y}{N}\right) \cdot \left(\frac{N-1}{N}\right)^{\frac{y(y+1)}{2}+2r-3}.$$

Proof: For $r \geq y + 2$, the event

$$\left((S_r[S_r[y]] = f_y) \wedge (S_r[y] \leq r-1)\right)$$

can occur in two mutually exclusive and exhaustive ways:

Case I: $\left((S_r[S_r[y]] = f_y) \wedge (S_r[y] \leq r-2)\right)$ and

Case II: $\left((S_r[S_r[y]] = f_y) \wedge (S_r[y] = r-1)\right).$

Let us now calculate the contribution of each one separately.

In the r-th round, $i = r - 1 \notin \{0, \ldots, r-2\}$. Thus, Case I occurs if we already had

Case I.a $\left((S_{r-1}[S_{r-1}[y]] = f_y) \wedge (S_{r-1}[y] \leq r-2)\right)$ and

Case I.b $j_r \notin \{y, S_{r-1}[y]\}$.

Hence, the contribution of this part is $q_{r-1}(y) \cdot (\frac{N-2}{N})$.

Case II occurs if after the $(r-1)$-th round, $S_{r-1}[r-1] = r-1$, $S_{r-1}[y] = f_y$ and in the r-th round (i.e., when $i = r-1$), $j_r = y$ causing a swap between the elements at the indices y and $r - 1$. The probabilities for these three events are as follows.

Case II.a We would have $S_{r-1}[r-1] = r-1$, if the index $r-1$ is not touched during the rounds $i = 0, \ldots, r-2$. This happens with probability $(\frac{N-1}{N})^{r-1}$.

Case II.b Following Theorem 3.1.5, The event $S_{r-1}[y] = f_y$ happens with a probability $(\frac{N-y}{N}) \cdot (\frac{N-1}{N})^{[\frac{y(y+1)}{2}+r-1]} + \frac{1}{N}$. For small values of y, this is approximately equal to $(\frac{N-y}{N}) \cdot (\frac{N-1}{N})^{[\frac{y(y+1)}{2}+r-2]}$ (obtained by considering the event A_y only in the proof of Lemma 3.1.3), which gives a simpler expression. For good approximation, we restrict y in $[0, 31]$.

Case II.c In the r-th round (when $i = r-1$), j_r becomes y with probability $\frac{1}{N}$.

Thus,

$$P\big((S_r[S_r[y]] = f_y) \wedge (S_r[y] = r-1)\big)$$

$$= \left(\frac{N-1}{N}\right)^{r-1} \cdot \left(\frac{N-y}{N}\right) \left(\frac{N-1}{N}\right)^{[\frac{y(y+1)}{2}+r-2]} \cdot \frac{1}{N}$$

$$= \frac{1}{N} \cdot \left(\frac{N-y}{N}\right) \cdot \left(\frac{N-1}{N}\right)^{\frac{y(y+1)}{2}+2r-3}.$$

Adding the contributions of Cases I and II, we get

$$q_r(y) = \left(\frac{N-2}{N}\right) q_{r-1}(y) + \frac{1}{N} \cdot \left(\frac{N-y}{N}\right) \cdot \left(\frac{N-1}{N}\right)^{\frac{y(y+1)}{2}+2r-3}.$$

∎

The recurrence in Lemma 3.1.16 and the base case in Lemma 3.1.15 completely specify the probabilities $q_r(y)$ for all y in $[0, \ldots, 31]$ and r in $[y+1, \ldots, N]$.

Now let us see what happens after the complete KSA.

Theorem 3.1.17. *For* $0 \le y \le 31$,

$$P(S_N[S_N[y]] = f_y) \approx \frac{y}{N} \cdot \left(\frac{N-1}{N}\right)^{\frac{y(y+1)}{2}+2(N-2)}$$

$$+ \frac{1}{N} \cdot \left(\frac{N-1}{N}\right)^{\frac{y(y+1)}{2}-y+2(N-1)}$$

$$+ \left(\frac{N-y-1}{N}\right) \cdot \left(\frac{N-y}{N}\right) \cdot \left(\frac{N-1}{N}\right)^{\frac{y(y+1)}{2}+2N-3}$$

Proof: Approximating $(\frac{N-2}{N})$ by $(\frac{N-1}{N})^2$, the recurrence in Lemma 3.1.16 can be rewritten as

$$q_r(y) = \left(\frac{N-1}{N}\right)^2 q_{r-1}(y) + \frac{1}{N} \left(\frac{N-y}{N}\right) \cdot \left(\frac{N-1}{N}\right)^{\frac{y(y+1)}{2}+2r-3},$$

i.e.,

$$q_r(y) = aq_{r-1}(y) + a^{r-1}b, \tag{3.2}$$

where

$$a = \left(\frac{N-1}{N}\right)^2 \text{ and } b = \frac{1}{N}\left(\frac{N-y}{N}\right) \cdot \left(\frac{N-1}{N}\right)^{\frac{y(y+1)}{2}-1}.$$

The solution to the recurrence in (3.2) is

$$q_r(y) = a^{r-y-1}q_{y+1}(y) + (r-y-1)a^{r-1}b, \qquad r \geq y+1.$$

Substituting the values of $q_{y+1}(y)$ (from Lemma 3.1.15), a and b, one can obtain

$$
\begin{aligned}
q_r(y) \;=\;& \frac{y}{N} \cdot \left(\frac{N-1}{N}\right)^{\frac{y(y+1)}{2}+2(r-2)} \\[2mm]
&+ \frac{1}{N} \cdot \left(\frac{N-1}{N}\right)^{\frac{y(y+1)}{2}-y+2(r-1)} \\[2mm]
&+ \left(\frac{r-y-1}{N}\right) \cdot \left(\frac{N-y}{N}\right) \cdot \left(\frac{N-1}{N}\right)^{\frac{y(y+1)}{2}+2r-3},
\end{aligned}
$$

for $y+1 \leq r \leq N$ and for initial values of y ($0 \leq y \leq 31$). Substituting $r = N$ and noting the fact that

$$P\big((S_N[S_N[y]] = f_y) \wedge (S_N[y] \leq N-1)\big) = P(S_N[S_N[y]] = f_y),$$

we get the result. ∎

The theoretical formula matches closely with the experimental results for $0 \leq y \leq 31$. Item 4 in the proof of Lemma 3.1.15 connects the bias of $S_r[y]$ toward f_y (empirically observed by Roos [146] and theoretically analyzed in Section 3.1.1 of this chapter) with that of $S_r[S_r[y]]$ toward f_y.

3.2   Non-Randomness of Permutation

RC4 KSA attempts to randomize the permutation S which is initialized to an identity permutation. How to generate a random permutation is a well-studied problem. Let us present one well known result from [88]. Given a permutation of n elements

$$\Pi = (\pi_0, \pi_1, \ldots, \pi_{n-1}),$$

a series of $n-1$ transpositions can produce a random permutation as follows:

Here, $rand(0, i)$ produces uniformly distributed random integers in the range 0 through i. We reproduce the following result from [143, Page 171].

Input: Any permutation $\Pi = (\pi_0, \pi_1, \ldots, \pi_{n-1})$.
Output: Random permutation $\Pi = (\pi_0, \pi_1, \ldots, \pi_{n-1})$.

for $i = n - 1$ *down to* 1 **do**
 | $\pi_i \leftrightarrow \pi_{rand(0,i)}$;
end

Algorithm 3.2.1: RandPerm

Theorem 3.2.1. *Algorithm 3.2.1 produces a random permutation.*

Proof: Use induction on n. For $n = 1$, the result is obvious. Suppose that the result holds for $n = k - 1$, i.e., each of the $(k-1)!$ factorial permutations are equally likely. Let

$$\Sigma = (\sigma_0, \sigma_1, \ldots, \sigma_{k-1})$$

be any permutation of $\{0, 1, \ldots, k-1\}$. We have

$$
P(\Pi = \Sigma) = P(\pi_{k-1} = \sigma_{k-1}) \cdot
$$
$$
P\Big(\big((\pi_0, \ldots, \pi_{k-2}) = (\sigma_0, \ldots, \sigma_{k-2})\big) \mid (\pi_{k-1} = \sigma_{k-1})\Big).
$$

By construction, the first probability is $\frac{1}{k}$ and by inductive hypothesis, the second probability is $\frac{1}{(k-1)!}$, giving

$$P(\Pi = \Sigma) = \frac{1}{k!}.$$

∎

The RC4 KSA and the *RandPerm* algorithm are not similar, on account of the following issues [110, Chapter 6].

1. The index i increases from 0 to $N-1$ in the RC4 KSA (number of transpositions is N), but the index i decreases from $N-1$ to 1 in *RandPerm* (number of transpositions is $N-1$).

2. The index j can take any value between 0 to $N-1$ at any step and the pseudo-random value of j is based on the secret key bytes in the KSA. On the other hand, in *RandPerm*, $j = rand(0, i)$, that implies that it can take values only in the range 0 to i at the i-th step.

In [110, Chapter 6], it has also been pointed out that the RC4 KSA does not produce permutations uniformly at random. On the other hand, the *RandPerm* algorithm cannot be used for key scheduling using the secret key bytes instead of $rand(0, i)$. In that case, the permutation after the key scheduling algorithm would have revealed the secret key completely (e.g., π_{N-1} will contain the first accessed secret key byte, π_{N-2} will contain the second accessed secret key byte, and so on). Thus, it is necessary to design a key scheduling algorithm that will produce a random looking permutation with certain cryptographic properties.

3.2.1   Biased Sign of RC4 Permutation

The sign of a permutation $\Pi = (\pi_0, \pi_1, \ldots, \pi_{n-1})$ is defined as

$$sgn(\Pi) = (-1)^{|I(\Pi)|},$$

where

$$I(\Pi) = \{(u, v) : u < v \ \& \ \pi_u > \pi_v\}$$

is the set of *inversions*. Alternatively, if Π can be decomposed into a product of m transpositions, then the sign is given by

$$sgn(\Pi) = (-1)^m = \begin{cases} +1, & \text{if } m \text{ is even;} \\ -1, & \text{if } m \text{ is odd.} \end{cases}$$

Although such a decomposition is not unique, the parity (i.e., oddness or evenness) of the number of transpositions in all decompositions of a permutation is the same, implying that the sign of a permutation is unique.

In [122, Section 4], the theory of random shuffles has been used to establish the following result on the sign of the RC4 permutation after the KSA is over.

Theorem 3.2.2. *Assume that the index j takes its value from $\mathbb{Z}_N$ independently and uniformly at random at each round of the KSA. Then the sign of the permutation after the KSA can take the two possible values $(-1)^N$ and $(-1)^{N-1}$ with the following probabilities.*

$$P\left(sgn(S_N) = (-1)^N\right) \ \approx \ \frac{1}{2}(1 + e^{-2})$$

$$and \quad P\left(sgn(S_N) = (-1)^{N-1}\right) \ \approx \ \frac{1}{2}(1 - e^{-2}).$$

Proof: The RC4 KSA begins with an identity permutation, i.e., $sgn(S_0) = +1$. At any step, if $i = j$, the probability of which is $\frac{1}{N}$, there is no change of sign. On the other hand, if $i \neq j$, which happens with a probability of $1 - \frac{1}{N}$, there is a sign change. Thus, the probability that the sign changes exactly k times and does not change the other $N - k$ times is given by $\binom{N}{k}(1 - \frac{1}{N})^k \frac{1}{N^{N-k}}$. Hence,

$$
\begin{aligned}
P\left(sgn(S_N) = (-1)^N\right) &= \sum_{k=N,N-2,N-4,\ldots} \binom{N}{k}\left(1 - \frac{1}{N}\right)^k \frac{1}{N^{N-k}} \\
&\approx e^{-1}\left(1 + \frac{1}{2!} + \frac{1}{4!} + \cdots\right) \\
&= \frac{1}{2}(1 + e^{-2}).
\end{aligned}
$$

In the above derivation, it is assumed that N is sufficiently large. The other probability is given by $1 - P\left(sgn(S_N) = (-1)^N\right)$.   ∎

The implication of this result is that one can predict the sign of the permutation after the key scheduling with an advantage of $\frac{1}{2}e^{-2} \approx 6.7\%$ over random guess.

3.2.2 Bias in Each Permutation Byte

If the permutation after the KSA is perfectly random, then $P(S_N[u] = v)$ should be $\frac{1}{N}$ for $0 \le u, v \le N - 1$. However, in [110, Chapter 6 and Appendix C] it has been shown that this bivariate distribution is not uniform.

Theorem 3.2.3. *At the end of KSA, for $0 \le u \le N - 1$, $0 \le v \le N - 1$,*

$$P(S_N[u] = v) = \begin{cases} \frac{1}{N}\left(\left(\frac{N-1}{N}\right)^v + \left(1 - \left(\frac{N-1}{N}\right)^v\right)\left(\frac{N-1}{N}\right)^{N-u-1}\right) & \text{if } v \le u; \\ \frac{1}{N}\left(\left(\frac{N-1}{N}\right)^{N-u-1} + \left(\frac{N-1}{N}\right)^v\right) & \text{if } v > u. \end{cases}$$

Later, the same distribution has been independently studied in [122] and [135]. Here, we present the proof approach of [135].

In Figure 3.3, we present a few graphs for $P(S_N[u] = v)$ versus v, $0 \le v \le N - 1$, for a few values of u as motivating examples. These graphs are plotted using 10 million trials over randomly chosen keys of 32 bytes. It is clear that the probabilities $P(S_N[u] = v)$ are not uniform.

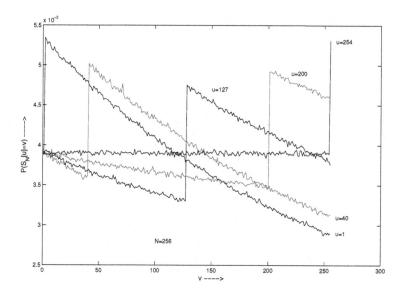

FIGURE 3.3: $P(S_N[u] = v)$ versus v for some specific u's.

Below we present the proof technique of [135] for estimating the probabilities $P(S_N[u] = v)$. It turns out that each permutation byte after the KSA is significantly biased (either positive or negative) toward many values in the range $0, \dots, N - 1$. These biases are independent of the secret key and thus present an evidence that the permutation after the KSA can be distinguished from random permutation without any assumption on the secret key.

Due to the small keys (say 5 to 32 bytes) generally used in RC4, some of the assumptions differ from practice and hence the theoretical formulae do not match with the experimental results for some values of u and v. We also discuss this issue later.

Lemma 3.2.4. $P(S_2[0] = 1) = \frac{2(N-1)}{N^2}$.

Proof: In the first round, we have $i = 0$, and

$$
\begin{aligned}
j_1 &= 0 + S[0] + K[0] \\
    &= K[0].
\end{aligned}
$$

In the second round, $i = 1$ and

$$j_2 = j_1 + S_1[1] + K[1].$$

Consider two mutually exclusive and exhaustive cases, namely, $K[0] = 1$ and $K[0] \neq 1$.

Case I: Take $K[0] = 1$. So, after the first swap, $S_1[0] = 1$ and $S_1[1] = 0$. Now,

$$
\begin{aligned}
j_2 &= K[0] + 0 + K[1] \\
    &= K[0] + K[1].
\end{aligned}
$$

After the second swap, $S_2[0]$ will remain 1, if $K[0] + K[1] \neq 0$. Hence the contribution of this case to the event $(S_2[0] = 1)$ is

$$
\begin{aligned}
P(K[0] = 1) \cdot P(K[0] + K[1] \neq 0) &= \frac{1}{N} \cdot \frac{N-1}{N} \\
&= \frac{N-1}{N^2}.
\end{aligned}
$$

Case II: Now consider $K[0] \neq 1$. After the first swap, $S_1[1]$ remains 1. Now,

$$
\begin{aligned}
j_2 &= K[0] + 1 + K[1] \\
    &= K[0] + K[1] + 1.
\end{aligned}
$$

Thus, after the second swap, $S_2[0]$ will obtain the value 1, if $K[0]+K[1]+1 = 0$. Hence the contribution of this case to the event $(S_2[0] = 1)$ is

$$
\begin{aligned}
P(K[0] \neq 1) \cdot P(K[0] + K[1] + 1 = 0) &= \frac{N-1}{N} \cdot \frac{1}{N} \\
&= \frac{N-1}{N^2}.
\end{aligned}
$$

Adding the two contributions, we get the total probability as $\frac{2(N-1)}{N^2}$. ∎

Here $P(S_{v+1}[u] = v)$ is calculated for the special case $u = 0$, $v = 1$. Note that the form of $P(S_{v+1}[u] = v)$ for $v \geq u + 1$ in general (see Lemma 3.2.7 later) does not work for the case $u = 0, v = 1$ only. This will be explained more clearly in Remark 3.2.8.

Proposition 3.2.5. *For $v \geq 0$, $P(S_v[v] = v) = \left(\frac{N-1}{N}\right)^v$.*

Proof: In the rounds 1 through v, the deterministic index i touches the permutation indices $0, 1, \ldots, v-1$. Thus, after round v, $S_v[v]$ will remain the same as $S_0[v] = v$, if v has not been equal to any of the v many pseudo-random indices $j_1, j_2, \ldots, j_v$. The probability of this event is $\left(\frac{N-1}{N}\right)^v$. So the result holds for $v \geq 1$. Furthermore,

$$P(S_0[0] = 0) = 1 = \left(\frac{N-1}{N}\right)^0.$$

Hence, for any $v \geq 0$, we have

$$P(S_v[v] = v) = \left(\frac{N-1}{N}\right)^v.$$

∎

Proposition 3.2.6. *For $v \geq u + 1$,*

$$P(S_v[u] = v) = \frac{1}{N} \cdot \left(\frac{N-1}{N}\right)^{v-u-1}.$$

Proof: In round $u + 1$, the permutation index u is touched by the deterministic index i for the first time and the value at index u is swapped with the value at a random location based on j_{u+1}. Hence, $P(S_{u+1}[u] = v) = \frac{1}{N}$.

The probability that the index u is not touched by any of the subsequent $v - u - 1$ many j values, namely, $j_{u+2}, \ldots, j_v$, is given by $\left(\frac{N-1}{N}\right)^{v-u-1}$. So, after the end of round v,

$$P(S_v[u] = v) = \frac{1}{N} \cdot \left(\frac{N-1}{N}\right)^{v-u-1}.$$

∎

In Proposition 3.2.6, we have considered that the value v occupies the location u of the permutation just before the deterministic index i becomes v. In the following lemma, we analyze the same event (i.e., v occupying location u) after the swap in the next round when i has already touched the location v.

Lemma 3.2.7. *For $v \geq u + 1$, except for the case "$u = 0$ and $v = 1$,"*

$$P(S_{v+1}[u] = v) = \frac{1}{N} \cdot \left(\frac{N-1}{N}\right)^{v-u} + \frac{1}{N} \cdot \left(\frac{N-1}{N}\right)^v - \frac{1}{N^2} \cdot \left(\frac{N-1}{N}\right)^{2v-u-1}.$$

Proof: In round $v+1$, $i = v$ and $j_{v+1} = j_v + S_v[v] + K[v]$. The event $(S_{v+1}[u] = v)$ can occur in two ways.

1. $S_v[u]$ already had the value v and the index u is not involved in the swap in round $v+1$.

2. $S_v[u] \neq v$ and the value v comes into the index u from the index v (i.e., $S_v[v] = v$) by the swap in round $v+1$.

From Proposition 3.2.5, we have

$$P(S_v[v] = v) = \left(\frac{N-1}{N}\right)^v$$

and also from Proposition 3.2.6, we get

$$P(S_v[u] = v) = \frac{1}{N} \cdot \left(\frac{N-1}{N}\right)^{v-u-1}.$$

Hence, except for the case "$u = 0$ and $v = 1$" (see Remark 3.2.8),

$$
\begin{aligned}
P(S_{v+1}[u] = v) &= P(S_v[u] = v) \cdot P(j_v + S_v[v] + K[v] \neq u) \\
&\quad + P(S_v[u] \neq v) \cdot P(S_v[v] = v) \cdot P(j_v + S_v[v] + K[v] = u) \\
&= \left(\frac{1}{N} \cdot \left(\frac{N-1}{N}\right)^{v-u-1}\right) \cdot \left(\frac{N-1}{N}\right) \\
&\quad + \left(1 - \frac{1}{N} \cdot \left(\frac{N-1}{N}\right)^{v-u-1}\right) \cdot \left(\frac{N-1}{N}\right)^v \cdot \frac{1}{N} \\
&= \frac{1}{N} \cdot \left(\frac{N-1}{N}\right)^{v-u} + \frac{1}{N} \cdot \left(\frac{N-1}{N}\right)^v \\
&\quad - \frac{1}{N^2} \cdot \left(\frac{N-1}{N}\right)^{2v-u-1}.
\end{aligned}
$$

∎

Remark 3.2.8. *Case 1 in the proof of Lemma 3.2.7 applies to Lemma 3.2.4 also. In case 2, i.e., when $S_v[u] \neq v$, in general we may or may not have $S_v[v] = v$. However, for $u = 0$ and $v = 1$,*

$$(S_1[0] \neq 1) \iff (S_1[1] = 1),$$

the probability of each of which is $\frac{N-1}{N}$ (note that there has been only one swap involving the indices 0 and $K[0]$ in round 1). Hence the contribution of case 2 except for "$u = 0$ and $v = 1$" would be

$$P(S_v[u] \neq v) \cdot P(S_v[v] = v) \cdot P(j_v + S_v[v] + K[v] = u),$$

and for "$u = 0$ and $v = 1$" it would be

$$P(S_1[0] \neq 1) \cdot P(j_1 + S_1[1] + K[1] = 0)$$

or, equivalently,

$$P(S_1[1] = 1) \cdot P(j_1 + S_1[1] + K[1] = 0).$$

Let $p_r^{u,v}$ denote the probability $P(S_r[u] = v)$, for $1 \leq r \leq N$. The following lemma connects the probabilities between rounds τ and $r \geq \tau$.

Lemma 3.2.9. *Suppose, $p_\tau^{u,v}$ is given for any intermediate round τ, $max\{u, v\} < \tau \leq N$. Then, for $\tau \leq r \leq N$, we have*

$$p_r^{u,v} = p_\tau^{u,v} \cdot \left(\frac{N-1}{N}\right)^{r-\tau} + (1 - p_\tau^{u,v}) \cdot \frac{1}{N} \left(\frac{N-1}{N}\right)^v \cdot \left(1 - \left(\frac{N-1}{N}\right)^{r-\tau}\right).$$

Proof: After round τ ($> max\{u, v\}$), there may be two different cases: $S_\tau[u] = v$ and $S_\tau[u] \neq v$. Both of these can contribute to the event $(S_r[u] = v)$ in the following ways.

Case I: $S_\tau[u] = v$ and the index u is not touched by any of the subsequent $r - \tau$ many j values. The contribution of this part is

$$P(S_\tau[u] = v) \cdot \left(\frac{N-1}{N}\right)^{r-\tau} = p_\tau^{u,v} \cdot \left(\frac{N-1}{N}\right)^{r-\tau}.$$

Case II: $S_\tau[u] \neq v$ and for some x in the interval $[\tau, r-1]$, $S_x[x] = v$ which comes into the index u from the index x by the swap in round $x+1$, and after that the index u is not touched by any of the subsequent $r - 1 - x$ many j values. So the contribution of the second part is given by

$$P(S_\tau[u] \neq v) \cdot \left(\sum_{x=\tau}^{r-1} P(S_x[x] = v) \cdot P(j_{x+1} = u) \cdot \left(\frac{N-1}{N}\right)^{r-1-x}\right).$$

Suppose, the value v remains in location v after round v. By Proposition 3.2.5, this probability, i.e., $P(S_v[v] = v)$, is $\left(\frac{N-1}{N}\right)^v$. The swap in the next round moves the value v to a random location $x = j_{v+1}$. Thus,

$$\begin{aligned}
P(S_{v+1}[x] = v) &= P(S_v[v] = v) \cdot P(j_{v+1} = x) \\
&= \left(\frac{N-1}{N}\right)^v \cdot \frac{1}{N}.
\end{aligned}$$

For all $x > v$, until x is touched by the deterministic index i, i.e., until round $x + 1$, v will remain randomly distributed. Hence, for all $x > v$,

$$P(S_x[x] = v) = P(S_{v+1}[x] = v) = \frac{1}{N} \left(\frac{N-1}{N}\right)^v \quad \text{and}$$

$$P(S_\tau[u] \neq v) \cdot \left(\sum_{x=\tau}^{r-1} P(S_x[x] = v) \cdot P(j_{x+1} = u) \cdot \left(\frac{N-1}{N}\right)^{r-1-x} \right)$$

$$= (1 - p_\tau^{u,v}) \cdot \left(\sum_{x=\tau}^{r-1} \frac{1}{N} \left(\frac{N-1}{N}\right)^v \cdot \frac{1}{N} \cdot \left(\frac{N-1}{N}\right)^{r-1-x} \right)$$

$$= (1 - p_\tau^{u,v}) \cdot \frac{1}{N^2} \left(\frac{N-1}{N}\right)^v \cdot \left(\sum_{x=\tau}^{r-1} \left(\frac{N-1}{N}\right)^{r-1-x} \right)$$

$$= (1 - p_\tau^{u,v}) \cdot \frac{1}{N^2} \left(\frac{N-1}{N}\right)^v \cdot \left(\frac{1 - a^{r-\tau}}{1 - a} \right),$$

where $a = \frac{N-1}{N}$. Substituting the value of a and simplifying, we get the above probability as $(1 - p_\tau^{u,v}) \cdot \frac{1}{N} (\frac{N-1}{N})^v \cdot \left(1 - (\frac{N-1}{N})^{r-\tau} \right)$.

Combining the contributions of Cases I and II, we get

$$p_r^{u,v} = p_\tau^{u,v} \cdot \left(\frac{N-1}{N}\right)^{r-\tau} + (1 - p_\tau^{u,v}) \cdot \frac{1}{N} \left(\frac{N-1}{N}\right)^v \cdot \left(1 - \left(\frac{N-1}{N}\right)^{r-\tau} \right).$$

∎

Corollary 3.2.10. *Suppose, $p_\tau^{u,v}$ is given for any intermediate round τ, $max\{u, v\} < \tau \leq N$. Then, $P(S_N[u] = v)$, i.e., the probability after the complete KSA is given by*

$$p_\tau^{u,v} \cdot \left(\frac{N-1}{N}\right)^{N-\tau} + (1 - p_\tau^{u,v}) \cdot \frac{1}{N} \left(\frac{N-1}{N}\right)^v \cdot \left(1 - \left(\frac{N-1}{N}\right)^{N-\tau} \right).$$

Proof: Substitute $r = N$ in Lemma 3.2.9. ∎

Theorem 3.2.11.
(1) *For $0 \leq u \leq N - 2$, $u + 1 \leq v \leq N - 1$,*

$$P(S_N[u] = v) = p_{v+1}^{u,v} \cdot \left(\frac{N-1}{N}\right)^{N-1-v}$$

$$+ (1 - p_{v+1}^{u,v}) \cdot \frac{1}{N} \cdot \left(\left(\frac{N-1}{N}\right)^v - \left(\frac{N-1}{N}\right)^{N-1} \right),$$

where

$$p_{v+1}^{u,v} = \begin{cases} \frac{2(N-1)}{N^2} & \text{if } u = 0 \ \& \ v = 1; \\ \frac{1}{N} \cdot (\frac{N-1}{N})^{v-u} + \frac{1}{N} \cdot (\frac{N-1}{N})^v - \frac{1}{N^2} \cdot (\frac{N-1}{N})^{2v-u-1} & \text{otherwise.} \end{cases}$$

(2) *For $0 \leq v \leq N - 1$, $v \leq u \leq N - 1$,*

$$P(S_N[u] = v) = \frac{1}{N} \cdot \left(\frac{N-1}{N}\right)^{N-1-u} + \frac{1}{N} \cdot \left(\frac{N-1}{N}\right)^{v+1}$$

$$- \frac{1}{N} \cdot \left(\frac{N-1}{N}\right)^{N+v-u}.$$

Proof: First we prove item (1). Since $v > u$, so for any $\tau > v$, we will have $\tau > max\{u, v\}$. Substituting $\tau = v + 1$ in Corollary 3.2.10, we have

$$
\begin{aligned}
P(S_N[u] = v) &= p_{v+1}^{u,v} \cdot \left(\frac{N-1}{N}\right)^{N-1-v} \\
&\quad + (1 - p_{v+1}^{u,v}) \cdot \frac{1}{N} \left(\frac{N-1}{N}\right)^{v} \cdot \left(1 - \left(\frac{N-1}{N}\right)^{N-1-v}\right) \\
&= p_{v+1}^{u,v} \cdot \left(\frac{N-1}{N}\right)^{N-1-v} \\
&\quad + (1 - p_{v+1}^{u,v}) \cdot \frac{1}{N} \cdot \left(\left(\frac{N-1}{N}\right)^{v} - \left(\frac{N-1}{N}\right)^{N-1}\right).
\end{aligned}
$$

Now, from Lemma 3.2.7, we get

$$
p_{v+1}^{u,v} = \frac{1}{N} \cdot \left(\frac{N-1}{N}\right)^{v-u} + \frac{1}{N} \cdot \left(\frac{N-1}{N}\right)^{v} - \frac{1}{N^2} \cdot \left(\frac{N-1}{N}\right)^{2v-u-1},
$$

except for "$u = 0$ and $v = 1$". Also, Lemma 3.2.4 gives

$$
p_2^{0,1} = \frac{2(N-1)}{N^2}.
$$

Substituting these values of $p_{v+1}^{u,v}$, one can get the result.

Now let us prove item (2). Here we have $u \geq v$. So for any $\tau > u$, we will have $\tau > max\{u, v\}$. Substituting $\tau = u + 1$ in Corollary 3.2.10, we have

$$
\begin{aligned}
P(S_N[u] = v) &= p_{u+1}^{u,v} \cdot \left(\frac{N-1}{N}\right)^{N-1-u} \\
&\quad + (1 - p_{u+1}^{u,v}) \cdot \frac{1}{N} \left(\frac{N-1}{N}\right)^{v} \cdot \left(1 - \left(\frac{N-1}{N}\right)^{N-1-u}\right).
\end{aligned}
$$

As $p_{u+1}^{u,v} = P(S_{u+1}[u] = v) = \frac{1}{N}$ (see the proof of Proposition 3.2.6), substituting this in the above expression, we get

$$
\begin{aligned}
P(S_N[u] = v) &= \frac{1}{N} \cdot \left(\frac{N-1}{N}\right)^{N-1-u} \\
&\quad + \left(1 - \frac{1}{N}\right) \cdot \frac{1}{N} \left(\frac{N-1}{N}\right)^{v} \cdot \left(1 - \left(\frac{N-1}{N}\right)^{N-1-u}\right) \\
&= \frac{1}{N} \cdot \left(\frac{N-1}{N}\right)^{N-1-u} + \frac{1}{N} \cdot \left(\frac{N-1}{N}\right)^{v+1} \\
&\quad - \frac{1}{N} \cdot \left(\frac{N-1}{N}\right)^{N+v-u}.
\end{aligned}
$$

■

The final formulae in Theorem 3.2.11 are very close to the results presented in [110] apart from some minor differences as terms with N^2 in the denominator or a difference in 1 in the power. For $N = 256$, the maximum absolute difference between these results and the results of [110] is 0.000025 as well as the average of absolute differences is 0.000005.

However, the proof approach of [135] is different from that of [110]. In [110], the idea of relative positions is introduced. If the current deterministic index is i, then relative position a means the position $(i+1+a) \bmod N$. The transfer function $T(a, b, r)$, which represents the probability that value in relative position a in S will reach relative position b in the permutation generated from S by executing r RC4 rounds, has the following explicit form by [110, Claim C.3.3]:

$$T(a, b, r) = \begin{cases} p(q^a + q^{r-(b+1)} - q^{a+r-(b+1)}) & \text{if } a \le b; \\ p(q^a + q^{r-(b+1)}) & \text{if } a > b, \end{cases}$$

where $p = \frac{1}{N}$ and $q = (\frac{N-1}{N})$. This solution is obtained by solving a recurrence [110, Equation C.3.1] which expresses $T(a, b, r)$ in terms of $T(a - 1, b - 1, r - 1)$. Instead, the proof idea here uses the probabilities $P(S_\tau[u] = v)$ in order to calculate the probabilities $P(S_r[u] = v)$ which immediately gives $P(S_N[u] = v)$ with $r = N$. When $v > u$, we take $\tau = v + 1$ and when $v \le u$, we take $\tau = u+1$ (see Theorem 3.2.11). However, the values $u + 1$ and $v + 1$ are not special. If one happens to know the probabilities $P(S_\tau[u] = v)$ at any round τ between $max\{u, v\}+1$ and N, then it is possible to arrive at the probabilities $P(S_r[u] = v)$ using Lemma 3.2.9. The recurrence relation in [110] is over three variables a, b and r, and at each step each of these three variables is reduced by one. On the other hand, the model of [135] has the following features.

1. It connects four variables u, v, τ and r which respectively denote any index u in the permutation (analogous to b), any value $v \in [0, \ldots N - 1]$ (analogous to the value at a), any round $\tau > max\{u, v\}$ and a particular round $r \ge \tau$.

2. Though the formulation of [135] does not solve any recurrence and provides a direct proof, it can be considered analogous to a recurrence over a single variable r, the other two variables u and v remaining fixed.

For graphical comparison of the theoretical and the experimental values of $P(S_N[u] = v)$ versus $0 \le u, v \le N-1$, experiments have been performed with 100 million randomly chosen secret keys of 32 bytes. In Figure 3.4, the empirical values are plotted in the top part and the the numerical values obtained from the formula in Theorem 3.2.11 are plotted in the bottom part. Observe that the graph from experimental data has a few downward spikes. These actually correspond to the *anomaly pairs* described in Section 3.2.3 below. Had the permutation been random, the surface would have been flat at a height

$\frac{1}{N}$. However, the surface is not flat and this identifies that the permutation after the RC4 KSA can be distinguished from random permutation.

3.2.3 Anomaly Pairs

To evaluate how closely the theoretical formulae tally with the experimental results, we use average percentage absolute error $\bar{\epsilon}$. Let $p_N^{u,v}$ and $q_N^{u,v}$ respectively denote the theoretical and the experimental value of the probability $P(S_N[u] = v)$, $0 \le u \le N - 1$, $0 \le v \le N - 1$. We define

$$\epsilon_{u,v} = \left(\frac{|p_N^{u,v} - q_N^{u,v}|}{q_N^{u,v}} \right) \cdot 100\% \quad \text{and} \quad \bar{\epsilon} = \frac{1}{N^2} \sum_{u=0}^{N-1} \sum_{v=0}^{N-1} \epsilon_{u,v}.$$

We ran experiments for 100 million randomly chosen secret keys of 32 bytes and found that $\bar{\epsilon} = 0.22\%$. The maximum of the $\epsilon_{u,v}$'s was 35.37% and it occurred for $u = 128$ and $v = 127$. Though the maximum error is quite high, we find that out of $N^2 = 65536$ (with $N = 256$) many $\epsilon_{u,v}$'s, only 11 (< 0.02% of 65536) exceeded the 5% error margin. These cases are summarized Table 3.3 below. We call the pairs (u, v) for which $\epsilon_{u,v} > 5\%$ *anomaly pairs*.

u	v	$p_N^{u,v}$	$q_N^{u,v}$	$|p_N^{u,v} - q_N^{u,v}|$	$\epsilon_{u,v}$ (in %)
38	6	0.003846	0.003409	0.000437	12.82
38	31	0.003643	0.003067	0.000576	18.78
46	31	0.003649	0.003408	0.000241	7.07
47	15	0.003774	0.003991	0.000217	5.44
48	16	0.003767	0.003974	0.000207	5.21
66	2	0.003882	0.003372	0.000510	15.12
66	63	0.003454	0.002797	0.000657	23.49
70	63	0.003460	0.003237	0.000223	6.89
128	0	0.003900	0.003452	0.000448	12.98
128	127	0.003303	0.002440	0.000863	35.37
130	127	0.003311	0.003022	0.000289	9.56

TABLE 3.3: The anomaly pairs for key length 32 bytes.

The experimental values of $P(S_N[u] = v)$ match with the theoretical ones given by the formula of Theorem 3.2.11 except at these few anomaly pairs. As an illustration, we plot $q_N^{u,v}$ (calculated by running the KSA with 100 million random secret keys of length 32 bytes) and $p_N^{u,v}$ versus v for $u = 38$ in Figure 3.5. We see that $q_N^{38,v}$ follows the pattern predicted by $p_N^{38,v}$ for all v's, $0 \le v \le 255$ except at $v = 6$ and $v = 31$ as pointed out in Table 3.3.

Experimentation with different key lengths (100 million random keys for each key length) reveals that the location of the anomaly pairs and the total number of anomaly pairs vary with the key lengths in certain cases. Table 3.4 shows the number n_5 of anomaly pairs (when $\epsilon_{u,v} > 5\%$) for different key

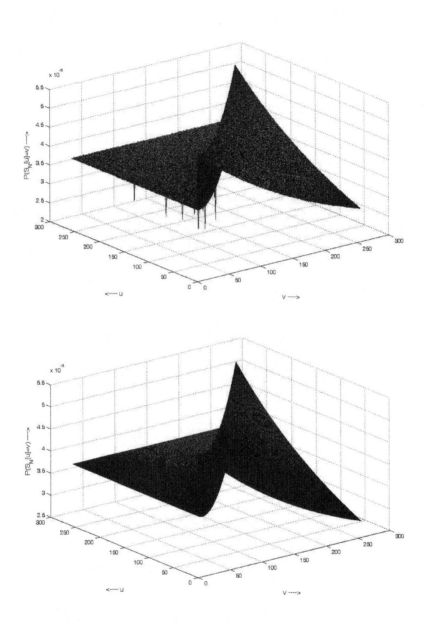

FIGURE 3.4: Experimental and theoretical $P(S_N[u] = v)$ versus $0 \leq u, v \leq N - 1$.

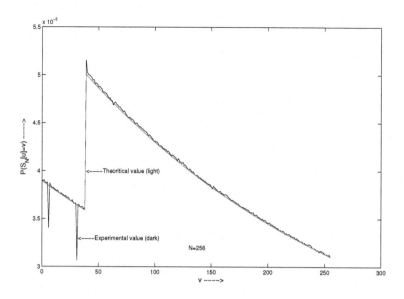

FIGURE 3.5: Comparing experimental and theoretical values of $P(S_N[38] = v)$ versus v.

lengths l (in bytes) along with the average $\bar{\epsilon}$ and the maximum ϵ_{max} of the $\epsilon_{u,v}$'s. u_{max} and v_{max} are the (u, v) values which correspond to ϵ_{max}. Though for some key lengths there are more than a hundred anomaly pairs, most of them have $\epsilon_{u,v} \leq 10\%$. To illustrate this, we look at the column n_{10} which shows how many of the anomaly pairs exceed the 10% error margin. The two rightmost columns show what percentage of $256^2 = 65536$ (total number of (u, v) pairs) are the numbers n_5 and n_{10}.

These results show that as the key length increases, the proportion of anomaly pairs tends to decrease. With 256-byte key, there is no anomaly pair with $\epsilon_{u,v} > 5\%$, i.e., $n_5 = 0$. It has also been pointed out in [110] that as the key length increases, the actual random behavior of the key is demonstrated and that is why the number of anomaly pairs decrease and experimental results match the theoretical formulae. In [110, Section 6.3.2] the anomalies are discussed for rows and columns 9, 19 and also for the diagonal given short keys as 5 bytes.

l	$\bar{\epsilon}$ (in %)	ϵ_{max} (in %)	u_{max}	v_{max}	n_5	n_{10}	n_5 (in %)	n_{10} (in %)
5	0.75	73.67	9	254	1160	763	1.770	1.164
8	0.48	42.48	15	255	548	388	0.836	0.592
12	0.30	21.09	23	183	293	198	0.447	0.302
15	0.25	11.34	44	237	241	2	0.368	0.003
16	0.24	35.15	128	127	161	7	0.246	0.011
20	0.20	5.99	30	249	3	0	0.005	0.000
24	0.19	4.91	32	247	0	0	0.000	0.000
30	0.19	6.54	45	29	1	0	0.002	0.000
32	0.22	35.37	128	127	11	6	0.017	0.009
48	0.18	4.24	194	191	0	0	0.000	0.000
64	0.26	35.26	128	127	6	4	0.009	0.006
96	0.21	4.52	194	191	0	0	0.000	0.000
128	0.34	37.00	128	127	3	2	0.005	0.003
256	0.46	2.58	15	104	0	0	0.000	0.000

TABLE 3.4: The number and percentage of anomaly pairs along with the average and maximum error for different key lengths.

3.3   Movement Frequency of Permutation Values

During the KSA, each index of the permutation is touched exactly once by i and some indices are touched by j additionally. Let us turn the view around. Instead of the permutation indices, let us investigate the actual entries in the permutation in terms of being touched by i and j. We are going to show that many values in the permutation are touched at least once with a very high probability by the indices i, j during the KSA.

Theorem 3.3.1. *The probability that a value v in the permutation is touched exactly once during the KSA by the indices i, j, is given by $\frac{2v}{N} \cdot (\frac{N-1}{N})^{N-1}$, $0 \le v \le N - 1$.*

Proof: Initially, v is located at index v in the permutation. It is touched exactly once in one of the following two ways.

Case I: The location v is not touched by any of $\{j_1, j_2, \ldots, j_v\}$ in the first v rounds. This happens with probability $(\frac{N-1}{N})^v$. In round $v + 1$, when i becomes v, the value v at index v is moved to the left by j_{v+1} due to the swap and remains there until the end of KSA. This happens with probability

$$P(j_{v+1} \in \{0, \ldots, v-1\}) \cdot P(j_\tau \ne j_{v+1}, v+1 \le \tau \le N)$$

$$= \frac{v}{N} \cdot \left(\frac{N-1}{N}\right)^{N-v-1}.$$

Thus, the probability contribution of this part is

$$\left(\frac{N-1}{N}\right)^v \cdot \frac{v}{N} \cdot \left(\frac{N-1}{N}\right)^{N-v-1} = \frac{v}{N} \cdot \left(\frac{N-1}{N}\right)^{N-1}.$$

Case II: For some τ, $1 \le \tau \le v$, it is not touched by any of $\{j_1, j_2, \ldots, j_{\tau-1}\}$; after that it is touched for the first time by $j_\tau = v$ in round τ and hence is moved to index $\tau - 1$; and it is not touched by any one of the subsequent $(N - \tau)$ many j values. The probability contribution of this part is

$$\sum_{\tau=1}^{v} \left(\frac{N-1}{N}\right)^{\tau-1} \cdot \frac{1}{N} \cdot \left(\frac{N-1}{N}\right)^{N-\tau} = \frac{v}{N} \cdot \left(\frac{N-1}{N}\right)^{N-1}.$$

Adding the above two contributions, we get the result. ∎

Using similar arguments one could compute the probability that a value is touched exactly twice, thrice and in general x times, during the KSA. However, the computation would be tedious and complicated for $x > 1$. Next, let us look into a more natural measure of this asymmetric behavior, namely, the expected number of times each value in the permutation is touched during the KSA. This is computed in the following theorem.

Theorem 3.3.2. *For $0 \le v \le N - 1$, the expected number of times a value v in the permutation is touched by the indices i, j during the KSA is given by*

$$E_v = 1 + \left(\frac{2N-v}{N}\right) \cdot \left(\frac{N-1}{N}\right)^v.$$

Proof: Let $x_{v,y} = 1$, if the value v is touched by the indices i, j in round $y + 1$ of the KSA (i.e., when $i = y$); otherwise, let $x_{v,y} = 0$, $0 \le v \le N - 1$, $0 \le y \le N - 1$. Then the number of times v is touched by i, j during the KSA is given by

$$X_v = \sum_{y=0}^{N-1} x_{v,y}.$$

In any round $y + 1$, any value v is touched by j with a probability $\frac{1}{N}$. To this, we need to add the probability of v being touched by i, in order to find $P(x_{v,y} = 1)$. Now, v is touched by the index i in round $y + 1$, if and only if $S_y[y] = v$. We consider three possible ways in which $S_y[y]$ can become v.

Case I: Consider $y < v$. Initially, the value v was situated in index v. In order for v to move from index v to index $y < v$, either v has to be touched by i and y has to be touched by j, or vice versa, during the first y rounds. However, this is not possible, since neither the index y nor the index v has been touched by the index i so far. Thus,

$$P(S_y[y] = v) = 0.$$

Case II: Consider $y = v$. We would have $S_v[v] = v$, if v is not touched by any of $\{j_1, j_2, \ldots, j_v\}$ in the first v rounds, the probability of which is $(\frac{N-1}{N})^v$.

Case III: Consider $y > v$. Once $S_v[v] = v$, the swap in the next round moves the value v to a random location j_{v+1}. Thus,

$$
\begin{aligned}
P(S_{v+1}[y] = v) &= P(S_v[v] = v) \cdot P(j_{v+1} = y) \\
&= \left(\frac{N-1}{N}\right)^v \cdot \frac{1}{N}.
\end{aligned}
$$

For all $y > v$, until y is touched by the deterministic index i, i.e., until round $y + 1$, v will remain randomly distributed. Hence, for all $y > v$,

$$
P(S_y[y] = v) = P(S_{v+1}[y] = v) = \frac{1}{N}\left(\frac{N-1}{N}\right)^v.
$$

Noting that

$$
E(x_{v,y}) = P(x_{v,y} = 1) = \frac{1}{N} + P(S_y[y] = v),
$$

we have

$$
\begin{aligned}
E_v &= E(X_v) \\
&= \sum_{y=0}^{N-1} E(x_{v,y}) \\
&= 1 + \sum_{y=0}^{N-1} P(S_y[y] = v) \\
&= 1 + \sum_{y=0}^{v-1} P(S_y[y] = v) + P(S_v[v] = v) + \sum_{y=v+1}^{N-1} P(S_y[y] = v) \\
&= 1 + 0 + \left(\frac{N-1}{N}\right)^v + (N-v) \cdot \frac{1}{N}\left(\frac{N-1}{N}\right)^v \\
&= 1 + \left(\frac{2N-v}{N}\right) \cdot \left(\frac{N-1}{N}\right)^v.
\end{aligned}
$$

∎

One can observe that E_v decreases from 3.0 to 1.37, as v increases from 0 to 255. To demonstrate how close the experimental values of the expectations match with the theoretical values, we have performed 100 million runs of the KSA, with random key of 16 bytes in each run. As depicted in Figure 3.6, The experimental data (left) correspond to the theoretical values (right). One may also refer to the first two rows of Table 9.1 in Chapter 9 for more details.

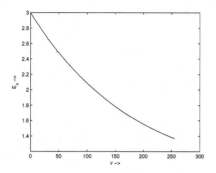 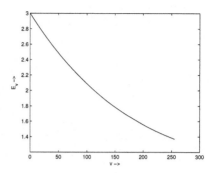

FIGURE 3.6: Experimental and theoretical E_v versus v, $0 \le v \le 255$.

In [5, 134], it is shown that the probabilities $P(j_{y+1} = S_N^{-1}[y])$ increase with increasing y. This is related to the above decreasing pattern in the expectations. In the first half of the KSA, i.e., when y is small, the values $v = S[y]$ are thrown more to the right of S with high probability by the index j_{y+1} due to the swaps and hence are touched again either by the deterministic index i or by the pseudo-random index j in the subsequent rounds. On the other hand, in the second half of the KSA, i.e., when $y \ge 128$, the values $v = S[y]$ are thrown more to the left of S by the index j_{y+1} due to the swap and hence are never touched by i in the subsequent rounds, and may be touched by j with a small probability.

An important requirement in designing a key scheduling algorithm in shuffle-exchange paradigm is that each value in the permutation should be touched (and therefore moved with probability almost one) sufficient number of times. This would make it harder to guess the values of j for which a permutation byte is swapped. However, in RC4 KSA, there are many permutation bytes which are swapped only once with a high probability, leading to information leakage from S_N regarding the secret key bytes.

3.4 Key Collisions

Suppose $S^{(1)}$ and $S^{(2)}$ are the internal states and $j^{(1)}$ and $j^{(2)}$ are the j index when the RC4 KSA is executed with two different secret keys K_1 and K_2 respectively. The keys K_1 and K_2 are called *colliding keys*, if they yield the same internal state after the KSA (i.e., if $S_N^{(1)} = S_N^{(2)}$) and hence produce the same keystream sequence during the PRGA.

Construction of a colliding key pair of RC4 was first attempted in [14, Section 5.1]. Consider secret keys of size 32 bytes. From a given 32-byte key K_1, a second 32-byte key K_2 was formed as follows.

$$K_2[i] = \begin{cases} K_1[i] & \text{for } i = 0, \dots, 29; \\ K_1[i] + 1 & \text{for } i = 30; \\ K_1[i] - 1 & \text{for } i = 31. \end{cases}$$

With these two keys, for the first 30 rounds of the KSA, S_1 and S_2 would remain the same. In round 31 (when $i = 30$), if $j_{31}^{(1)} = 30$, then $j_{31}^{(2)} = 31$. Due to this, no swap happens in $S^{(1)}$ and a swap happens in $S^{(2)}$ between the indices 30 and 31. In the next round (i.e., in round 31), i becomes 31. Now if $j_{32}^{(1)} = 31$, then $j_{32}^{(2)} = 30$. Again no swap happens in $S^{(1)}$ and a swap happens in $S^{(2)}$ between the indices 30 and 31. Thus, after the first 32 rounds of the KSA,

$$\begin{aligned} P(S_{32}^{(1)} = S_{32}^{(2)}) &= P(j_{31}^{(1)} = 30) \cdot P(j_{32}^{(1)} = 31) \\ &= 2^{-16}. \end{aligned}$$

Since the 32-byte secret key is repeated 8 times before the KSA is complete, we have

$$P(S_N^{(1)} = S_N^{(2)}) = (2^{-16})^8 = 2^{-128}. \tag{3.3}$$

This collision finding method is based on the natural expectation that the differences in two byte positions would compensate each other. However, as Equation (3.3) points out, it is practically infeasible to generate a colliding key pair using this method for 32-byte key length. In general, since a key of length l bytes is repeated at least $\lfloor \frac{N}{l} \rfloor$ times, the probability of finding a colliding key pair using the above approach is less than or equal to $(2^{-16})^{\lfloor \frac{N}{l} \rfloor}$, which decreases when l decreases. This implies that it is more difficult to find collision for shorter keys than for longer keys.

Later, for the first time, a practical method of constructing colliding key pairs of RC4 was presented in [111]. However counter-intuitive it may appear, in [111], no attempt was made to compensate one difference with another. Rather, the keys were selected so that they differ in a single byte position only, and the initial differences in the permutation are automatically balanced at the end of the KSA.

The *distance of a key pair at step* r may be defined as the number of distinct bytes between $S_r^{(1)}$ and $S_r^{(2)}$. If the distance between a key pair is 0 at $r = N$, then they are a colliding key pair. A collision search algorithm is designed in [111], where the key pairs are constructed in such a way that the distance of a key pair is at most 2 in any step r ($1 \leq r \leq N$). Let us briefly discuss this search strategy.

For a given key K and two indices y and δ in $[0, N-1]$, define a modified key $K\langle y, \delta\rangle$ as

$$K\langle y, \delta\rangle[y] = K[y] + \delta,$$
$$K\langle y, \delta\rangle[y+1] = K[y+1] - \delta,$$
$$\text{and} \quad K\langle y, \delta\rangle[x] = K[x] + \delta, \quad \text{for } x \neq y, y+1.$$

A given key K and its modified key $K\langle y, \delta\rangle$ are expected to generate similar initial states. So when the collision check fails for the pair (K_1, K_2), one can try with the pair $(K_1\langle y, \delta\rangle, K_2\langle y, \delta\rangle)$, instead of restarting the search with another random pair. Which candidate pair is closer to collision can be measured by the function $MCS(K_1, K_2)$ that returns the maximal step r such that the distance between K_1 and K_2 is at most 2 at all steps up to r. For a colliding or near-colliding key pair, $MCS(K_1, K_2) = N$.

Algorithm *CollisionSearch* is invoked with a random key pair (K_1, K_2), each of length l, that are related as follows.

$$K_1[y] = \begin{cases} K_2[y] - 1, & \text{if } y = d; \\ K_2[y] = l - d - 1, & \text{if } y = d + 1; \\ K_2[y], & \text{otherwise,} \end{cases} \tag{3.4}$$

where $0 \leq d \leq l - 1$.

The variable *maxc* is assigned a predefined value.

Three colliding key pairs of lengths 64, 43 and 24 bytes respectively are reported in [111].

64-byte colliding key pairs:

$$
\begin{aligned}
K_1 = \quad & \text{45 3d 7d 3d c9 45 57 12 00 00 00 00 00 00 00 00} \\
& \text{00 00 00 00 00 00 00 00 00 00 00 00 00 00 00 00} \\
& \text{00 00 00 00 00 00 00 00 00 00 00 00 00 00 00 00} \\
& \text{00 00 00 00 00 00 00 00 00 00 00 00 00 00 00 00} \\
K_2 = \quad & \text{45 3d 7e 3d c9 45 57 12 00 00 00 00 00 00 00 00} \\
& \text{00 00 00 00 00 00 00 00 00 00 00 00 00 00 00 00} \\
& \text{00 00 00 00 00 00 00 00 00 00 00 00 00 00 00 00} \\
& \text{00 00 00 00 00 00 00 00 00 00 00 00 00 00 00 00}
\end{aligned}
$$

Input: Key pair (K_1, K_2) satisfying Equation (3.4).
Output: A colliding or near-colliding key pair.

$r = MCS(K_1, K_2)$;
if $r = N$ **then**
  | Stop (found a (near-)colliding pair) or return (to find more);
end
$maxr = \max\limits_{\substack{0 \le y \le N-1, y \ne d, d+1 \\ 1 \le \delta \le N-1}} MCS(K_1\langle y, \delta\rangle, K_2\langle y, \delta\rangle)$;
if $maxr \le r$ **then**
  | Return;
end
$c = 0$;
for *all y from 0 to $N-1$ except d and $d+1$ and for all δ from 1 to* $N-1$ **do**
  if $MCS(K_1\langle y, \delta\rangle, K_2\langle y, \delta\rangle) = maxr$ **then**
    | Call CollisionSearch($K_1\langle y, \delta\rangle, K_2\langle y, \delta\rangle$);
    | $c = c + 1$;
    | **if** $c = maxc$ **then**
    | | Return;
    | **end**
  end
end
Return;

Algorithm 3.4.1: CollisionSearch

43-byte colliding key pairs:

$$K_1 \;=\; \text{00 6d 41 8b 95 46 07 a4 87 8d 69 d7 bc bc c4 70}$$
$$\text{4a 3b ed 94 34 50 04 68 4d 4f 2e 30 c1 6e 20 a8}$$
$$\text{bf 80 b6 ae df ae 43 56 0a 80 e7}$$
$$K_2 \;=\; \text{00 6d 41 8b 95 46 07 a4 87 8d 69 d7 bc bc c4 70}$$
$$\text{4a 3b ed 94 34 50 04 68 4d 4f 2e 30 c1 6e 20 a8}$$
$$\text{bf 80 b6 ae df ae 43 56 0a 80 e8}$$

24-byte colliding key pairs:

$$K_1 \;=\; \text{00 42 CE D3 DF DD B6 9D 41 3D BD 3A B1 16 5A 33}$$
$$\text{ED A2 CD 1F E2 8C 01 76}$$
$$K_2 \;=\; \text{00 42 CE D3 DF DD B6 9D 41 3D BD 3A B1 16 5A 33}$$
$$\text{ED A2 CD 1F E2 8C 01 77}$$

Note that in each of the colliding key pairs, one key differs from the other in a single byte position only. The next example shows a 20-byte *near-colliding key pair*, i.e., a key pair whose corresponding states after the KSA differ in exactly two positions.

20-byte near-colliding key pairs:

$$K_1 \;=\; \text{00 73 2F 6A 01 37 89 C5 15 49 9A 55 98 54 D7 53 4E F6 4F DC}$$
$$K_2 \;=\; \text{00 73 2F 6A 01 37 89 C5 15 49 9A 55 98 54 D7 53 4E F6 4F DD.}$$

Apart from *maxc*, the depth of recursion is also a parameter of Algorithm 3.4.1. The experiments described in [111] have considered a *maxc* value around 10 and the maximum depth of recursion less than 20. Since it is a heuristic search, further improvement is possible and is an open research topic.

Research Problems

Problem 3.1 The proportion of *anomaly pairs* tends to decrease with increasing key length. Theoretical investigation into the nature of anomaly pairs and their relation with the key length is an open problem.

Problem 3.2 Can you propose modifications to the RC4 KSA such that there is no permutation-key correlation and no bias in the permutation after the KSA? You are supposed to run it N times and the modification should be minimum and the increase in the number of steps should be minimized.

Problem 3.3 Does there exist a collide key pair with length below 24 bytes?

Chapter 4

Key Recovery from State Information

In the previous chapter, we presented a detailed analysis of the RC4 KSA and demonstrated different types of biases present in the permutation bytes. Here we discuss different algorithms for recovering the secret key of RC4 from the state information.

In a shuffle-exchange kind of stream cipher, for proper cryptographic security, one may expect that after the key scheduling algorithm one should not be able to get any information on the secret key bytes from the random looking permutation in time complexity less than the exhaustive key search. The KSA of RC4 is weak in this aspect.

There are three primary motivations for studying the possibility of key recovery from a known RC4 state.

1. An important class of attacks on RC4 is the class of state recovery attacks [87, 116, 181]. Key recovery from the internal state is useful to turn a state recovery attack into a key recovery attack. If the complexity of recovering the secret key from the permutation is less than that of recovering RC4 permutation from the keystream, then by cascading the techniques of the latter with those of the former, recovering the secret key from the keystream is possible at the same complexity as the latter.

2. In many practical applications, a secret key is combined with a known IV to form a *session key*. Generally, recovering the permutation is enough for cryptanalysis of a single session only. However, there are many applications (such as WEP [92]), where the key and the IV are combined to form the session key in such a way that the secret key can be easily extracted from the session key. For such an application, if the session key can be recovered from the permutation, then the secret key is immediately broken. Moreover, for subsequent sessions, where the same secret key would be used with different known IVs, the RC4 encryption would be rendered completely insecure.

3. Apart from cryptanalytic significance, the different methods of key retrieval from state information provide guidelines in designing improved versions or modes of operations of RC4 with better security.

The first section of this chapter shows that the steps of RC4 PRGA are reversible and hence explains why the final permutation after the KSA should be the starting point of key recovery from state information.

Mantin mentioned in [52, Appendix D] as well as in [110, Appendix A.2] how to recover the first A bytes of the secret key from an "Early Permutation State" after round A. Apart from this, key recovery from RC4 permutation has earlier been considered in various forms [86, 180, 184] (see Chapter 7 for details). However, the first algorithm to recover the complete key from the final permutation after the KSA, without any assumption on the key or IV, appeared in [133, Section 3]. This work is described in Section 4.2. Subsequently, Biham and Carmeli [15] refined the idea of [133] to devise an improved algorithm. We summarize this improvement in Section 4.3. The following section, i.e., Section 4.4, outlines further improvement in the sequel made by Akgün et al. [5]. Around the same time as [5], two other independent techniques for the secret key recovery were developed. One takes a byte-by-byte approach [134, Sections 2.3, 3.4] and the other a bit-by-bit approach [83]. These two works are described in Sections 4.5 and 4.6 respectively. Finally, Section 4.7 concludes this chapter by describing the recently developed bidirectional key recovery algorithm [10] based on certain sequences of the index j.

4.1 Reversibility of RC4 PRGA

The RC4 state information consists of

- the entire permutation S,

- the number of keystream output bytes generated (which is related to the index i) and

- the value of the index j.

If this state information at any instant during the PRGA is available, then it is possible to run the PRGA backwards to get back to the permutation after the KSA. Once the final permutation after the KSA is retrieved, the secret key can be reconstructed using any one of the techniques described in the subsequent sections.

Suppose we know the RC4 internal state after τ rounds of PRGA. We can get back the permutation S_N after the KSA using the *PRGAreverse* algorithm described below. Even when the value j_τ^G of the index j is not known, one can make an exhaustive search over $\mathbb{Z}_N$ and for each value perform the state-reversing loop in PRGAreverse. If, for a value of j_τ^G, the final value of j after the completion of the loop equals zero, the resulting final permutation is a candidate for S_N. On average, one expects to get one such candidate.

Note that we do not need the keystream bytes themselves. The algorithm requires the value of τ only. From τ, we can get the current value of i as

$i_\tau^G = \tau \bmod N$. All subtractions except $r = r - 1$ in Algorithm *PRGAreverse* are performed modulo N.

Input:

    1. Number of rounds τ.

    2. The permutation S_τ^G.

Output: Few candidates for S_N, the final permutation after the KSA.

$i_\tau^G = \tau \bmod N$;
for $j_\tau^G = 0$ *to* $N - 1$ **do**
    $i = i_\tau^G$; $j = j_\tau^G$; $S = S_\tau^G$; $r = \tau$;
    repeat
        Swap($S[i]$, $S[j]$);
        $j = j - S[i]$; $i = i - 1$; $r = r - 1$;
    until $r = 0$;
    if $j = 0$ **then**
        Report S as a candidate for S_N;
    end
end

Algorithm 4.1.1: PRGAreverse

Once S_N is derived from the current state information, the next step of the attacker would be to invert S_N to get back the secret key. The technique for recovering the secret key from the permutation after the KSA or at any stage during the KSA is the main theme of the following sections.

4.2 Recovery through Solving Simultaneous Equations

This was the first strategy developed for secret key recovery of RC4 from the permutation after the KSA [133]. The complexity is of the order of square root of exhaustive search complexity, i.e., for a secret key of size $8l$ bits ($40 \leq 8l \leq 128$), the key can be recovered in $O(2^{\frac{8l}{2}})$ effort with a constant probability of success.

To illustrate the idea, let us start with an example.

Example 4.2.1. *Consider a 5-byte secret key $K[0\ldots4]$ with $K[0] = 106, K[1] = 59, K[2] = 220, K[3] = 65$, and $K[4] = 34$. If one runs the KSA with this key, then the first 16 bytes of the final permutation are as follows.*

y	0	1	2	3	4	5	6	7
f_y	106	166	132	200	238	93	158	129
$S_{256}[y]$	230	166	87	48	238	93	68	239

y	8	9	10	11	12	13	14	15
f_y	202	245	105	175	151	229	21	142
$S_{256}[y]$	202	83	105	147	151	229	35	142

The key recovery strategy is as follows. Consider all possible systems of 5 equations chosen from the 16 equations of the form $S_N[y] = f_y$, $0 \leq y \leq 15$. A solution to such a system would give one candidate key. The correctness of a candidate key can be verified by running the KSA with this key and comparing the permutation thus obtained with the permutation in hand. Of course, some of the systems may not be solvable at all.

The correct solution for this example correspond to the choices $y = 1, 4, 5, 8$ and 12, and the corresponding equations are:

$$
\begin{aligned}
K[0] + K[1] + (1 \cdot 2)/2 &= 166 \\
K[0] + K[1] + K[2] + K[3] + K[4] + (4 \cdot 5)/2 &= 238 \\
K[0] + \ldots + K[5] + (5 \cdot 6)/2 &= 93 \\
K[0] + \ldots + K[8] + (8 \cdot 9)/2 &= 202 \\
K[0] + \ldots + K[12] + (12 \cdot 13)/2 &= 151
\end{aligned}
$$

The correctness of a solution depends on the correctness of the selected equations. The probability that we would indeed get a correct solution from a system is equal to the joint probability of $S_r[y] = f_y$ for the set of chosen y-values. However, we do not need the assumption that the majority of the equations are correct. Whether a system of equations is correct or not can be cross-checked by running the KSA again. Moreover, empirical results show that in a significant proportion of the cases one might obtain enough correct equations to have the correct key as one of the solutions.

The exhaustive search for a 5-byte key requires a complexity of 2^{40}. Whereas in the above approach, one needs to consider at the most $\binom{16}{5} = 4368 < 2^{13}$ sets of 5 equations. Since the equations are triangular in form, solving each system of 5 equations would take approximately $5^2 = 25$ (times a small constant) $< 2^5$ many additions/subtractions. Thus the improvement over exhaustive search is almost by a factor of $\frac{2^{40}}{2^{13} \cdot 2^5} = 2^{22}$.

Let us now describe the approach in general. Theorem 3.1.5 gives us how $S_r[y]$ is biased to different combinations of the keys, namely, with

$$
f_y = \frac{y(y+1)}{2} + \sum_{x=0}^{y} K[x].
$$

Let us denote

$$
P(S_r[y] = f_y) = p_{r,y}, \qquad \text{for } 0 \leq y \leq r - 1, \, 1 \leq r \leq N.
$$

Let m be a parameter, such that one wants to recover exactly m out of the l secret key bytes by solving equations and the other $l - m$ bytes by exhaustive key search. Let n be another parameter ($m \leq n \leq r$) of our choice, that denotes how many equations of the form $S_r[y] = f_y$, $y = 0, 1, \ldots, n - 1$, in l variables (the key bytes) are to be considered.

Let EI_τ denote the set of all systems of τ equations that are independent. Equivalently, EI_τ may be thought of as the collection of the subsets of indices

$$\{y_1, y_2, \ldots, y_\tau\} \subseteq \{0, 1, \ldots, n - 1\},$$

corresponding to all systems of τ independent equations (selected from the above system of n equations).

In general, we need to check whether each of the $\binom{n}{m}$ systems of m equations is independent or not. The next result establishes the criteria for independence of such a system of equations and also the total number of such systems.

Theorem 4.2.2. *Let $l \geq 2$ be the RC4 key length in bytes. Suppose we want to select systems of m independent equations, $2 \leq m \leq l$, from the n equations of the form $S_r[y] = f_y$, $0 \leq y \leq n - 1$, involving the permutation bytes after round r of the KSA, $m \leq n \leq r \leq N$.*

Part 1. *The system $S_r[y_q] = f_{y_q}$, $1 \leq q \leq m$, of m equations corresponding to $y = y_1, y_2, \ldots, y_m$, is independent if and only if any one of the following two conditions hold:*

(i) *$y_q \bmod l$, $1 \leq q \leq m$, yields m distinct values, or*

(ii) *$y_q \bmod l \neq (l - 1)$, $1 \leq q \leq m$, and there is exactly one pair $y_a, y_b \in \{y_1, y_2, \ldots, y_m\}$ such that $y_a = y_b \pmod{l}$, and all other $y_q \bmod l$, $q \neq a$, $q \neq b$, yields $m - 2$ distinct values different from $y_a, y_b \pmod{l}$.*

Part 2. *The total number of independent systems of m (≥ 2) equations is given by*

$$|EI_m| = \sum_{x=0}^{m} \binom{n \bmod l}{x} \binom{l - n \bmod l}{m - x} (\lfloor \tfrac{n}{l} \rfloor + 1)^x (\lfloor \tfrac{n}{l} \rfloor)^{m-x}$$

$$+ \binom{n \bmod l}{1} \binom{\lfloor \frac{n}{l} \rfloor + 1}{2} \sum_{x=0}^{m-2} \binom{n \bmod l-1}{x} \binom{l - n \bmod l-1}{m-2-x} (\lfloor \tfrac{n}{l} \rfloor + 1)^x (\lfloor \tfrac{n}{l} \rfloor)^{m-2-x}$$

$$+ \binom{l - n \bmod l-1}{1} \binom{\lfloor \frac{n}{l} \rfloor}{2} \sum_{x=0}^{m-2} \binom{n \bmod l}{x} \binom{l - n \bmod l-2}{m-2-x} (\lfloor \tfrac{n}{l} \rfloor + 1)^x (\lfloor \tfrac{n}{l} \rfloor)^{m-2-x},$$

where the binomial coefficient $\binom{u}{v}$ has the value 0, if $u < v$.

Proof: Part 1. First, we will show that any one of the conditions (i) and (ii) is sufficient. Suppose that the condition (i) holds, i.e., $y_q \bmod l$ ($1 \leq q \leq m$) yields m distinct values. Then each equation involves a different key byte as a variable, and hence the system is independent.

Now, suppose that the condition (ii) holds. Then there exists exactly one pair $a, b \in \{1, \ldots, m\}$, $a \neq b$, where $y_a = y_b \bmod l$. Without loss of generality, let $y_a < y_b$. So we can subtract $S_r[y_a] = f_{y_a}$ from $S_r[y_b] = f_{y_b}$ to get one equation involving some multiple of the sum $s = \sum_{x=0}^{l-1} K[x]$ of the key bytes. Thus, one can replace exactly one equation involving either y_a or y_b with the new equation involving s, which will become a different equation with a new variable $K[l-1]$, since $l - 1 \notin \{y_1 \bmod l, y_2 \bmod l, \ldots, y_m \bmod l\}$. Hence the resulting system is independent.

Next, we show that the conditions are necessary. Suppose that neither condition (i) nor condition (ii) holds. Then either we will have a triplet a, b, c such that $y_a = y_b = y_c = \bmod\ l$, or we will have a pair a, b with $y_a = y_b \bmod l$ and $l - 1 \in \{y_1 \bmod l, y_2 \bmod l, \ldots, y_m \bmod l\}$. In the first case, subtracting two of the equations from the third one would result in two equations involving s and the same key bytes as variables. Thus the resulting system will not be independent. In the second case, subtracting one equation from the other will result in an equation which is dependent on the equation involving the key byte $K[l-1]$.

Part 2. We know that $n = (\lfloor \frac{n}{l} \rfloor)l + (n \bmod l)$. If we compute $y \bmod l$, for $y = 0, 1, \ldots n - 1$, then we will have the following residue classes:

$$
[0] \;=\; \left\{ 0, l, 2l, \ldots, \left(\left\lfloor \frac{n}{l} \right\rfloor\right) l \right\}
$$

$$
[1] \;=\; \left\{ 1, l+1, 2l+1, \ldots, \left(\left\lfloor \frac{n}{l} \right\rfloor\right) l + 1 \right\}
$$

$$
\vdots \quad \vdots \quad \vdots
$$

$$
[n \bmod l - 1] \;=\; \left\{ n \bmod l - 1, l + (n \bmod l - 1), 2l + (n \bmod l - 1), \ldots, \right.
$$
$$
\left. \left(\left\lfloor \frac{n}{l} \right\rfloor\right) l + (n \bmod l - 1) \right\}
$$

$$
[n \bmod l] \;=\; \left\{ n \bmod l, l + (n \bmod l), 2l + (n \bmod l), \ldots, \left(\left\lfloor \frac{n}{l} \right\rfloor - 1\right) l \right.
$$
$$
\left. + (n \bmod l) \right\}
$$

$$
\vdots \quad \vdots \quad \vdots
$$

$$
[l-1] \;=\; \left\{ l-1, l + (l-1), 2l + (l-1), \ldots, \left(\left\lfloor \frac{n}{l} \right\rfloor - 1\right) l + (l-1) \right\}
$$

The set of these l many residue classes can be split into two mutually exclusive subsets, namely $A = \{[0], \ldots, [n \bmod l - 1]\}$ and $B = \{[n \bmod l], \ldots, [l-1]\}$, such that each residue class $\in A$ has $\lfloor \frac{n}{l} \rfloor + 1$ members and each residue class $\in B$ has $\lfloor \frac{n}{l} \rfloor$ members. Note that $|A| = n \bmod l$ and $|B| = l - (n \bmod l)$.

Now, the independent systems of m equations can be selected in three mutually exclusive and exhaustive ways. Case I corresponds to the condition (i) and Cases II & III correspond to the condition (ii) stated in the theorem.

<u>Case I</u>: *Select m different residue classes from $A \cup B$ and choose one y-value (the equation number) from each of these m residue classes.* Now, x of the m residue classes can be selected from the set A in $\binom{n \bmod l}{x}$ ways and the remaining $m - x$ can be selected from the set B in $\binom{l-n \bmod l}{m-x}$ ways. Again, corresponding to each such choice, the first x residue classes would give $\lfloor \frac{n}{l} \rfloor + 1$ choices for y (the equation number) and each of the remaining $m - x$ residue classes would give $\lfloor \frac{n}{l} \rfloor$ choices for y. Thus, the total number of independent equations in this case is given by

$$\sum_{x=0}^{m} \binom{n \bmod l}{x} \binom{l - n \bmod l}{m - x} \left(\left\lfloor \frac{n}{l} \right\rfloor + 1\right)^x \left(\left\lfloor \frac{n}{l} \right\rfloor\right)^{m-x}.$$

<u>Case II</u>: *Select two y-values from any residue class in A. Then select $m-2$ other residue classes except $[l-1]$ and select one y-value from each of those $m-2$ residue classes.* One can pick one residue class $a \in A$ in $\binom{n \bmod l}{1}$ ways and subsequently two y-values from a in $\binom{\lfloor \frac{n}{l} \rfloor + 1}{2}$ ways. Of the remaining $m-2$ residue classes, x can be selected from $A \setminus \{a\}$ in $\binom{n \bmod l-1}{x}$ ways and the remaining $m - 2 - x$ can be selected from $B \setminus \{[l-1]\}$ in $\binom{l-n \bmod l-1}{m-2-x}$ ways. Again, corresponding to each such choice, the first x residue classes would give $\lfloor \frac{n}{l} \rfloor + 1$ choices for y (the equation number) and each of the remaining $m - 2 - x$ residue classes would give $\lfloor \frac{n}{l} \rfloor$ choices for y. Thus, the total number of independent equations in this case is given by

$$\binom{n \bmod l}{1} \binom{\lfloor \frac{n}{l} \rfloor + 1}{2} s_m,$$

where

$$s_m = \sum_{x=0}^{m-2} \binom{n \bmod l - 1}{x} \binom{l - n \bmod l - 1}{m - 2 - x} \left(\left\lfloor \frac{n}{l} \right\rfloor + 1\right)^x \left(\left\lfloor \frac{n}{l} \right\rfloor\right)^{m-2-x}.$$

<u>Case III</u>: *Select two y-values from any residue class in $B \setminus \{[l-1]\}$. Then select $m - 2$ other residue classes and select one y-value from each of those $m - 2$ residue classes.* This case is similar to case II, and the total number of independent equations in this case is given by

$$\binom{l - n \bmod l - 1}{1} \binom{\lfloor \frac{n}{l} \rfloor}{2} s'_m,$$

where

$$s'_m = \sum_{x=0}^{m-2} \binom{n \bmod l}{x} \binom{l - n \bmod l - 2}{m - 2 - x} \left(\left\lfloor \frac{n}{l} \right\rfloor + 1\right)^x \left(\left\lfloor \frac{n}{l} \right\rfloor\right)^{m-2-x}.$$

Adding the counts for the above three cases, we get the result for Part 2. ∎

Proposition 4.2.3. *Given n and m, it takes $O(m^2 \cdot \binom{n}{m})$ time to generate the set EI_m using Theorem 4.2.2.*

Proof: One needs to check a total of $\binom{n}{m}$ many m tuples $\{y_1, y_2, \ldots, y_m\}$, and using the independence criteria of Theorem 4.2.2, it takes $O(m^2)$ amount of time to determine if each tuple belongs to EI_m or not.   ■

Proposition 4.2.4. *Suppose we have an independent system of equations of the form $S_r[y_q] = f_{y_q}$ involving the l key bytes as variables corresponding to the tuple $\{y_1, y_2, \ldots, y_m\}$, $0 \leq y_q \leq n-1$, $1 \leq q \leq m$. If there is one equation in the system involving $s = \sum_{x=0}^{l-1} K[x]$, then there are at most $\lfloor \frac{n}{l} \rfloor$ many solutions for the key.*

Proof: If the coefficient of s is a, then by Linear Congruence Theorem [169, Page 56], we would have at most $gcd(a, N)$ many solutions for s, each of which would give a different solution for the key. To find the maximum possible number of solutions, we need to find an upper bound of $gcd(a, N)$.

Since the key is of length l, the coefficient a of s would be $\lfloor \frac{y_s}{l} \rfloor$, where y_s is the y-value $\in \{y_1, y_2, \ldots, y_m\}$ corresponding to the equation involving s. Thus,

$$gcd(a, N) \leq a = \left\lfloor \frac{y_s}{l} \right\rfloor \leq \left\lfloor \frac{n}{l} \right\rfloor.$$

   ■

Let us provide an example to demonstrate the case when we have two y-values (equation numbers) from the same residue class in the selected system of m equations, but still the system is independent and hence solvable.

Example 4.2.5. *Assume that the secret key is of length 5 bytes. Consider 16 equations of the form $S_N[y] = f_y$, $0 \leq y \leq 15$. One can study all possible sets of 5 equations chosen from the above 16 equations and then try to solve them. One such set would correspond to $y = 0, 1, 2, 3$ and 13. Let the corresponding $S_N[y]$ values be 246, 250, 47, 204 and 185 respectively. Then we can form the following equations:*

$$K[0] = 246 \tag{4.1}$$
$$K[0] + K[1] + (1 \cdot 2)/2 = 250 \tag{4.2}$$
$$K[0] + K[1] + K[2] + (2 \cdot 3)/2 = 47 \tag{4.3}$$
$$K[0] + K[1] + K[2] + K[3] + (3 \cdot 4)/2 = 204 \tag{4.4}$$
$$K[0] + \ldots + K[13] + (13 \cdot 14)/2 = 185. \tag{4.5}$$

From the first four equations, one readily gets $K[0] = 246, K[1] = 3, K[2] = 51$ and $K[3] = 154$. Since the key is 5 bytes long, $K[5] = K[0], \ldots, K[9] = K[4], K[10] = K[0], \ldots, K[13] = K[3]$. Denoting the sum of the key bytes $K[0] + \ldots + K[4]$ by s, we can rewrite Equation (4.5) as:

$$2s + K[0] + K[1] + K[2] + K[3] + 91 = 185 \tag{4.6}$$

Subtracting Equation (4.4) from Equation (4.6), and solving for s, we get $s = 76$ *or* 204. *Taking the value* 76, *we get*

$$K[0] + K[1] + K[2] + K[3] + K[4] = 76 \qquad (4.7)$$

Subtracting Equation (4.4) from Equation (4.7), we get $K[4] = 134$. *Taking* $s = 204$ *does not give the correct key, as can be verified by running the KSA and observing the permutation obtained.*

We now present the general algorithm for recovering the secret key bytes from the permutation at any stage of the KSA.

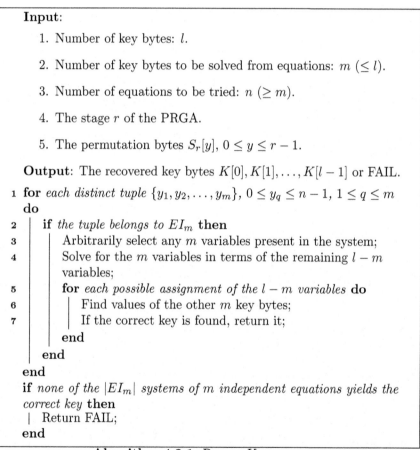

Input:

    1. Number of key bytes: l.

    2. Number of key bytes to be solved from equations: m $(\leq l)$.

    3. Number of equations to be tried: n $(\geq m)$.

    4. The stage r of the PRGA.

    5. The permutation bytes $S_r[y]$, $0 \leq y \leq r - 1$.

Output: The recovered key bytes $K[0], K[1], \ldots, K[l-1]$ or FAIL.

1 **for** *each distinct tuple* $\{y_1, y_2, \ldots, y_m\}$, $0 \leq y_q \leq n-1$, $1 \leq q \leq m$ **do**
2   **if** *the tuple belongs to* EI_m **then**
3   Arbitrarily select any m variables present in the system;
4   Solve for the m variables in terms of the remaining $l - m$ variables;
5   **for** *each possible assignment of the* $l - m$ *variables* **do**
6   Find values of the other m key bytes;
7   If the correct key is found, return it;
        end
    end
end
if *none of the* $|EI_m|$ *systems of* m *independent equations yields the correct key* **then**
    | Return FAIL;
end

Algorithm 4.2.1: RecoverKey

Note that the correctness of a key can be verified by running the KSA and comparing the resulting permutation with the permutation at hand.

If one does not use the independence criteria (Theorem 4.2.2), all $\binom{n}{m}$ sets of equations need to be checked. However, the number of independent systems is $|EI_m|$, which is much smaller than $\binom{n}{m}$. Table 4.1 shows that $|EI_m| < \frac{1}{2}\binom{n}{m}$ for most values of l, n, and m. Thus, the independence criteria in Step 2 reduces the number of trials inside the "if" block of Step 2 by a substantial factor.

The following Theorem quantifies the amount of time required to recover the key due to the algorithm.

Theorem 4.2.6. *The time complexity of the RecoverKey algorithm is given by*

$$O\left(m^2 \cdot \binom{n}{m} + |EI_m| \cdot \left(m^2 + \left\lfloor \frac{n}{l} \right\rfloor \cdot 2^{8(l-m)}\right)\right),$$

where $|EI_m|$ is given by Theorem 4.2.2.

Proof: According to Proposition 4.2.3, for a complete run of the algorithm, checking the condition at Step 2 has $O(m^2 \cdot \binom{n}{m})$ time complexity.

Further, the Steps 3, 4 and 5 are executed $|EI_m|$ times. Among them, finding the solution in Step 4 involves $O(m^2)$ many addition/subtraction operations (the equations being triangular in form). By Proposition 4.2.4, each system can yield at the most $O(\lfloor \frac{n}{l} \rfloor)$ many solutions for the key. After the solution is found, Step 5 involves $2^{8(l-m)}$ many trials. Thus, the total time consumed by the Steps 3, 4 and 5 for a complete run would be

$$O\left(|EI_m| \cdot \left(m^2 + \left\lfloor \frac{n}{l} \right\rfloor \cdot 2^{8(l-m)}\right)\right).$$

Hence, the time complexity of the RecoverKey algorithm is given by

$$O\left(m^2 \cdot \binom{n}{m} + |EI_m| \cdot \left(m^2 + \left\lfloor \frac{n}{l} \right\rfloor \cdot 2^{8(l-m)}\right)\right).$$

■

Next, we estimate the probability of getting a set of independent correct equations when we run the above algorithm. Suppose for the given system of equations $S_r[y] = f_y$, $y = 0, 1, \ldots, n-1$, $m \leq n \leq r \leq N$, $c_{r,n}$ denotes the number of independent correct equations and $p_r(y_1, y_2, \ldots, y_\tau)$ denotes the joint probability that the τ equations corresponding to the indices $\{y_1, y_2, \ldots, y_\tau\}$ are correct and the other $n - \tau$ equations corresponding to the indices $\{0, 1, \ldots, n-1\} \setminus \{y_1, y_2, \ldots, y_\tau\}$ are incorrect. Then one can immediately state the following result.

Proposition 4.2.7. $P(c_{r,n} \geq m) = \sum\limits_{\tau=m}^{n} \sum\limits_{\{y_1, y_2, \ldots, y_\tau\} \in EI_\tau} p_r(y_1, y_2, \ldots, y_\tau).$

Proof: One needs to sum $|EI_\tau|$ number of terms of the form $p_r(y_1, y_2, \ldots, y_\tau)$ to get the probability that exactly τ equations are correct, i.e.,

$$P(c_{r,n} = \tau) = \sum_{\{y_1, y_2, \ldots, y_\tau\} \in EI_\tau} p_r(y_1, y_2, \ldots, y_\tau).$$

Hence,

$$P(c_{r,n} \geq m) = \sum_{\tau=m}^{n} P(c_{r,n} = \tau)$$

$$= \sum_{\tau=m}^{n} \sum_{\{y_1,y_2,\ldots,y_\tau\}\in EI_\tau} p_r(y_1, y_2, \ldots, y_\tau).$$

∎

Note that $P(c_{r,n} \geq m)$ gives the success probability with which one can recover the secret key from the permutation after the r-th round of the KSA.

In Theorem 3.1.5, we observe that as the number r of rounds increase, the probabilities $P(S_r[y] = f_y)$ decrease. Finally, after the KSA, when $r = N$, (see Corollary 3.1.6) the probabilities settle to the values as given in Table 3.2. However, as the events $(S_r[y] = f_y)$ are not independent for different y's, deriving the formulae for the joint probability $p_r(y_1, y_2, \ldots, y_\tau)$ seems to be extremely tedious.

Consider the first n equations $S_N[y] = f_y, 0 \leq y \leq n-1$, after the complete KSA (i.e., $r = N$). In Table 4.1 (taken from [136]), we provide experimental results on the probability of having at least m independent correct equations for different values of n, m, and the key length l, satisfying $m \leq l \leq n$.

For each probability computation, the complete KSA (with $N = 256$ rounds) is repeated a million times, each time with a randomly chosen key. We also compare the values of the exhaustive search complexity and the reduction due to the above algorithm. Let

$$e = \log_2 \left(m^2 \cdot \binom{n}{m} + |EI_m| \cdot \left(m^2 + \left\lfloor \frac{n}{l} \right\rfloor \cdot 2^{8(l-m)} \right) \right).$$

The time complexity of exhaustive search is $O(2^{8l})$ and that of the *RecoverKey* algorithm, according to Theorem 4.2.6, is given by $O(2^e)$. Thus, the reduction in search complexity due to the algorithm is by a factor $O(2^{8l-e})$. By suitably choosing the parameters (see Table 4.1), one can achieve the search complexity $O(2^{\frac{8l}{2}})$, i.e., $O(2^{4l})$, which is the square root of the exhaustive key search complexity.

The results in Table 4.1 show that the probabilities (i.e., the empirical values of $P(c_{N,n} \geq m)$) in most of the cases are greater than 10%. However, the algorithm does not use the probabilities to recover the key. For certain keys the algorithm would be able to recover the keys and for certain other keys the algorithm will not. The success probability can be interpreted as the proportion of keys for which the algorithm would be able to successfully recover the key. The keys, that can be recovered from the permutation after the KSA using the *RecoverKey* algorithm, may be considered as weak keys in RC4.

The last 8 entries corresponding to 16-byte key in Table 4.1 give a flavor about the relationship between the complexity and the success probability.

l	n	m	$\binom{n}{m}$	$|EI_m|$	$8l$	e	$P(c_{N,n} \geq m)$
5	48	5	1712304	238500	40	25.6	0.431
5	24	5	42504	7500	40	20.3	0.385
5	16	5	4368	810	40	17.0	0.250
8	22	6	74613	29646	64	31.9	0.414
8	16	6	8008	3472	64	28.8	0.273
8	20	7	77520	13068	64	23.4	0.158
10	16	7	11440	5840	80	36.5	0.166
10	24	8	735471	130248	80	34.0	0.162
12	24	8	735471	274560	96	51.1	0.241
12	24	9	1307504	281600	96	43.1	0.116
12	21	10	352716	49920	96	31.6	0.026
16	24	9	1307504	721800	128	75.5	0.185
16	32	10	64512240	19731712	128	73.2	0.160
16	32	11	129024480	24321024	128	65.5	0.086
16	40	12	5586853480	367105284	128	61.5	0.050
16	27	12	17383860	2478464	128	53.2	0.022
16	26	12	9657700	1422080	128	52.4	0.019
16	44	14	114955808528	847648395	128	46.9	0.006
16	24	14	1961256	69120	128	32.2	0.0006

TABLE 4.1: Running the *RecoverKey* algorithm using different parameters for the final permutation after the complete KSA (with $N = 256$ rounds).

For example, when the success probability increases from 0.0006 to 0.006 (i.e. increases by 10 times), the complexity increases by a factor of $2^{14.7}$.

4.3   Improvement by Difference Equations

The above algorithm was improved in [15] to achieve faster recovery of the secret key at the same success probability.

Let

$$C_y = S_N[y] - \frac{y(y+1)}{2}.$$

The equations of Section 4.2 can be rewritten as

$$K[0 \uplus y_1] \;=\; C_{y_1}, \tag{4.8}$$
$$K[0 \uplus y_2] \;=\; C_{y_2}, \tag{4.9}$$

etc., where $K[a \uplus b]$ has the usual meaning of $\displaystyle\sum_{x=a}^{b} K[x]$. One can subtract the equations of the above form to generate more equations of the form

$$K[0 \uplus y_2] - K[0 \uplus y_1] \;=\; K[y_1 + 1 \uplus y_2]$$
$$= \; C_{y_2} - C_{y_1}. \tag{4.10}$$

The approach in [15] makes use of the fact that Equation (4.10) holds with more probability than the product of the probabilities of Equations (4.8) and (4.9). For example, according to Corollary 3.1.6, $P(K[0 \uplus 50] = C_{50}) = 0.0059$ and $P(K[0 \uplus 52] = C_{52}) = 0.0052$, but $P(K[51 \uplus 52] = C_{52} - C_{50}) = 0.0624$. Hence the use of the difference equations can give a more efficient key reconstruction. We summarize this technique in the following discussion.

Among the sums $K[a \uplus b]$ for different a, b, the sum

$$s = K[0 \uplus l - 1]$$

of all the key bytes is guessed first. Plugging in the value of s reduces all the remaining equations to sums of fewer than l variables, of the form

$$K[y_1 \uplus y_2], \quad 0 \le y_1 < l, y_1 \le y_2 < y_1 + l - 1.$$

At each stage, the value for a new sum of key bytes is guessed. Equations linearly dependent on prior guesses are no longer considered. After l such guesses, the resulting set of equations reveals the key. Below we enumerate additional techniques used in [15] toward improvement.

1. Several equations may suggest different values of the same sum of the key bytes. To resolve this issue, each equation with a specific sum is associated with a set of N counters for storing the weight of each possible value in $\mathbb{Z}_N$. However, each value is not given the same weight. The weight of a particular value depends on the frequency of its occurrence. A weight of 2 is assigned to values with probability > 0.05, a weight of 1 is assigned to values with probability between 0.008 and 0.05 and a weight of 0 to all other values (these weights are tuned empirically).

2. For each equation, the value with the highest weight is considered first. If this fails to retrieve the correct key, backtracking is performed and the value with the second highest weight is considered, and so on. The algorithm is parameterized by the number of attempts λ_t to be tried on the t-th guess, $0 \le t < l$. These parameters are tuned empirically.

3. An equivalence relation is defined over the set of all possible sums. Two sums are said to be in the same equivalence class, if and only if the value of each of them can be computed from the value of the other and the values of the known sums. Counters of all sums in an equivalence class are merged together and exactly one representative of each equivalence class is kept.

4. If the sum $K[y_1 + 1 \uplus y_2]$ is correct, then it is expected that all the following three events occurred with high probability.

   - $S_r[r] = r$, where $r \in [y_1 + 1, y_2]$.
   - $S[y_1] = j_{y_1+1}$.
   - $S[y_2] = j_{y_2+1}$.

   This information is utilized in two ways.

   (a) When considering a value for a sum of of key bytes $K[y_1 + 1 \uplus y_2]$ which is still unknown, if the known sums indicate that the above three events are likely to have occurred for y_1, y_2, then the weight associated with the value for the sum $K[y_1 + 1 \uplus y_2]$ is increased.

   (b) All suggestions passing over some r, $y_1 < r < y_2$, for which $S_r[r] \ne r$ must have the same error

   $$\Delta = C_{y_2} - C_{y_1} - K[y_1 + 1 \uplus y_2].$$

   So, if several suggestions passing over some r have the same error Δ, then other suggestions passing over r are corrected by Δ.

5. Suppose $S[y] = v$. Two cases may arise.

   - If $v < y$, then the equation derived from $S[y]$ is discarded, as it is expected that v has already been swapped when $i = v$ had occurred and so is not likely to be in location y after y iterations of the KSA.

- If $v > y$, then there are two ways in which the assignment $S[y] \leftarrow v$ might have occurred: (a) when $i = y$ and $j = S[j] = v$, or (b) when $i = S[i] = v$ and $j = y$. The first case yields an equation of the typical form $K[0 \uplus v] = C_v$. In the second case, the event $(y = K[0 \uplus v] + \frac{v(v+1)}{2})$ occurs with high probability and so

$$\overline{C}_v = S^{-1}[v] - \frac{v(v+1)}{2}$$

can be considered as an alternative suggestion (in addition to C_v) for $K[0 \uplus v]$.

6. If the already-made guesses are correct, then after merging the counters, it is expected that the weight of the highest counter would be significantly higher than the other counters. When considering candidates for the t-th guess, only the ones with a weight not less than some threshold μ_t are considered, $0 \leq t < l$. The optimal values of these thresholds are determined empirically so as to increase the chances of eliminating the wrong guesses.

Table 4.2 below presents a comparison of the times T_1 and T_2 (in seconds) required by Algorithm 4.2.1 and its improvement due to the difference equations respectively to achieve the same success probabilities P_{suc}. As reported in [15, Section 7], the data were generated on a 3GHz Pentium IV CPU. The running times are averaged over 10000 random keys.

l	P_{suc}	T_1	T_2
5	0.8640	366	0.02
8	0.4058	2900	0.60
10	0.0786	183	1.46
10	0.1290	2932	3.93
12	0.0124	100	3.04
12	0.0212	1000	7.43
16	0.0005	500	278

TABLE 4.2: Improvement over Algorithm 4.2.1 in execution time due to using the difference equations.

4.4 Group of Key Bytes Approach

Subsequent to [15], the work [5] has revisited the key reconstruction from S_N with further improvements. They have accumulated the earlier ideas along with some additional new results to devise a more efficient algorithm for key recovery.

The key retrieval algorithm in [5] considers 6 types of events for guessing one j value:

1. $j_{y+1} = S_N[y]$.

2. $j_{y+1} = S_N^{-1}[y]$.

3. $j_{y+1} = S_N[S_N[[y]]$.

4. $j_{y+1} = S_N^{-1}[S_N^{-1}[y]]$.

5. $j_{y+1} = S_N[S_N[S_N[y]]]$.

6. $j_{y+1} = S_N^{-1}[S_N^{-1}[S_N^{-1}[y]]]$.

From two successive j values j_y and j_{y+1}, $6 \times 6 = 36$ candidates for the key byte $K[y]$ are obtained. They are weighted according to their probabilities. In addition, the Equations (4.10) of Section 4.3 [15] are also used. The sum s of the key bytes is guessed first. Then, for all m-byte combinations of l key bytes, the selected m key bytes under consideration are assigned values with the highest weight.

The remaining $l - m$ key bytes are solved as follows. For each group of four bytes, some of them are already guessed (being part of the selected m-combination). New candidates for the remaining (being part of the other $l-m$ key bytes) ones are found using the sum information (i.e., C_y values) in these 4-byte sequence. The values for these are then fixed one by one, by trying all possible candidates through a selected depth h.

The correctness of a key can be verified by the same technique as discussed for Algorithm 4.2.1, i.e., by running the KSA and comparing the resulting permutation with the permutation at hand.

In Table 4.3, we present some experimental data from [5], that corresponds to the best success probabilities obtained by running Algorithm 4.4.1. The implementation was performed on a 2.67GHz Intel CPU with 2GB RAM. The success probabilities are derived from 10000 randomly generated key-state table pairs.

4.5   Byte by Byte Recovery

This technique, reported in [134, Sections 2.3, 3.4], exploits all entries of both the permutations S_N and S_N^{-1} to narrow down the range of possible values of j in each round of the KSA to only two values in $\mathbb{Z}_N$ with a very good and constant success probability (> 0.37). The estimates of the two successive

Input:

1. Number l of key bytes.

2. The parameter m ($\leq l$).

3. The parameter h.

4. The permutation bytes $S_N[y]$, $0 \leq y \leq N - 1$, after the KSA.

Output: The recovered key bytes $K[0], K[1], \ldots, K[l-1]$ or FAIL.

Compute all C_i values for i and obtain all suggestions for sum such as $K[0 \uplus l] = C_{l+1} - C_1$;

Among the suggestions select the one with the highest weight as sum value;

Reduce all C_is in which $i > l$ to suggestions for sequences in which $i < l$;

for *all m-byte combinations of l* **do**

    Fix the specific bytes of the key that are declared to be true;

    for *each 4-byte group* **do**

        Start the update process which chooses the first h candidates that have been sorted according to the weights for the unknown key bytes;

    end

    Try all combinations of resulting candidates obtained for 4-byte groups;

    Return the correct key when it is found;

end

Return FAIL;

Algorithm 4.4.1: KeyGroupRetrieval

l	m	h	P_{suc}	Time (seconds)
5	3	256	0.998	0.008
8	5	64	0.931	8.602
12	4	6	0.506	54.390
16	6	4	0.745	1572

TABLE 4.3: Selected experimental results corresponding to Algorithm 4.4.1.

pairs of j's give four possible values of a key byte with good probability. Since each key is repeated at least $\lfloor \frac{N}{T} \rfloor$ times, a frequency table can be constructed for each secret key byte, by accumulating many suggestive values for the key.

4.5.1   Related Theoretical Results

Each value y in the permutation is touched at least once during the KSA by the indices i, j, $0 \leq y \leq N - 1$. Initially, y is located at index y. In round $y + 1$, when i equals y, either the value y is still in index y, or it has been moved due to swaps in one of the previous y rounds. In the former case, i touches it in round $y + 1$ and in the latter case, one of $\{j_1, j_2, \ldots, j_y\}$ has touched it already.

Theorem 4.5.1. *For* $0 \leq y \leq N - 1$,

$$P\left(j_{y+1} = S_N^{-1}[y] \text{ or } j_{y+1} = S_N[y]\right) = \frac{N-2}{N} \cdot \left(\frac{N-1}{N}\right)^{N-1} + \frac{2}{N}.$$

Proof: The event $(j_{y+1} = S_N^{-1}[y])$, i.e., the event $(S_N[j_{y+1}] = y)$ occurs, if $A_1(y)$, which is a combination of the following two events, holds.

1. y is not touched by any of $\{j_1, j_2, \ldots, j_y\}$ in the first y rounds. This happens with probability $(\frac{N-1}{N})^y$.

2. In round $y + 1$, when i becomes y, j_{y+1} moves y to one of the indices in $\{0, \ldots, y\}$ due to the swap and y remains there until the end of KSA. This happens with probability

$$P(j_{y+1} \in \{0, \ldots, y\}) \cdot P(j_\tau \neq j_{y+1}, y + 2 \leq \tau \leq N)$$
$$= \frac{y+1}{N} \cdot \left(\frac{N-1}{N}\right)^{N-y-1}.$$

Thus,

$$P\left(A_1(y)\right) = \left(\frac{N-1}{N}\right)^y \cdot \frac{y+1}{N} \cdot \left(\frac{N-1}{N}\right)^{N-y-1}$$
$$= \frac{y+1}{N} \cdot \left(\frac{N-1}{N}\right)^{N-1}.$$

Again, the event $(j_{y+1} = S_N[y])$ occurs, if $A_2(y)$, which is a combination of the following two events, holds. (Note that the event $A_2(y)$ is taken from Lemma 3.1.4. We here outline the proof of the probability of $A_2(y)$ for easy reference.)

1. $S_y[j_{y+1}] = j_{y+1}$ and therefore after the swap in round $y + 1$, $S_{y+1}[y] = j_{y+1}$. This happens if $j_{y+1} \in \{y, \ldots, N - 1\}$ and had not been touched by any of $\{j_1, j_2, \ldots, j_y\}$ in the first y rounds. The probability of this is $\frac{N-y}{N} \cdot (\frac{N-1}{N})^y$.

2. Once j_{y+1} sits in index y due to the above, it is not touched by any of the remaining $N - y - 1$ many j values until the end of the KSA. The probability of this is $(\frac{N-1}{N})^{N-y-1}$.

Thus,

$$
\begin{aligned}
P\left(A_2(y)\right) &= \frac{N-y}{N} \cdot \left(\frac{N-1}{N}\right)^{y} \cdot \left(\frac{N-1}{N}\right)^{N-y-1} \\
&= \frac{N-y}{N} \cdot \left(\frac{N-1}{N}\right)^{N-1}.
\end{aligned}
$$

Now, both $A_1(y)$ and $A_2(y)$ hold, if y is not touched by any of $\{j_1, j_2, \ldots, j_y\}$ in the first y rounds, and then $j_{y+1} = y$ so that y is not moved due to the swap, and subsequently y is not touched by any of the remaining $N - y - 1$ many j values until the end of the KSA. Thus,

$$
\begin{aligned}
P\left(A_1(y) \cap A_2(y)\right) &= \left(\frac{N-1}{N}\right)^{y} \cdot \frac{1}{N} \cdot \left(\frac{N-1}{N}\right)^{N-y-1} \\
&= \frac{1}{N}\left(\frac{N-1}{N}\right)^{N-1}.
\end{aligned}
$$

Hence,

$$
\begin{aligned}
P\left(A_1(y) \cup A_2(y)\right) &= P\left(A_1(y)\right) + P\left(A_2(y)\right) - P\left(A_1(y) \cap A_2(y)\right) \\
&= \frac{y+1}{N} \cdot \left(\frac{N-1}{N}\right)^{N-1} + \frac{N-y}{N} \cdot \left(\frac{N-1}{N}\right)^{N-1} \\
&\quad - \frac{1}{N}\left(\frac{N-1}{N}\right)^{N-1} \\
&= \left(\frac{N-1}{N}\right)^{N-1}.
\end{aligned}
$$

One way the event $\left(j_{y+1} = S_N^{-1}[y] \text{ or } j_{y+1} = S_N[y]\right)$ occurs is through $A_1(y) \cup A_2(y)$. Another way is that neither $A_1(y)$ nor $A_2(y)$ holds, yet $j_{y+1} \in \{S_N^{-1}[y], S_N[y]\}$ due to random association, whose probability contribution is

$$
\left(1 - \left(\frac{N-1}{N}\right)^{N-1}\right) \cdot \frac{2}{N}.
$$

Adding these two contributions, we get the result. ∎

For $N = 256$, the probability turns out to be > 0.37, which conforms to experimental observation.

Two events, namely $A_1(y)$ and $A_2(y)$, are introduced in the above proofs. These events would subsequently be referred in the proof of Theorem 4.5.2.

Theorem 4.5.1 implies that the permutation S_N and its inverse S_N^{-1} reveal information about the secret index j in each byte. This result can be used to reveal the secret key in the following manner.

Let

$$\mathcal{G}_0 = \{S_N[0], S_N^{-1}[0]\}$$

and for $1 \le y \le N - 1$, let

$$\mathcal{G}_y = \left\{ u - v - y \mid u \in \{S_N[y]\} \cup \{S_N^{-1}[y]\}, v \in \{S_N[y-1]\} \cup \{S_N^{-1}[y-1]\} \right\}.$$

Once more we like to remind that in $(u - v - y)$, the operations are modulo N. For $0 \le y \le N - 1$, $\mathcal{G}_y$ represents the set of possible values that the key byte $K[y]$ can take.

It is highly likely that $S_N[y] \ne S_N^{-1}[y]$ and $S_N[y-1] \ne S_N^{-1}[y-1]$. Hence we consider $|\mathcal{G}_0| = 2$ and $|\mathcal{G}_y| = 4$, $1 \le y \le N - 1$. We write

$$\mathcal{G}_0 = \{g_{01}, g_{02}\},$$

where $g_{01} = S_N^{-1}[0]$, and $g_{02} = S_N[0]$; and for $1 \le y \le N - 1$,

$$\mathcal{G}_y = \{g_{y1}, g_{y2}, g_{y3}, g_{y4}\},$$

where

$$
\begin{aligned}
g_{y1} &= S_N^{-1}[y] - S_N^{-1}[y-1] - y, \\
g_{y2} &= S_N[y] - S_N[y-1] - y, \\
g_{y3} &= S_N^{-1}[y] - S_N[y-1] - y, \\
g_{y4} &= S_N[y] - S_N^{-1}[y-1] - y.
\end{aligned}
$$

Further, let

$$p_{0x} = P(K[0] = g_{0x}), \qquad 1 \le x \le 2,$$

and for $1 \le y \le N - 1$, let

$$p_{yx} = P(K[y] = g_{yx}), \qquad 1 \le x \le 4.$$

We have the following result.

Theorem 4.5.2.
(1) $p_{01} = \frac{1}{N} \cdot \left(\frac{N-1}{N}\right)^N + \frac{1}{N}$ and $p_{02} = \left(\frac{N-1}{N}\right)^N + \frac{1}{N}$.
(2) *For* $1 \le y \le N - 1$,

$$
\begin{aligned}
p_{y1} &= \frac{y(y+1)}{N^2} \cdot \left(\frac{N-1}{N}\right)^{2N-1} + \frac{1}{N}, \\
p_{y2} &= \frac{(N-y)(N-y+1)}{N^2} \cdot \left(\frac{N-1}{N}\right)^{2N-1+y} + \frac{1}{N}, \\
p_{y3} &= \frac{(y+1)(N-y+1)}{N^2} \cdot \left(\frac{N-1}{N}\right)^{2N-1+y} + \frac{1}{N}, \\
p_{y4} &= \frac{y(N-y)}{N^2} \cdot \left(\frac{N-1}{N}\right)^{2N-1+y} + \frac{1}{N}.
\end{aligned}
$$

Proof: We would be referring to the events $A_1(y)$ and $A_2(y)$ in the proof of Theorem 4.5.1. From Theorem 4.5.1, we have

$$P(A_1(y)) = \frac{y+1}{N} \cdot \left(\frac{N-1}{N}\right)^{N-1}$$

$$\text{and} \quad P(A_2(y)) = \frac{N-y}{N} \cdot \left(\frac{N-1}{N}\right)^{N-1}.$$

For each probability p_{yx} in items (1) and (2), we would consider two components. The component which comes due to the contributions of the events $A_1(y)$, $A_2(y)$ etc, would be called α_{yx}. The other component is due to random association and is given by $(1 - \alpha_{yx}) \cdot \frac{1}{N}$. So for each probability p_{yx}, deriving the part α_{yx} suffices, as the total probability can be computed as

$$\alpha_{yx} + (1 - \alpha_{yx}) \cdot \frac{1}{N} = \frac{N-1}{N} \cdot \alpha_{yx} + \frac{1}{N}.$$

Consider the update rule in the KSA:

$$j_{y+1} = j_y + S_y[y] + K[y], \quad 0 \le y \le N - 1,$$

where $j_0 = 0$.

First, we prove item (1). Since $S_0[0] = 0$, we can write $j_1 = K[0]$. Considering $j_1 = S_N^{-1}[0]$, we have

$$\alpha_{01} = P(A_1(0))$$

and considering $j_1 = S_N[0]$, we have

$$\alpha_{02} = P(A_2(0)).$$

Substituting 0 for y in the expressions for $P(A_1(y))$ and $P(A_2(y))$, we get the results.

Now, we come to item (2). In the update rule, $S_y[y]$ can be replaced by y, assuming that it has not been touched by any one of $j_1, j_2, \ldots, j_y$ in the first y rounds of the KSA. This happens with a probability

$$\left(\frac{N-1}{N}\right)^y, \quad 0 \le y \le N - 1.$$

Assuming $S_y[y] = y$, we can write

$$K[y] = j_{y+1} - j_y - y.$$

When considering the contribution of $A_1(y)$ to j_{y+1}, the factor $\left(\frac{N-1}{N}\right)^y$ need not be taken into account, as the event $(S_y[y] = y)$ is already contained in

$A_1(y)$. Thus, the components α_{yx}'s for the probabilities p_{y1}, p_{y2}, p_{y3} and p_{y4} are respectively given by

$$
\begin{aligned}
\alpha_{y1} &= P\left(A_1(y)\right) \cdot P\left(A_1(y-1)\right), \\
\alpha_{y2} &= P\left(A_2(y)\right) \cdot P\left(A_2(y-1)\right) \cdot \left(\frac{N-1}{N}\right)^y, \\
\alpha_{y3} &= P\left(A_1(y)\right) \cdot P\left(A_2(y-1)\right), \\
\alpha_{y4} &= P\left(A_2(y)\right) \cdot P\left(A_1(y-1)\right) \cdot \left(\frac{N-1}{N}\right)^y.
\end{aligned}
$$

Substituting the probability expressions for $A_1(y)$, $A_1(y-1)$, $A_2(y)$ and $A_2(y-1)$, we get the results.   ∎

Corollary 4.5.3.
(1) $P(K[0] \in \mathcal{G}_0) = 1 - (1 - p_{01})(1 - p_{02})$.
(2) *For* $1 \le y \le N - 1$,

$$
P(K[y] \in \mathcal{G}_y) = 1 - (1 - p_{y1})(1 - p_{y2})(1 - p_{y3})(1 - p_{y4}).
$$

Substituting the values for y yields

$$
P(K[0] \in \mathcal{G}_0) \approx 0.37
$$

and for $1 \le y \le N - 1$, $P(K[y] \in \mathcal{G}_y)$ varies between 0.12 and 0.15. Experimental results also confirm these theoretical estimates.

Based on Theorems 4.5.1 and 4.5.2, we build the framework of retrieving individual key bytes.

The RC4 key κ of l bytes is repeated in key length boundaries to fill the N bytes of the key array K. The number of places in K where the same key byte $\kappa[w]$ is repeated is given by

$$
n_w = \begin{cases} \lfloor \frac{N}{l} \rfloor + 1 & \text{for } 0 \le w < N \bmod l; \\ \lfloor \frac{N}{l} \rfloor & \text{for } N \bmod l \le w < l. \end{cases}
$$

Thus, when considering a key byte $\kappa[w]$, we are interested in the set

$$
T_w = \bigcup_{y \in \mathbb{Z}_N,\, y \bmod l = w} \mathcal{G}_y,
$$

which is a union of n_w many $\mathcal{G}_y$'s, say, $\mathcal{G}_{w_1}, \mathcal{G}_{w_2}, \ldots, \mathcal{G}_{w_{n_w}}$. Denote the corresponding p_{yx}'s by $p_{w_1 x}, p_{w_2 x}, \ldots, p_{w_{n_w} x}$. Also, for notational convenience in representing the formula, we denote the size of $\mathcal{G}_y$ by M_y. We have already argued that M_0 can be taken to be 2 and for $1 \le y \le N - 1$, M_y can be considered to be 4.

Theorem 4.5.4. *For* $0 \le w \le l - 1$, *let* $freq_w$ *be the frequency (i.e., number of occurrences) of* $\kappa[w]$ *in the set* T_w. *Then for* $0 \le w \le l - 1$, *we have the* *following.*

(1) $P(\kappa[w] \in T_w) = 1 - \prod_{\tau=1}^{n_w} \prod_{x=1}^{M_{w_\tau}} (1 - p_{w_{tau}x})$.

(2) $P(freq_w = c) =$

$$\sum_{\substack{\{\tau_1, \tau_2, \ldots, \tau_c\} \\ \subseteq \{1, 2, \ldots, n_w\}}} \left(\prod_{r=1}^{c} \sum_{x=1}^{M_{w_{\tau_r}}} p_{w_{\tau_r}x} \prod_{x' \neq x} (1 - p_{w_{\tau_r}x'}) \right) \left(\prod_{\substack{r \in \{1, 2, \ldots, n_w\} \setminus \\ \{\tau_1, \tau_2, \ldots, \tau_c\}}} \prod_{x=1}^{M_{w_r}} (1 - p_{w_r x}) \right).$$

(3) $E(freq_w) = \sum_{\tau=1}^{n_w} \sum_{x=1}^{M_{w_\tau}} p_{w_\tau x}$.

Proof: First, we prove item (1). We know that $\kappa[w] \notin T_w$, if and only if $\kappa[w] \notin \mathcal{G}_{w_\tau}$ for all $\tau \in \{1, \ldots, n_w\}$. Again, $\kappa[w] \notin \mathcal{G}_{w_\tau}$, if and only if for each $x \in \{1, \ldots, M_{w_\tau}\}$, $\kappa[w] \neq g_{w_\tau x}$, the probability of which is $(1 - p_{w_\tau x})$. Hence the result follows.

Next, we prove item (2). Item (2) is a generalization of item (1), since

$$P(\kappa[w] \in T_w) = 1 - P(freq_w = 0).$$

For an arbitrary c, $0 \leq c \leq n_w$, $\kappa[w]$ occurs exactly c times in T_w, if and only if it occurs once in exactly c out of n_w many $\mathcal{G}_w$'s, say once in each of $\mathcal{G}_{w_{\tau_1}}, \mathcal{G}_{w_{\tau_2}}, \ldots, \mathcal{G}_{w_{\tau_c}}$, and it does not occur in any of the remaining $n_w - c$ many G_w's. We call such a division of the $\mathcal{G}_w$'s a *c-division*. Again, for $1 \leq r \leq c$, $\kappa[w]$ occurs exactly once in $\mathcal{G}_{w_{\tau_r}}$, if and only if $\kappa[w]$ equals exactly one of the $M_{w_{\tau_r}}$ members of $\mathcal{G}_{w_{\tau_r}}$ and it does not equal any of the remaining $(M_{w_{\tau_r}} - 1)$ members of $\mathcal{G}_{w_{\tau_r}}$. Thus, the probability that $\kappa[w]$ occurs exactly once in $\mathcal{G}_{w_{\tau_r}}$ is given by

$$\sum_{x=1}^{M_{w_{\tau_r}}} p_{w_{\tau_r}x} \prod_{x' \neq x} (1 - p_{w_{\tau_r}x'}).$$

Also, for any $r \in \{1, 2, \ldots, n_w\} \setminus \{\tau_1, \tau_2, \ldots, \tau_c\}$, $\kappa[w]$ does not occur in $\mathcal{G}_{w_r}$ with probability $\prod_{x=1}^{M_{w_r}} (1 - p_{w_r x})$. Adding the contributions of all $\binom{n_w}{c}$ many *c-division*'s, we get the result.

Finally, we come to item (3). For $1 \leq \tau \leq n_w$, $1 \leq x \leq M_{w_\tau}$ let $u_{\tau,x} = 1$, if $\kappa[w] = g_{w_\tau x}$; otherwise, let $u_{\tau,x} = 0$. Thus, the number of occurrences of $\kappa[w]$ in T_w is

$$freq_w = \sum_{\tau=1}^{n_w} \sum_{x=1}^{M_{w_\tau}} u_{\tau,x}.$$

Then, by linearity of expectation, we have

$$
\begin{aligned}
E(freq_w) &= \sum_{\tau=1}^{n_w} \sum_{x=1}^{M_{w\tau}} E(u_{\tau,x}) \\
&= \sum_{\tau=1}^{n_w} \sum_{x=1}^{M_{w\tau}} P(u_{\tau,x} = 1) \\
&= \sum_{\tau=1}^{n_w} \sum_{x=1}^{M_{w\tau}} p_{w_{\tau}x}.
\end{aligned}
$$

∎

Corollary 4.5.5. *For* $0 \le w \le l-1$, *given a threshold* H, $P(freq_w > TH)$ *can be estimated as* $1 - \sum_{c=0}^{TH} P(freq_w = c)$, *where* $P(freq_w = c)$*'s are as given in Theorem 4.5.4, item 2.*

Define the following notation.

$$
q_{yx} = \frac{1 - p_{yx}}{N-1}, \qquad \text{for } 0 \le x \le M_y, 0 \le y \le N-1.
$$

Theorem 4.5.6. *For* $0 \le w \le l-1$, *we have the following.*
(1) *The expected number of distinct values in* $\mathbb{Z}_N$ *occurring in* T_w *is given by*

$$
E_{dist} = N - \prod_{\tau=1}^{n_w} \prod_{x=1}^{M_{w\tau}} (1 - p_{w_{\tau}x}) - (N-1) \prod_{\tau=1}^{n_w} \prod_{x=1}^{M_{w\tau}} (1 - q_{w_{\tau}x}).
$$

(2) *The expected number of distinct values in* $\mathbb{Z}_N$, *each occurring exactly* c *times in* T_w, *is given by*

$$
E_c = \sum_{\substack{\{\tau_1,\tau_2,\ldots,\tau_c\} \\ \subseteq \{1,2,\ldots,n_w\}}} p(\tau) + (N-1) \sum_{\substack{\{\tau_1,\tau_2,\ldots,\tau_c\} \\ \subseteq \{1,2,\ldots,n_w\}}} q(\tau),
$$

where

$$
p(\tau) = \left(\prod_{r=1}^{c} \sum_{x=1}^{M_{w\tau_r}} p_{w_{\tau_r}x} \prod_{x'\neq x} (1 - p_{w_{\tau_r}x'}) \right) \left(\prod_{\substack{r\in\{1,2,\ldots,n_w\}\setminus \\ \{\tau_1,\tau_2,\ldots,\tau_c\}}} \prod_{x=1}^{M_{w r}} (1 - p_{w_r x}) \right)
$$

and

$$
q(\tau) = \left(\prod_{r=1}^{c} \sum_{x=1}^{M_{w\tau_r}} q_{w_{\tau_r}x} \prod_{x'\neq x} (1 - q_{w_{\tau_r}x'}) \right) \left(\prod_{\substack{r\in\{1,2,\ldots,n_w\}\setminus \\ \{\tau_1,\tau_2,\ldots,\tau_c\}}} \prod_{x=1}^{M_{w r}} (1 - q_{w_r x}) \right).
$$

Proof: First, we prove item (1). For $0 \leq u \leq N - 1$, let $x_u = 1$, if u does not occur in T_w; otherwise, let $x_u = 0$. Hence, the number of values in $\mathbb{Z}_N$ that do not occur at all in T_w is given by $X = \sum_{u=0}^{N-1} x_u$. A value u does not occur in T_w, if and only if it does not occur in any $\mathcal{G}_{w_\tau}$. For $u = \kappa[w]$, according to item 1 of Theorem 4.5.4,

$$P(x_u = 1) = \prod_{\tau=1}^{n_w} \prod_{x=1}^{M_{w_\tau}} (1 - p_{w_\tau x}).$$

Assuming that each $g_{w_\tau x}$, $1 \leq x \leq M_{w_\tau}$, takes each value $u \in \mathbb{Z}_N \setminus \{\kappa[w]\}$ with equal probabilities, we have

$$
\begin{aligned}
P(g_{w_\tau x} = u) &= \frac{1 - p_{w_\tau x}}{N - 1} \\
&= q_{w_\tau x}.
\end{aligned}
$$

So, for $u \in \mathbb{Z}_N \setminus \{\kappa[w]\}$,

$$P(x_u = 1) = \prod_{\tau=1}^{n_w} \prod_{x=1}^{M_{w_\tau}} (1 - q_{w_\tau x}).$$

We compute

$$
\begin{aligned}
E(X) &= \sum_{u=0}^{N-1} E(x_u) \\
&= E(x_{\kappa[w]}) + \sum_{u \in \mathbb{Z}_N \setminus \{\kappa[w]\}} E(x_u),
\end{aligned}
$$

where $E(x_u) = P(x_u = 1)$. The expected number of distinct values $\in \mathbb{Z}_N$ occurring in T_w is then given by $N - E(X)$.

Next, we come to item (2). Here, let $x'_u = 1$, if u occurs exactly c times in T_w; otherwise, let $x'_u = 0$. Hence, the number of values from $\mathbb{Z}_N$ that occurs exactly c times in T_w is given by

$$X' = \sum_{u=0}^{N-1} x'_u.$$

Now, u occurs exactly c times in T_w, if and only if it occurs once in exactly c out of n_w many $\mathcal{G}_w$'s and it does not occur in any of the remaining $n_w - c$ many $\mathcal{G}_w$'s. For $u = \kappa[w]$, $P(x'_u = 1)$ is given by item (2) of Theorem 4.5.4. In the same expression, $p_{w_{\tau_r} x}$ would be replaced by $q_{w_{\tau_r} x}$, when $u \neq \kappa[w]$. Since $E(x'_u) = P(x'_u = 1)$, and $E(X') = \sum_{u=0}^{N-1} E(x'_u)$, we get the result by adding the individual expectations. ∎

Corollary 4.5.7. *For* $0 \leq w \leq l-1$, *given a threshold* H, *the expected number of distinct values* $\in \mathbb{Z}_N$, *each occurring* $> H$ *times in* T_w, *can be estimated as* $N - \displaystyle\sum_{c=0}^{TH} E_c$, *where* E_c *is the expected number of distinct values* $\in \mathbb{Z}_N$, *each occurring exactly* c *times in* T_w, *as given in Theorem 4.5.6, item 2.*

Now let us see how close the theoretical estimates are to the experimental ones. Table 4.4 depicts some results corresponding to Theorems 4.5.4, 4.5.6. For all key lengths, the estimates are given for $\kappa[3]$ (i.e. for $w = 3$ and $c = 2$) only as a representative. The experimental results are obtained by averaging over 1 million runs, each with a randomly chosen key.

Similar results hold for all key bytes $\kappa[w]$, $0 \leq w \leq l - 1$, for each key length $l = 5, 8, 10, 12$ and 16, and for different values of c.

		l	5	8	10	12	16
$P(\kappa[w] \in T_w)$	Theory		0.9991	0.9881	0.9729	0.9532	0.8909
	Exp.		0.9994	0.9902	0.9764	0.9578	0.8970
$P(freq_w = c)$	Theory		0.0225	0.1210	0.1814	0.2243	0.2682
	Exp.		0.0186	0.1204	0.1873	0.2353	0.2872
$E(freq_w)$	Theory		6.8	4.3	3.5	3.0	2.1
	Exp.		6.8	4.3	3.5	2.9	2.1
E_{dist}	Theory		138.5	99.2	84.2	73.4	55.9
	Exp.		138.2	98.9	84.0	73.2	55.8
E_c	Theory		34.7	18.1	13.1	10.0	5.8
	Exp.		35.2	18.6	13.6	10.4	6.2

TABLE 4.4: Theoretical vs. empirical estimates with $w = 3$ and $c = 2$ for Theorems 4.5.4, 4.5.6.

There are a few exceptions where the theoretical and the empirical values are not close (e.g., the values in the second row and the column corresponding to $l = 5$). These cases arise because the idealistic assumption of independence of events does not always hold in practice. However, in general the theoretical formula fits the empirical data to a very good approximation.

4.5.2   A Set of Recovery Methods

The above results can be used to devise an algorithm *BuildKeyTable* for building a frequency table for each key byte and one can use the table to extract information about the key.

Arrays *jarr*1 and *jarr*2 contain the values of j_y's estimated from S_N and S_N^{-1} respectively. For each key byte $\kappa[w]$, $0 \leq w \leq l - 1$, $karr[w][y]$ stores the frequency of the value $y \in \mathbb{Z}_N$. "$x \mathrel{+}= 1$" is a short form for "$x \longleftarrow x + 1$."

When guessing a specific key byte $\kappa[w]$, Algorithm 4.5.1 considers all values in $\mathbb{Z}_N$ that have occurred at least once. However, other alternatives may

---

Input: The permutation S_N after the KSA.
Output: A frequency table for each key byte.
Data: Arrays $jarr1[N+1], jarr2[N+1], karr[l][N]$.

$jarr1[0] = jarr2[0] = 0$;
for $y = 0$ **to** $N - 1$ **do**
$\quad |\quad jarr1[y+1] = S[y]$ and $jarr2[y+1] = S_N^{-1}[y]$;
end
for $w = 0$ **to** $l - 1$ *and* $y = 0$ **to** $N - 1$ **do**
$\quad |\quad karr[w][y] = 0$;
end
$karr\big[0\big]\big[jarr1[1]\big] += 1$;
$karr\big[0\big]\big[jarr2[1]\big] += 1$;
for $y = 1$ **to** $N - 1$ **do**
$\quad |\quad karr\big[y \bmod l\big]\big[jarr1[y+1] - jarr1[y] - y\big] += 1$;
$\quad |\quad karr\big[y \bmod l\big]\big[jarr1[y+1] - jarr2[y] - y\big] += 1$;
$\quad |\quad karr\big[y \bmod l\big]\big[jarr2[y+1] - jarr1[y] - y\big] += 1$;
$\quad |\quad karr\big[y \bmod l\big]\big[jarr2[y+1] - jarr2[y] - y\big] += 1$;
end

Algorithm 4.5.1: BuildKeyTable

also be attempted, such as considering only those values in in $\mathbb{Z}_N$ that have frequency above a certain threshold c.

Experiments reveal that the *BuildKeyTable* algorithm recovers each individual key byte with a very high probability. Table 4.5 shows data for the first two bytes of secret keys with key lengths 5, 8, 10, 12 and 16. The results are obtained by considering 1 million randomly chosen secret keys of different key lengths. In Table 4.5, "Succ." denotes the success probability and "Search" denotes the number of values $\in \mathbb{Z}_N$ with frequency $> c$. Theoretical estimates of these values are given in Theorem 4.5.2 and Theorem 4.5.6 respectively.

As an example, each key byte for $l = 5$ can be guessed with a probability > 0.97 among only around 12 values from $\{0, 1, \ldots, 255\}$, each having frequency at least 3; whereas, for random guess, one need to consider at least 248 ($\approx 256 \times 0.97$) values to achieve the same probability.

For any key length, the success probability and search complexity of the key bytes from $\kappa[2]$ onwards are almost the same as those of $\kappa[1]$. Thus the data for $\kappa[1]$ is a representative of the other key bytes. Note that the success probability for $\kappa[0]$ is always little more than that for the other key bytes. This happens because $\kappa[0]$ is estimated as $j_1 - j_0 - 0$, where the event $j_0 = 0$ occurs with probability 1. This is also consistent with item (1) of Theorem 4.5.2.

For the same threshold c, the probability as well as the search complexity for each key byte decreases with increasing key length. This is due to the fact that the number of repetitions corresponding to each key byte also decreases with increasing key length.

l	Key byte	Threshold c = 0		Threshold c = 1		Threshold c = 2	
		Succ.	Search	Succ.	Search	Succ.	Search
5	$\kappa[0]$	0.9997	138.9	0.9967	47.9	0.9836	12.4
	$\kappa[1]$	0.9995	138.2	0.9950	47.4	0.9763	12.2
8	$\kappa[0]$	0.9927	97.6	0.9526	22.2	0.8477	4.1
	$\kappa[1]$	0.9902	98.9	0.9400	22.9	0.8190	4.2
10	$\kappa[0]$	0.9827	82.5	0.9041	15.7	0.7364	2.6
	$\kappa[1]$	0.9761	84.0	0.8797	16.3	0.6927	2.7
12	$\kappa[0]$	0.9686	71.6	0.8482	11.8	0.6315	1.8
	$\kappa[1]$	0.9577	73.2	0.8118	12.3	0.5763	1.8
16	$\kappa[0]$	0.9241	54.1	0.7072	6.7	0.4232	0.9
	$\kappa[1]$	0.8969	55.8	0.6451	7.1	0.3586	0.9

TABLE 4.5: Experimental results for first two key bytes using different thresholds.

Based on Algorithm 4.5.1, several variants of complete key recovery strategy exist. By *Method* 1, we refer to the simple strategy of guessing the complete key using different thresholds on the basic table obtained from *BuildKeyTable* algorithm.

Method 1A improves upon *Method* 1. It updates the basic frequency table obtained from the *BuildKeyTable* algorithm by considering the values obtained from the $S[S[y]]$ types of biases. In Section 3.1.3, it is demonstrated that there exist biases toward

$$f_y = \frac{y(y+1)}{2} + \sum_{x=0}^{y} K[y]$$

at $S_N[y]$, $S_N[S_N[y]]$, $S_N[S_N[S_N[y]]]$, $S_N[S_N[S_N[S_N[y]]]]$, and so on. In *Method* 1A, we guess $K[0]$ first. Then, we increment y one by one starting from 1 and given the values of $K[0], K[1], \ldots, K[y-1]$, we compute the value of $K[y]$ from the equations $S_N^I[y] = f_y$, where I denotes the level of indirections considered. For frequency updates of $\kappa[0]$, the first four levels of indirections (i.e., $I = 1, 2, 3$ and 4) are used. For frequency updates of other four key bytes, only the first two levels of indirections are used. Further, for frequency updates of $\kappa[1]$, only the values of $\kappa[0]$ with frequency > 2 are considered. Similarly, for $\kappa[2]$, the values of $\kappa[0]$ with frequency > 3 and the values of $\kappa[1]$ with frequency > 4 respectively are considered. These values (e.g., 2, 3, 4 above) are called the *threshold frequencies*. For $\kappa[3]$, the threshold frequencies of $\kappa[0], \kappa[1]$ and $\kappa[2]$ are 4 and 5 and 6 respectively. Finally, for $\kappa[4]$, the threshold frequencies of $\kappa[0], \kappa[1], \kappa[2]$ and $\kappa[3]$ are taken to be 4, 5, 6 and 7 respectively. Observe that while updating the table for the key byte $\kappa[w]$ for a fixed w, the thresholds for $\kappa[0], \ldots, \kappa[w-1]$ are increased as w increases. Small thresholds for $\kappa[0], \ldots, \kappa[w-1]$ substantially increase the number of choices for $\kappa[w]$ without significantly increasing the probability for obtaining the correct value of $\kappa[w]$. The thresholds mentioned above are tuned empirically.

Experimental Results for 5 Byte Keys

In Table 4.6, we present the search complexity and success probability related to all the bytes for a 5 byte secret key using both *Method* 1 and *Method* 1A. Here "Succ." denotes the success probability and "Comp." denotes the search complexity. The complexity of retrieving the entire key is calculated by multiplying the average number of values that need to be searched for individual key bytes. The success probability of recovering the entire key is computed empirically. This value is typically more than the product of the individual probabilities, indicating that the values of the key bytes are not independent given the final permutation bytes. Observe that just applying *Method* 1 helps to achieve 89.46% success rate in a complexity $12.4 \times (12.2)^4 \approx 2^{18.1}$.

Key byte	Method	Threshold $c = 0$		Threshold $c = 1$		Threshold $c = 2$	
		Succ.	Comp.	Succ.	Comp.	Succ.	Comp.
$\kappa[0]$	1	0.9997	138.9	0.9967	47.9	0.9836	12.4
	1A	0.9998	140.2	0.9980	49.2	0.9900	13.0
$\kappa[1]$	1	0.9995	138.2	0.9950	47.4	0.9763	12.2
	1A	0.9997	149.2	0.9971	56.6	0.9857	16.1
$\kappa[2]$	1	0.9995	138.2	0.9949	47.4	0.9764	12.2
	1A	0.9996	142.2	0.9965	50.7	0.9834	13.6
$\kappa[3]$	1	0.9995	138.2	0.9950	47.4	0.9761	12.2
	1A	0.9996	138.7	0.9958	47.8	0.9796	12.4
$\kappa[4]$	1	0.9995	138.2	0.9950	47.4	0.9766	12.2
	1A	0.9996	138.7	0.9958	47.8	0.9796	12.4
Entire Key	1	0.9976	$2^{35.6}$	0.9768	$2^{27.8}$	0.8946	$2^{18.1}$
	1A	0.9983	$2^{35.7}$	0.9830	$2^{28.3}$	0.9203	$2^{18.7}$

TABLE 4.6: Experimental results for all key bytes using different thresholds for $l = 5$.

Next, we discuss some enhancements of the basic technique for a 5-byte key. For each key byte, first the value with the maximum frequency is tried, then the second maximum and so on. The search is performed in an *iterative deepening* manner. If d_w is the depth (starting from the most frequent guess) of the correct value for $\kappa[w]$, then the search never goes beyond depth $d_{max} = max\{d_w, 0 \leq w \leq 4\}$ for any key byte. The complexity is determined by finding the average of d_{max}^5 over 1 million trials, each with a different key. A *depth limit D* is also set, which denotes the maximum number of different values that are to be tried for each key in descending order of their frequencies, starting from the most frequent value. We denote this strategy as *Method* 2. If the frequency table is updated as in *Method* 1A before performing the search in descending order of frequencies, then we name it *Method* 2A.

The experimental results for the above enhancements are presented in Table 4.7. The decimal fractions and the powers of 2 denote the success probabilities and the search complexities respectively.

D	10	16	32	48	64	80	96	160
2	0.8434 $2^{14.3}$	0.9008 $2^{17.0}$	0.9383 $2^{21.3}$	0.9719 $2^{23.5}$	0.9800 $2^{24.8}$	0.9841 $2^{26.0}$	0.9870 $2^{27.1}$	0.9980 $2^{29.0}$
2A	0.8678 $2^{14.0}$	0.9196 $2^{16.7}$	0.9503 $2^{20.9}$	0.9772 $2^{23.1}$	0.9855 $2^{24.5}$	0.9876 $2^{25.7}$	0.9906 $2^{26.6}$	0.9985 $2^{28.7}$

TABLE 4.7: Experimental results for $l = 5$ using the most frequent guesses of the key bytes.

Note that for a 5-byte key, [15] reports a success probability of 0.8640 in 0.02 seconds and [5] reports a success probability of 0.998 in 0.008 seconds, whereas *Method 2A* achieves a success probability of 0.9985 in complexity $2^{28.7}$.

Experimental Results for Other Key Lengths

For key lengths $l = 8, 10, 12$ and 16, we report results of simple experiments using *Method* 1A, where the technique of updating the frequency table is applied for the first 5 key bytes only. In [5,15], the time estimates are presented in seconds, whereas time complexity estimates of searching the number of keys are presented in other cases.

In Table 4.8, three sets of results are presented, namely, selected entries from Table 4.1 and data generated from Exp. I and II, which are explained below.

- In Exp. I, we report the exact time complexities for the experiments in which one can just exceed the probabilities given in [15, Table 4]. In Exp. II, we report the success probabilities for each key length that is achievable in complexity around 2^{40}, which should take less than a day using a few state of the art machines.

- In Exp. I, for $l = 8, 10$ and 12, a threshold of 2 for each of the first 4 bytes and for $l = 16$, a threshold of 2 for each of the first 2 bytes are used. A threshold of 1 is used for the remaining bytes for each key length. In Exp. II, for $l = 16$, the thresholds are the same as those in Exp. I. For other key lengths, the thresholds are as follows. For $l = 8$, a threshold of 1 is used for each of the first 6 bytes and a threshold of 0 for each of the last two bytes. A threshold of 1 is used for each key byte for the case $l = 10$. For $l = 12$, a threshold of 2 is used for the first byte and a threshold of 1 is used for each of the rest.

In the first and the last row of Table 4.8, the best probability values reported in [15, Table 4] and [5, Table 5] are presented respectively.

		8	10	12	16
Data from [15]	Probability	0.4058	0.1290	0.0212	0.0005
	Time in Seconds	0.60	3.93	7.43	278
Results from Table 4.1	Probability	0.414	0.162	0.026	0.0006
	Complexity	$2^{31.9}$	$2^{34.0}$	$2^{31.6}$	$2^{32.2}$
Exp. I	Probability	0.4362	0.1421	0.0275	0.0007
	Complexity	$2^{26.8}$	$2^{29.9}$	$2^{32.0}$	$2^{40.0}$
Exp. II	Probability	0.7250	0.2921	0.0659	0.0007
	Complexity	$2^{40.6}$	$2^{40.1}$	$2^{40.1}$	$2^{40.0}$
Data from [5]	Probability	0.931	–	0.506	0.0745
	Time in Seconds	8.602	–	54.390	1572

TABLE 4.8: Experimental results for $l = 8, 10, 12, 16$.

4.6 Bit by Bit Recovery

The work of Khazaei and Meier [83] starts with the equations

$$K[y_1 + 1 \ldots y_2] = C_{y_2} - C_{y_1}$$

of [15] and considers a bit-by-bit approach to key recovery. We briefly describe their work below.

The event

$$K[y_1 + 1 \ldots y_2] = C_{y_2} - C_{y_1}$$

holds with certain probability, say p_{y_1,y_2}. In practice, one can model the above equation as

$$K[y_1 + 1 \ldots y_2] = C_{y_2} - C_{y_1} + e_{y_1,y_2},$$

where e_{y_1,y_2} is the noise component noise whose distribution depends on p_{y_1,y_2}. The range of y_1, y_2 can be written as $-1 \leq y_1 < y_2 \leq N - 1$, with the choice $C_{-1} = 0$. Let E_{y_1,y_2} denote the random variables corresponding to the noises e_{y_1,y_2}. In this setup, the key can be recovered in a correlation-based attack by using a hypothesis-testing approach. Similar to [168], the authors of [83] assume that for wrong guess $\bar{K}$ of the key K, the values $C_{y_2} - C_{y_1}$ and $\bar{K}[y_1 + 1 \ldots y_2]$ are uncorrelated. Given $\frac{N(N+1)}{2}$ samples of

$$e_{y_1,y_2} = \bar{K}[y_1 + 1 \ldots y_2] - (C_{y_2} - C_{y_1})$$

as a realization of the random variables E_{y_1,y_2}, the problem is to decide whether the guess $\bar{K}$ is correct or not. Thus, one needs to test the null hypothesis

$$H_0 : \bar{K} = K, P(E_{y_1,y_2} = e \mid H_0) = \begin{cases} p_{y_1,y_2} & \text{for } e = 0; \\ \frac{1-p_{y_1,y_2}}{N-1} & \text{for } 1 \leq e \leq N - 1. \end{cases}$$

against the alternative hypothesis

$$H_1 : \bar{K} \neq K, P(E_{y_1,y_2} = e \mid H_1) = \frac{1}{N}, \quad 0 \leq e \leq N-1.$$

The optimum decision rule is given by the Neyman-Pearson Lemma [31] as follows: select H_0 if $M(S_N, \bar{K}) > T$; otherwise, select H_1, where

$$M(S_N, \bar{K}) = \sum_{-1 \leq y_1 < y_2 \leq N-1} log \frac{(N-1)p_{y_1,y_2}}{1 - p_{y_1,y_2}} \cdot \delta(e_{y_1,y_2})$$

$$\text{and} \quad \delta(e) = \begin{cases} 1 & \text{if } e = 0; \\ 0 & \text{otherwise.} \end{cases}$$

According to [83], the implementation requires $(\frac{l(l+1)}{2} - 1)N + 1$ memory and $\frac{l(l+1)}{2}N^l$ table look-ups. The false positive and negative rates depend on the threshold T.

The above idea can be extended to a bit-level decision-making process, assuming that N is a power of 2. In this case, one needs to test the null hypothesis

$$H_0^r : \bar{K} = K \bmod 2^r,$$

against the alternative hypothesis

$$H_1^r : \bar{K} \neq K \bmod 2^r,$$

where

$$P(E_{y_1,y_2} = e \bmod 2^r \mid H_0) = \begin{cases} p_{y_1,y_2}^r & \text{for } e = 0; \\ \frac{1-p_{y_1,y_2}^r}{2^r-1} & \text{for } 1 \leq e \leq 2^r - 1, \end{cases}$$

$$P(E_{y_1,y_2} = e \bmod 2^r \mid H_1) = \frac{1}{2^r}, \quad 0 \leq e \leq 2^r - 1, \text{ and}$$

$$p_{y_1,y_2}^r = p_{y_1,y_2} + \frac{N - 2^r}{2^r(N-1)}(1 - p_{y_1,y_2}).$$

The measure for comparison against the threshold T is given by

$$M^r(S_N, \bar{K}) = \sum_{-1 \leq y_1 < y_2 \leq N-1} log \frac{(2^r - 1)p_{y_1,y_2}^r}{1 - p_{y_1,y_2}^r} \cdot \delta(e_{y_1,y_2} \bmod 2^r).$$

The key recovery technique is to search over all 2^l possible values of the r-th LSB of the key bytes, $1 \leq r \leq 8$, assuming that the first $(r-1)$ LSBs of the key bytes are known, and choose only N_r out of them with the highest correlation measure $M^r(S_N, \bar{K})$. The complexity of this tree-based search algorithm can be shown to be $2^l(1 + \sum_{i=0}^{R-2} \prod_{j=0}^{i} N_j)$, where $R = \lceil log_2 N \rceil$.

4.7 Bidirectional Key Search

The main idea behind this approach [10] is to guess $K[0], K[1], \ldots$ using the left end of the permutation S and guess $K[l-1], K[l-2], \ldots$ using the right end of S simultaneously. After a number key bytes are guessed, the KSA may be run (in the forward or in the backward direction, depending on the position of the guessed key bytes) until a known j_{y+1} is encountered. If this matches with the computed j_{y+1}, then the guessing continues, otherwise the partial key is discarded. The indices y for which j_{y+1}'s are known act as "filters" for separating out the wrong candidate keys from the correct ones. This algorithm can recover secret keys of length 16 bytes with a success probability of 0.1409, which is almost two times the earlier best known value of 0.0745 reported in [5].

4.7.1 Sequences of Filter Indices

Here we present the structure and algebraic properties of the sequence of indices that act as filters for validating the secret key guesses. For simplicity, l is assumed to be a factor of N.

The secret index j is pseudo-random and may be considered to follow uniform distribution over the domain $[0, N-1]$. Hence, if the values of m consecutive j's are guessed randomly, the success probability is N^{-m}. For $N = 256$ and $m = 8$, this probability turns out to be $(256)^{-8} = 2^{-64}$. Whereas, using theoretical framework of this section, one can guess a sequence of j values with better probability (see Table 4.9).

Definition 4.7.1. (Event A_y) *For $0 \leq y \leq N-1$, event A_y occurs if and only if the following three conditions hold.*

1. $S_y[y] = y$ and $S_y[j_{y+1}] = j_{y+1}$.

2. $j_{y+1} \geq y$.

3. $S_N[y] = S_{y+1}[y]$.

Proposition 4.7.2. *For $0 \leq y \leq N-1$,*

$$p_y = P(A_y) = \left(\frac{N-y}{N}\right) \cdot \left(\frac{N-2}{N}\right)^y \cdot \left(\frac{N-1}{N}\right)^{N-1-y}.$$

Proof: Let us analyze the three cases mentioned in the definition of A_y.

1. $S_y[y] = y$ and $S_y[j_{y+1}] = j_{y+1}$ occur if $j_t \notin \{y, j_{y+1}\}$, for $1 \leq t \leq y$, the probability of which is $(\frac{N-2}{N})^y$.

2. $P(j_{y+1} \geq y) = \frac{N-y}{N}$.

3. $S_N[y] = S_{y+1}[y]$, if $j_t \neq y$, $\forall t \in [y+2, N]$. This happens with probability $(\frac{N-1}{N})^{N-y-1}$.

Multiplying the above three probabilities, we get the result.   ∎

Note that the event A_y implies $j_{y+1} = S_N[y]$ and Proposition 4.7.2 is a variant of Lemma 3.1.4 that relates j_{y+1} and $S_N[y]$.

Definition 4.7.3. (Event B_y) *For $0 \leq y \leq N - 1$, event B_y occurs if and only if the following three conditions hold.*

1. $S_y[y] = y$.

2. $j_{y+1} \leq y$.

3. $S_N[j_{y+1}] = S_{y+1}[j_{y+1}]$.

Proposition 4.7.4. *For $0 \leq y \leq N - 1$,*

$$p'_y = P(B_y) = \left(\frac{y+1}{N}\right) \cdot \left(\frac{N-1}{N}\right)^{N-1}.$$

Proof: Let us analyze the items mentioned in the definition of B_y.

1. $S_y[y] = y$ occurs if $j_t \neq y$ for $1 \leq t \leq y$, the probability of which is $(\frac{N-1}{N})^y$.

2. $P(j_{y+1} \leq y) = \frac{y+1}{N}$.

3. $S_N[j_{y+1}] = S_{y+1}[j_{y+1}]$, if $j_t \neq j_{y+1}$, $\forall t \in [y + 2, N]$. This happens with probability $(\frac{N-1}{N})^{N-y-1}$.

Multiplying the above three probabilities, we get the result.   ∎

Note that the event B_y implies $j_{y+1} = S_N^{-1}[y]$, $0 \leq y \leq N - 1$. B_y is the same as the event $A_1(y)$ defined in the proof of Theorem 4.5.1.

Definition 4.7.5. (Filter) *An index y in $[0, N - 1]$ is called a filter if either of the following two holds.*

1. $0 \leq y \leq \frac{N}{2} - 1$ and event A_y occurs.

2. $\frac{N}{2} \leq y \leq N - 2$ and event B_y occurs.

Definition 4.7.6. (Bisequence) *Suppose j_N is known. Then a sequence of at least $(t + t' + 1)$ many filters is called a (t, t')-bisequence if the following three conditions hold.*

1. *Exactly t many filters*

$$0 \leq i_1 < i_2 < \ldots < i_t \leq \frac{l}{2} - 1$$

exist in the interval $\left[0, \frac{l}{2} - 1\right]$.

2. *Exactly t' many filters*

$$N - 1 - \frac{l}{2} \le i'_{t'} < \ldots < i'_1 \le N - 2$$

exist in the interval $\left[N - 1 - \frac{l}{2}, N - 2\right]$.

3. *Either a filter i_{t+1} exist in the interval* $\left[\frac{l}{2}, l - 1\right]$ *or a filter $i'_{t'+1}$ exist in the interval* $\left[N - 1 - l, N - 2 - \frac{l}{2}\right]$.

Lemma 4.7.7. *Given a set F_t of t many indices in* $\left[0, \frac{l}{2} - 1\right]$, *a set $B_{t'}$ of t' many indices in* $\left[N - 1 - \frac{l}{2}, N - 2\right]$ *and an index x in* $\left[\frac{l}{2}, l - 1\right] \cup \left[N - 1 - l, N - 2 - \frac{l}{2}\right]$, *the probability of the sequence of indices $F_t \cup B_{t'} \cup \{x\}$ to be a (t, t')-bisequence is*

$$\left(\prod_{i_u \in F_t} p_{i_u}\right) \left(\prod_{i_v \in [0, l-1] \setminus (F_t \cup \{x\})} q_{i_v}\right) \left(\prod_{i'_{u'} \in B_{t'}} p'_{i'_{u'}}\right).$$

$$\left(\prod_{i'_{v'} \in [N-1-l, N-2] \setminus (B_{t'} \cup \{x\})} q'_{i'_{v'}}\right) \tilde{p}_x,$$

where $q_y = 1 - p_y$, $q'_y = 1 - p'_y$ for $0 \le y \le N - 1$ and $\tilde{p}_x = p_x$ or p'_x according as $x \in \left[\frac{l}{2}, l - 1\right]$ or $\left[N - 1 - l, N - 2 - \frac{l}{2}\right]$ respectively.

Proof: According to Definition 4.7.6, $F_t \cup B_{t'} \cup \{x\}$ would be an (t, t')-bisequence, if the indices in F_t and $B_{t'}$ and the index x are filters and the indices in $([0, l - 1] \cup [N - 1 - l, N - 2]) \setminus (F_t \cup B_{t'} \cup \{x\})$ are non-filters. Hence, the result follows from Propositions 4.7.2 and 4.7.4. ∎

Definition 4.7.8. (Critical Filter) *The last filter i_t within the first $\frac{l}{2}$ indices and the first filter i'_t within the last $\frac{l}{2}$ indices for an (t, t')-bisequence are called the left critical and the right critical filters respectively. Together, they are called the critical filters.*

Definition 4.7.9. (Favorable Bisequence) *For $d \le \frac{l}{2}$, a (t, t')-bisequence is called d-favorable, if the following seven conditions hold.*

1. $i_1 + 1 \le d$.

2. $i_{u+1} - i_u \le d, \ \forall u \in [1, t - 1]$.

3. $\frac{l}{2} - 1 - i_t \le d$.

4. $N - 1 - i'_1 \le d$.

5. $i'_v - i'_{v+1} \le d, \ \forall v \in [1, t' - 1]$.

6. $i'_{t'} - \left(N - 1 - \frac{l}{2}\right) \le d$.

7. $i_{t+1} - i_t \le d$ or $i'_{t'} - i'_{t'+1} \le d$.

Let us briefly explain the rationale behind Definition 4.7.9.

- For ease of analysis, a dummy index -1 is introduced to the left of index 0.

- Conditions 1 and 4 ensure that the first filter to the right of index -1 and the first filter to the left of index $N-1$ are at most at a distance d from indices -1 and $N-1$ respectively.

- Conditions 2 and 5 guarantee that the distance between any two consecutive filters to the left of the left critical filter and to the right of the right critical filter is at most d.

- Conditions 3 and 6 ascertain that the left and right critical filters are at most at a distance d from indices $\frac{l}{2} - 1$ and $N - 1 - \frac{l}{2}$ respectively. Consider the first filter i_{t+1} to the right of the left critical filter and the first filter $i'_{t'+1}$ to the left of the right critical filter. At least one of i_{t+1} and $i'_{t'+1}$ must exist for a (t, t')-bisequence.

- Finally, Condition 7 ensures that whichever exist (and at least one of the two, if both exist) is located at most at a distance d from the corresponding critical filter.

Lemma 4.7.10. *The number of distinct sequences of indices in*

$$\left[0, \frac{l}{2} - 1\right] \cup \left[N - 1 - \frac{l}{2}, N - 2\right]$$

satisfying the conditions 1 through 6 of Definition 4.7.9 is

$$\sum_{\delta \le d, \delta' \le d} \sum_{t \le \frac{l}{2} - \delta, t' \le \frac{l}{2} - \delta'} c(\delta, t) c(\delta', t'),$$

where $\delta = \frac{l}{2} - 1 - i_t$, $\delta' = i'_{t'} - \left(N - 1 - \frac{l}{2}\right)$, *and* $c(\delta, t)$ *is the coefficient of* $x^{\frac{l}{2} - \delta - t}$ *in* $(1 + x + \ldots + x^{d-1})^t$.

Proof: Let $x_1 = i_1$ and $x_u = i_u - i_{u-1} - 1$ for $2 \le u \le t$ be the number of non-filters between two consecutive filters in $[0, i_t]$. The total number of non-filters in the interval $[0, i_t]$ is

$$
\begin{aligned}
i_t - (t - 1) &= \left(\frac{l}{2} - 1 - \delta\right) - (t - 1) \\
&= \frac{l}{2} - \delta - t.
\end{aligned}
$$

Thus, the number of distinct sequences of indices in $\left[0, \frac{l}{2} - 1\right]$ satisfying conditions 1, 2 and 3 of Definition 4.7.9 is the same as the number of non-negative integral solutions of

$$x_1 + x_2 + \ldots + x_t = \frac{l}{2} - \delta - t,$$

where $0 \le x_u \le d - 1, \forall u \in [1,t]$. The number of solutions is given by $c(\delta, t)$. Similarly, the number of distinct sequences of indices in

$$\left[N - 1 - \frac{l}{2}, N - 2\right]$$

satisfying conditions 4, 5 and 6 of Definition 4.7.9 is $c(\delta', t')$. Hence, the number of distinct sequences of indices in

$$\left[0, \frac{l}{2} - 1\right] \cup \left[N - 1 - \frac{l}{2}, N - 2\right]$$

satisfying the conditions 1 through 6 is

$$\sum_{\delta \le d, \delta' \le d} \sum_{t \le \frac{l}{2} - \delta, t' \le \frac{l}{2} - \delta'} c(\delta, t) c(\delta', t').$$

∎

In Lemma 4.7.10, we have considered all the items except item 7 of Definition 4.7.9. Subsequent results related to Definition 4.7.9 would consider item 7 separately and would combine Lemma 4.7.10 with that computation.

Theorem 4.7.11. *The probability of existence of a d-favorable (t, t')-bisequence in*

$$[0, l - 1] \cup [N - 1 - l, N - 2]$$

is

$$\pi_d = \sum_{t,t'} \sum_{F_t, B_{t'}} \left(\prod_{i_u \in F_t} p_{i_u}\right) \left(\prod_{i_v \in [0, \frac{l}{2} - 1] \setminus F_t} q_{i_v}\right) \left(\prod_{i'_{u'} \in B_{t'}} p'_{i'_{u'}}\right) \cdot$$

$$\left(\prod_{i'_{v'} \in [N - 1 - \frac{l}{2}, N - 2] \setminus B_{t'}} q'_{i'_{v'}}\right) \left(1 - \prod_{\substack{y \in [\frac{l}{2}, i_t + d] \cup \\ [i_{t'} - d, N - 2 - \frac{l}{2}]}} \tilde{q}_y\right),$$

where the sum is over all $t, t', F_t, B_{t'}$ such that the sequence of indices $F_t \cup B_{t'}$ satisfy the conditions 1 through 6 in Definition 4.7.9 and $\tilde{q}_y = q_y$ or q'_y according as $y \in [\frac{l}{2}, i_t + d]$ or $[i_{t'} - d, N - 2 - \frac{l}{2}]$ respectively.

Proof: Immediately follows from Lemma 4.7.7. The term

$$\left(1 - \prod_{y \in [i_t + 1, i_t + d] \cup [i_{t'} - d, i_{t'} - 1]} \tilde{q}_y\right)$$

accounts for condition 7 of Definition 4.7.9. ∎

Using the definitions of $c(\delta, t)$, $c(\delta', t')$ introduced in Theorem 4.7.10, one can approximate the probability expression presented in Theorem 4.7.11 as follows.

Corollary 4.7.12. *If $l \leq 16$ (i.e., the key length is small), then we have*

$$\pi_d \approx \sum_{\delta \leq d, \delta' \leq d} \sum_{t \leq \frac{l}{2} - \delta, t' \leq \frac{l}{2} - \delta'} c(\delta, t) p^t q^{\frac{l}{2} - t} c(\delta', t') p'^{t'} q'^{\frac{l'}{2} - t'} (1 - q^{d-\delta} q'^{d-\delta'}),$$

where $p = \frac{2}{l} \sum_{y=0}^{\frac{l}{2}-1} p_y$, $p' = \frac{2}{l} \sum_{y=N-1-\frac{l}{2}}^{N-2} p'_y$, $q = 1 - p$, $q' = 1 - p'$.

Proof: Approximating each p_y in the left half by the average p of the first $\frac{l}{2}$ many p_y's and each p'_y in the right half by the average p' of the last $\frac{l}{2}$ many p'_y's, we get the result. The ranges of the variables in the summation account for conditions 3 and 6 of Definition 4.7.9. The term $c(\delta, t) p^t q^{\frac{l}{2} - t}$ accounts for the conditions 1 and 2, the term $c(\delta', t') p'^t q'^{\frac{l'}{2} - t'}$ accounts for the conditions 4 and 5, and the term $(1 - q^{d-\delta} q'^{d-\delta'})$ accounts for the condition 7 of Definition 4.7.9. ∎

A comparison of the theoretical estimates of π_d from Corollary 4.7.12 with the experimental values obtained by running the RC4 KSA with 10 million randomly generated secret keys of length 16 bytes is presented in Table 4.9.

d	2	3	4	5	6
Theoretical	0.0055	0.1120	0.3674	0.6194	0.7939
Experimental	0.0052	0.1090	0.3626	0.6152	0.7909

TABLE 4.9: Theoretical and experimental values of π_d vs. d with $l = 16$.

The results indicate that the theoretical values match closely with the experimental values.

Theorem 4.7.13. *The number of distinct d-favorable (t, t')-bisequences containing exactly $f \leq 2l$ filters is*

$$\sum_{\delta \leq d, \delta' \leq d} \sum_{t \leq \frac{l}{2} - \delta, t' \leq \frac{l}{2} - \delta'} c(\delta, t) c(\delta', t') \sum_{s=1}^{2d-\delta-\delta'} \binom{2d - \delta - \delta'}{s} \binom{l - 2d + \delta + \delta'}{f - t - t' - s}.$$

Proof: By Lemma 4.7.10, the number of distinct sequences of indices in

$$\left[0, \frac{l}{2} - 1\right] \cup \left[N - 1 - \frac{l}{2}, N - 2\right]$$

satisfying the conditions 1 through 6 of Definition 4.7.9 is

$$\sum_{\delta \leq d, \delta' \leq d} \sum_{t \leq \frac{l}{2} - \delta, t' \leq \frac{l}{2} - \delta'} c(\delta, t) c(\delta', t').$$

The justification for the two binomial coefficients with a sum over s is as follows. For condition 7 to be satisfied, at least one out of $2d - \delta - \delta'$ indices in

$$\left[\frac{l}{2}, \frac{l}{2} - 1 + d - \delta\right] \cup \left[N - 1 - \frac{l}{2} - (d - \delta'), N - 2 - \frac{l}{2}\right]$$

must be a filter. Let the number of filters in this interval be $s \geq 1$. So the remaining $f - t - t' - s$ many filters must be among the remaining $l - 2d + \delta + \delta'$ many indices in

$$\left[\frac{l}{2} + d - \delta, l - 1\right] \cup \left[N - 1 - l, N - 2 - \frac{l}{2} - (d - \delta')\right].$$

∎

Corollary 4.7.14. *The number of distinct d-favorable (t, t')-bisequences containing at most $F \leq 2l$ filters is*

$$C_{d,F} = \sum_{\delta \leq d, \delta' \leq d} \sum_{t \leq \frac{l}{2} - \delta, t' \leq \frac{l}{2} - \delta'} c(\delta, t)c(\delta', t')$$

$$\sum_{s=1}^{2d - \delta - \delta'} \binom{2d - \delta - \delta'}{s} \sum_{r=0}^{F - t - t' - s} \binom{l - 2d + \delta + \delta'}{r}.$$

Proof: In addition to $s \geq 1$ filters in

$$\left[\frac{l}{2}, \frac{l}{2} - 1 + d - \delta\right] \cup \left[N - 1 - \frac{l}{2} - (d - \delta'), N - 2 - \frac{l}{2}\right],$$

we need r more filters in

$$\left[\frac{l}{2} + d - \delta, l - 1\right] \cup \left[N - 1 - l, N - 2 - \frac{l}{2} - (d - \delta')\right],$$

where $t + t' + s + r \leq F$. ∎

Corollary 4.7.15. *The number of distinct d-favorable (t, t')-bisequences in $[0, l - 1] \cup [N - 1 - l, N - 2]$ (containing at most $2l$ filters) is*

$$C_{d,2l} = \sum_{\delta \leq d, \delta' \leq d} \sum_{t \leq \frac{l}{2} - \delta, t' \leq \frac{l}{2} - \delta'} c(\delta, t)c(\delta', t')(1 - 2^{-2d + \delta + \delta'})2^l.$$

Proof: Substitute $F = 2l$ in Corollary 4.7.14 and simplify.

Alternatively, the number of distinct sequences satisfying the conditions 1 through 6 is

$$\sum_{\delta, \delta'} \sum_{t, t'} c(\delta, t)c(\delta', t')2^l$$

and out of these the number of sequences violating the condition 7 is

$$\sum_{\delta, \delta'} \sum_{t, t'} c(\delta, t)c(\delta', t')2^{l - 2d + \delta + \delta'}.$$

Subtracting, we get the result. Note that the term 2^r stands for the number of ways in which a set of r indices may be a filter or a non-filter. ∎

As an illustration, we present $C_{d,F}$ for different values of d and F with $l = 16$ in Table 4.10.

F	32	20	16	12	8
$d = 2$	$2^{27.81}$	$2^{27.31}$	$2^{24.72}$	$2^{18.11}$	$2^{3.58}$
$d = 3$	$2^{30.56}$	$2^{30.38}$	$2^{29.02}$	$2^{24.86}$	$2^{15.95}$
$d = 4$	$2^{31.47}$	$2^{31.35}$	$2^{30.38}$	$2^{27.17}$	$2^{20.14}$

TABLE 4.10: $C_{d,F}$ for different F and d values with $l = 16$.

Though we have assumed that l divides N in the above analysis, the same idea works when l is not a factor of N. One only needs to map the indices from the right end of S to the appropriate key bytes.

4.7.2   Application of Filter Sequences in Bidirectional Key Search

The theoretical properties of filter sequences discussed in the previous subsection can be used to devise a novel key retrieval algorithm.

Suppose that after the RC4 KSA is over, the final permutation S_N and the final value j_N of the index j are known. The update rule for the KSA is

$$j_{y+1} = j_y + S_y[y] + K[y], \qquad 0 \le y \le N - 1.$$

Since S_0 is the identity permutation and $j_0 = 0$, we have $j_1 = K[0]$. Thus, if index 0 is a filter, then $K[0]$ is known. Again, if index $N - 2$ is a filter, then j_{N-1} is known and hence $K[l - 1] = K[N - 1]$ can be determined as

$$j_N - j_{N-1} - S_{N-1}[N - 1] = j_N - j_{N-1} - S_N[j_N].$$

Moreover, if two consecutive indices $y - 1$ and y happen to be filters, $1 \le y \le N - 2$, then j_y and j_{y+1} are known and $S_y[y] = y$ from the definition of filters. Thus, the correct value of the key byte $K[y \bmod l]$ can be extracted as

$$K[y \bmod l] = j_{y+1} - j_y - y.$$

For complete key recovery, apart from the knowledge of the final permutation S_N and the final value j_N of the pseudo-random index j, we also need to assume that a *d-favorable* (t, t')-*bisequence* exists. Given such a sequence, our algorithm does not require one to guess more than d many key bytes at a time. For faster performance, one should keep d small (typically, 4).

The basic strategy for full key recovery is as follows. First, a partial key recovery is performed using the filters in the interval

$$[0, M - 1] \cup [N - 1 - M, N - 2],$$

where $l \leq M \leq 2l$, such that the correct values of at least n_{corr} out of l key bytes are uniquely determined and at least n_{in} out of n_{corr} are derived from the filters in

$$[0, l-1] \cup [N-1-l, N-2].$$

Let $\mathcal{P}$ be the set of the partially recovered key bytes, each of which is associated with a single value, namely, the correct value of that key byte.

Next, a frequency table is built for the $l - n_{corr}$ many unknown key bytes by guessing one j_{y+1} from the 8 values

$$S_N[y], S_N^{-1}[y], S_N[S_N[[y]], S_N^{-1}[S_N^{-1}[y]], S_N[S_N[S_N[y]]],$$
$$S_N^{-1}[S_N^{-1}[S_N^{-1}[y]]], S_N[S_N[S_N[S_N[y]]]] \text{ and } S_N^{-1}[S_N^{-1}[S_N^{-1}[S_N^{-1}[y]]]].$$

From two successive j values j_y and j_{y+1}, $8 \times 8 = 64$ candidates for the key byte $K[y]$ are found. These candidates are weighted according to their probabilities. The 256 possible values of each unknown key byte $K[y]$ are grouped into 4 sets of size 64 each. Let G_y be the set of values obtained from the above frequency table and let $H_{1,y}, H_{2,y}, H_{3,y}$ be the other sets, called *bad* sets. For $0 \leq y \leq l-1$, let $g_y = P(K[y] \in G_y)$. Since empirical results show that g_y is almost the same for each y, one can use the average $g = \frac{1}{l} \sum_{y=0}^{l-1} g_y$ in place of each g_y.

After this, a bidirectional search is performed from both ends several times, each time with exactly one of the sets $\{G_y, H_{1,y}, H_{2,y}, H_{3,y}\}$ for each unknown key byte. The bad sets are used for at most b of the unknown key bytes. The search algorithm, called *BidirKeySearch*, takes as inputs the final permutation S_N, the final value j_N, the partially recovered key set $\mathcal{P}$, a set $\mathcal{S}$ of $l - n_{corr}$ many candidate sets each of size 64 and a d-favorable (t, t')-bisequence $\mathcal{B}$.

For the left end, first one guesses the key bytes $K[0], \ldots, K[i_1]$ except those in $\mathcal{P}$. Starting with S_0 and j_0, the KSA is run until S_{i_1+1} and j_{i_1+1} are obtained. Since the correct value of j_{i_1+1} is known, the tuples leading to the correct j_{i_1+1} form a set T_{i_1} of candidate solutions for $K[0 \ldots i_1]$. These tuples are said to "pass the filter" i_1. Similarly, starting with j_{i_1+1} and each state S_{i_1+1} associated with each tuple in T_{i_1}, the key bytes $K[i_1+1], \ldots, K[i_2]$ excluding those in $\mathcal{P}$ are guessed. Thus, a set T_{i_2} of candidate solutions is obtained for $K[0 \ldots i_2]$ passing the filter i_2, and so on until a stage is reached where a set T_{i_t} of candidate solutions for $K[0 \ldots i_t]$ passing the *left critical filter* i_t would exist.

In the same manner, one starts with S_N and j_N and runs the KSA backwards until one has a set $T_{i'_{t'}}$ of candidate solutions for $K[i'_{t'} + 1 \ldots l - 1]$ passing the *right critical filter* $i'_{t'}$.

Among the set of remaining key bytes

$$L = \{K[i_t + 1], \ldots, K[i_{t+1}]\} \setminus \mathcal{P}$$

on the left, and the set of remaining key bytes

$$R = \{K[i'_{t'+1} + 1], \ldots, K[i'_{t'}]\} \setminus \mathcal{P}$$

on the right, some bytes are common. First, the smaller of these two sets is guessed and then only those key bytes from the larger set are guessed that are needed to complete the whole key. One can reduce the candidate keys by filtering the left half of the key using the right filters and the right half of the key using the left filters.

In the *BidirKeySearch* algorithm description, i_0 and i_0' denote the two default filters -1 (a dummy filter) and $N - 1$ respectively.

Let m_1 be the complexity and α be the probability of obtaining a d-favorable (t, t') bisequence along with the partially recovered key set $\mathcal{P}$, satisfying the input requirements of *BidirKeySearch*. We require a search complexity of

$$m_2 = \binom{2(M - l)}{n_{corr} - n_{in}}$$

for finding the $n_{corr} - n_{in}$ many correct key bytes from the filters in

$$[l, l + M - 1] \cup [N - 1 - l, N - 2 - M].$$

We need to run the *BidirKeySearch* algorithm a total of

$$m_3 = \sum_{r=0}^{b} \binom{n - n_{corr}}{r} 3^r$$

times, leading to a success probability

$$\beta = \sum_{r=n-n_{corr}-b}^{n-n_{corr}} \binom{n - n_{corr}}{r} g^r (1 - g)^{n - n_{corr} - r}.$$

If each run of the *BidirKeySearch* algorithm consumes τ time, the overall complexity for the full key recovery is given by $m_1 m_2 m_3 \tau 2^8$ and the success probability is given by $\alpha\beta$. The term 2^8 appears due to guessing the correct value of j_N.

For $l = 16$, $d = 4$, $M = 20$, $n_{corr} = 6$, $n_{in} = 4$, experiments with 10 million randomly generated secret keys reveal that $\alpha \approx 0.1537$ and $g \approx 0.8928$. Also, $m_1 \approx 2^{31}$ (see Table 4.10) and $m_2 \approx 2^5$. With $b = 2$, β and m_3 turn out to be 0.9169 and 2^9 (approx.) respectively. Thus, the success probability is $\alpha\beta \approx 0.1409$ and the complexity is approximately $2^{31+5+9+8}\tau = 2^{53}\tau$. Though the the exact estimation of τ is not feasible in theory, τ is expected to be small, since for wrong filter sequences the search is likely to terminate in a negligible amount of time. Before [10], the best known success probability for recovering a 16-byte secret key has been reported in [5] as 0.0745 and the corresponding time required by Algorithm 4.4.1 is 1572 seconds, but it is not clear how the complexity of this algorithm grows with increase in probability. Algorithm 4.7.1 [10] achieves better probability, but with very high time complexity.

Input: S_N, j_N, $\mathcal{P}$, $\mathcal{S}$ and $\mathcal{B}$.
Output: The recovered key bytes $K[0], K[1], \ldots, K[l-1]$ or FAIL.

Guess the Left Half-Key:
for $u = 1$ *to* t **do**
$\quad$ Iteratively guess $\{K[i_{u-1}+1], \ldots, K[i_u]\} \setminus \mathcal{P}$;
$\quad$ $T_{i_u} \leftarrow \{\{K[0], \ldots, K[i_u]\} : \text{the tuple pass the filter } i_u\}$;
end

Guess the Right Half-Key:
for $v = 1$ *to* t' **do**
$\quad$ Iteratively guess $\{K[i'_v+1], \ldots, K[i'_{v-1}]\} \setminus \mathcal{P}$;
$\quad$ $T_{i_v} \leftarrow \{\{K[i'_v+1], \ldots, K[l-1]\} : \text{the tuple pass the filter } i_v\}$;
end

Merge to Get the Full Key:
$L \leftarrow \{K[i_t+1], \ldots, K[i_{t+1}]\} \setminus \mathcal{P}$ and
$R \leftarrow \{K[i'_{t'+1}+1], \ldots, K[i'_{t'}]\} \setminus \mathcal{P}$;
if $|L| < |R|$ **then**
$\quad$ Guess L;
$\quad$ $T_{i_{t+1}} \leftarrow \{\{K[0], \ldots, K[i_{t+1}]\} : \text{the tuple pass the filter } i_{t+1}\}$;
$\quad$ Guess $R \setminus T_{i_{t+1}}$ using the filter $i'_{t'+1}$;
end
else
$\quad$ Guess R;
$\quad$ $T_{i'_{t'+1}} \leftarrow \{\{K[i'_{t'}+1], \ldots, K[l-1]\} : \text{the tuple pass the filter } i_v\}$;
$\quad$ $L \setminus T_{i'_{t'+1}}$ using the filter i_{t+1};
end

Cross-Filtration:
if $K[0 \ldots m-1]$ *is guessed from the left filters and* $K[m \ldots l-1]$ *is guessed from the right filters* **then**
$\quad$ then validate $K[m \ldots l-1]$ using the left filters up to $l-1$ and
$\quad$ validate $K[0 \ldots m-1]$ using the right filters up to $N-1-l$;
end

Algorithm 4.7.1: BidirKeySearch

Research Problems

Problem 4.1 Characterize the weak keys at the end of Section 4.2.

Problem 4.2 What is the relation between the time complexity and the success probability of each of the key recovery algorithms?

Problem 4.3 Find an estimate of τ in the complexity expression for bidirectional key search.

Chapter 5

Analysis of Keystream Generation

The previous two chapters focused on the RC4 KSA. In this chapter, we present a detailed analysis of the keystream generation component PRGA of RC4.

Certain impossible states of RC4 PRGA were discovered by Finney [50]. We begin the chapter with a discussion on these states in Section 5.1. Next, in Section 5.2, we discuss Glimpse Theorem [78,108], a very important result about the leakage of state information in keystream.

Most of the cryptanalytic materials on RC4 PRGA point out the weaknesses for the initial keystream bytes [52, 86, 103, 107, 108, 122, 131, 139, 146, 180, 184, 185] and many of these works can be used to attack RC4 in IV mode (e.g., WEP applications). There are comparatively fewer results when RC4 is considered after throwing away any amount of initial keystream bytes. One such work is Mantin's digraph distinguisher [109] described in the next chapter (in Section 6.3.1). Another important work that concentrates on studying the behavior of RC4 at any stage of keystream generation and makes no use of the RC4 initialization process is [9], which provides a theoretical investigation into the structure and evolution of RC4 PRGA. Section 5.4 is based on the analysis of [9]. In this section, a complete characterization of the RC4 PRGA for a single step is presented. The detailed analysis of the single step itself is quite involved and tedious, and so the complete analysis of two consecutive steps becomes harder. Thus, a particular case for two steps is studied, from which weaknesses of RC4 can be identified.

The next two sections deal with the biases of the keystream output bytes toward the secret key. Section 5.5 analyzes the first keystream byte of RC4. As part of this analysis, we present the proof of Roos' experimental observations [146] that appeared for the first time in [131]. Biases of several forms toward the secret key in many keystream bytes, that have been reported in [86, 103], are presented in Section 5.6. In addition to the biases in the initial keystream bytes, biases at much later stages such as in rounds 256, 257 and beyond are discussed in this section.

The last section gives an overview of exhaustive enumeration of all linear biases in RC4, that has been recently attempted in [158].

5.1   Finney Cycles

Shortly after the RC4 algorithm was known publicly, Finney [50] discovered an interesting set of states linked by a set of short cycles.

Theorem 5.1.1. *Suppose $j_r^G = i_r^G + 1$ and $S_r^G[j_r^G] = 1$ hold in some round r of RC4 PRGA. Then*
    (1) *the same relations continue to hold in all subsequent rounds,*
    (2) *a cycle of length $N(N-1)$ is formed in the state-space and*
    (3) *the output bytes $z_{r+(N-1)}, z_{r+2(N-1)}, \ldots, z_{r+N(N-1)}$ are a shift of S_r^G.*

Proof: Suppose that after round r of the RC4 PRGA, $j_r^G = i_r^G + 1$ and $S_r^G[j_r^G] = 1$. In the next round, $i_{r+1}^G = i_r^G + 1$ and

$$
\begin{aligned}
j_{r+1}^G &= j_r^G + S_r^G[i_{r+1}^G] \\
&= i_r^G + 1 + 1 \\
&= i_r^G + 2.
\end{aligned}
$$

Now, the swap yields $S_{r+1}^G[i_r^G + 2] = 1$ and $S_{r+1}^G[i_r^G + 1] = S_r^G[i_r^G + 2]$. Thus, the state at round $r + 1$ has the same relation among (S^G, i^G, j^G) as that at round r. The same logic, when applied to round $r + 1$, yields the same relations in round $r + 2$ and so on. This proves (1).

In each subsequent round, entries $S^G[y + 1]$ are moved back to $S^G[y]$ and the value "1" moves up. After N rounds, the "1" returns to its original position and the rest of the entries shift one position backwards. After $N - 1$ such sets of N rounds, i.e., at round $r + N(N - 1)$, all the entries will return to their original positions as in round r. Thus, a cycle of length $N(N - 1)$ is formed in the state-space. This proves (2).

For a fixed value of i^G, the entry $S^G[i^G]$ repeats it's value every $N - 1$ steps and the value of $S^G[j^G]$ remains 1 always. Thus the index $S^G[i^G] + S^G[j^G]$ of the output byte is repeated every $N - 1$ steps. As proved in (2) above, the value at this entry is rotated by one position to make room for the next value in the cyclic order. Hence the N keystream output bytes, each separated by $N - 1$ rounds from the previous, are nothing but a shift of the permutation. This proves (3).   ∎

The cycle thus formed in the state space is called a *Finney cycle*. Since each state in this cycle has the same relations (the preconditions of Lemma 5.1.1) and the PRGA round is reversible (see Section 4.1), it is not possible to move from a state not in this form to one that is. Such a state is called a *Finney state*. Fortunately, the initialization $i_0^G = j_0^G = 0$ in RC4 PRGA prevents the condition for a Finney cycle to occur in normal RC4 operation.

As an example, consider the permutation during the PRGA as shown in Table 5.1.

If $i^G = 0$ and $j^G = 1$, this is a Finney state, since $S^G[1] = 1$. Table 5.2

Permutation entry $S^G[y]$, where y varies from 0 to 255 in row major order															
205	1	142	124	195	47	132	187	113	200	109	106	54	65	75	241
190	183	159	157	244	34	115	31	140	138	76	216	251	20	93	180
52	66	119	71	154	72	103	230	55	135	186	172	51	168	82	69
114	253	139	169	161	26	178	44	39	57	121	247	27	77	38	131
133	218	111	42	207	136	53	137	229	217	228	9	40	84	194	48
3	60	107	182	101	203	96	238	243	234	214	45	86	211	173	130
50	78	226	64	146	163	156	14	170	202	174	41	0	36	134	73
246	112	184	49	11	17	160	175	141	13	68	248	151	15	120	108
2	117	25	61	224	250	83	185	166	89	21	32	143	62	59	16
233	221	118	88	199	122	100	7	193	123	179	43	215	254	240	176
94	104	129	209	6	144	58	110	196	125	92	191	164	90	220	149
210	208	192	24	8	225	46	171	213	19	74	242	4	197	237	150
162	127	177	70	167	10	249	67	219	80	212	147	188	18	153	35
152	223	198	181	126	28	95	79	22	99	189	97	206	148	245	102
81	239	231	63	85	91	98	255	158	227	252	56	155	30	222	12
145	5	128	29	235	37	201	236	204	23	87	33	165	232	105	116

TABLE 5.1: Permutation in a Finney state.

shows the keystream, where two consecutive values denote keystream bytes separated by 255 rounds of the PRGA.

Keystream byte z_r corresponding to $r = 255n$, for $n = 1, 2, \ldots, 256$ in row major order															
35	152	223	198	181	126	28	95	79	22	99	189	97	206	148	245
102	81	239	231	63	85	91	98	255	158	227	252	56	155	30	222
12	145	5	128	29	235	37	201	236	204	23	87	33	165	232	105
116	205	1	142	124	195	47	132	187	113	200	109	106	54	65	75
241	190	183	159	157	244	34	115	31	140	138	76	216	251	20	93
180	52	66	119	71	154	72	103	230	55	135	186	172	51	168	82
69	114	253	139	169	161	26	178	44	39	57	121	247	27	77	38
131	133	218	111	42	207	136	53	137	229	217	228	9	40	84	194
48	3	60	107	182	101	203	96	238	243	234	214	45	86	211	173
130	50	78	226	64	146	163	156	14	170	202	174	41	0	36	134
73	246	112	184	49	11	17	160	175	141	13	68	248	151	15	120
108	2	117	25	61	224	250	83	185	166	89	21	32	143	62	59
16	233	221	118	88	199	122	100	7	193	123	179	43	215	254	240
176	94	104	129	209	6	144	58	110	196	125	92	191	164	90	220
149	210	208	192	24	8	225	46	171	213	19	74	242	4	197	237
150	162	127	177	70	167	10	249	67	219	80	212	147	188	18	153

TABLE 5.2: Decimated (at every $N - 1$ rounds) keystream bytes corresponding to the Finney State in Table 5.1.

Observe that the output bytes shown in Table 5.2 are a shift of the permutation shown in 5.1. After 256×255 rounds the same state is repeated.

5.2 Glimpse Theorem

One of the important results on the weakness of RC4 PRGA is the *Glimpse Theorem*, also known as *Jenkins' Correlation* [78, 108]. It gives a glimpse of the hidden state in the keystream as follows.

Theorem 5.2.1. *[Glimpse Theorem] After the r-th round of the PRGA, $r \geq 1$,*

$$P(S_r^G[j_r^G] = i_r^G - z_r) = P(S_r^G[i_r^G] = j_r^G - z_r) \approx \frac{2}{N}.$$

Proof: We are going to prove the first probability, i.e.,

$$P(S_r^G[j_r^G] = i_r^G - z_r) \approx \frac{2}{N}.$$

The second one follows similarly. We consider two mutually exclusive and exhaustive cases:

$$i_r^G = S_r^G[i_r^G] + S_r^G[j_r^G]$$

and

$$i_r^G \neq S_r^G[i_r^G] + S_r^G[j_r^G].$$

Case I: When $i_r^G = S_r^G[i_r^G] + S_r^G[j_r^G]$,

$$
\begin{aligned}
z_r &= S_r^G\left[S_r^G[i_r^G] + S_r^G[j_r^G]\right] \\
&= S_r^G[i_r^G] \\
&= i_r^G - S_r^G[j_r^G].
\end{aligned}
$$

Thus, given the event

$$(i_r^G = S_r^G[i_r^G] + S_r^G[j_r^G]),$$

which occurs with probability $\frac{1}{N}$, the event

$$(S_r^G[j_r^G] = i_r^G - z_r)$$

happens with probability 1. Hence, the contribution of this case is

$$\frac{1}{N} \cdot 1 = \frac{1}{N}.$$

Case II: When $i_r^G \neq S_r^G[i_r^G] + S_r^G[j_r^G]$, which happens with probability $\frac{N-1}{N}$, the event

$$(S_r^G[j_r^G] = i_r^G - z_r)$$

occurs as random association with probability $\frac{1}{N}$. Hence, the contribution of this case is

$$\frac{N-1}{N} \cdot \frac{1}{N} = \frac{1}{N} - \frac{1}{N^2}.$$

The sum of the above two contributions gives the total probability of the event $(S_r^G[j_r^G] = i_r^G - z_r)$ to be $\frac{2}{N} - \frac{1}{N^2} \approx \frac{2}{N}$. ∎

We may rewrite the first relation between $S_r^G[j_r^G]$ and $i_r^G - z_r$ as follows.

Corollary 5.2.2. $P(z_r = r - S_{r-1}^G[r]) \approx \frac{2}{N}$, $r \geq 1$.

Proof: Since, $i_r^G = r$, the Glimpse Theorem gives

$$P(z_r = r - S_r^G[j_r^G]) \approx \frac{2}{N}, \qquad \text{for } r \geq 1.$$

Due to the swap in round r, $S_r^G[j_r^G]$ is assigned the value of $S_{r-1}^G[r]$. Thus, we get

$$P(z_r = r - S_{r-1}^G[r]) \approx \frac{2}{N}, \qquad \text{for } r \geq 1.$$

■

The motivation of rewriting the Glimpse Theorem in the form of Corollary 5.2.2 comes from the fact that relating "z_r to $S_{r-1}^G[r]$" will ultimately relate "z_r to the secret key bytes," as the permutations in the initial rounds of the PRGA have correlations with the secret key.

More details on Glimpse Theorem and its generalizations are discussed in [110, Chapter 7] and [108, Section 3].

5.3 Biased Permutation Index Selection for the First Keystream Byte

The index t_1 in S_1^G from where the first byte z_1 of RC4 keystream is chosen is biased. The probability of $t_1 = 2$ is approximately twice as large as that of t_1 taking any other value. This result first appeared in [131].

Theorem 5.3.1. *Assume that the permutation S_0^G after the KSA is chosen uniformly at random from the set of all possible permutations of $\mathbb{Z}_N$. Then the probability distribution of the output index t_1, that selects the first byte z_1 of the keystream output, is given by*

$$P(t_1 = x) = \begin{cases} \frac{1}{N} & \text{for odd } x; \\ \frac{1}{N} - \frac{2}{N(N-1)} & \text{for even } x \neq 2; \\ \frac{2}{N} - \frac{1}{N(N-1)} & \text{for } x = 2. \end{cases}$$

Proof: In the first round of the PRGA, i_1^G is set to 1, j_1^G is updated to $S_0^G[1]$. So after the swap in this round, $S_1^G[i_1^G]$ contains the value $S_0^G[j_1^G]$, i.e., $S_0^G[S_0^G[1]]$ and $S_1^G[j_1^G]$ contains the value $S_0^G[i_1^G]$, i.e., $S_0^G[1]$. Now,

$$
\begin{aligned}
S_1^G[j_1^G] = 1 &\iff S_0^G[1] = 1, && \text{since } S_1^G[j_1^G] = S_0^G[1] \\
&\iff j_1^G = 1, && \text{since } j_1^G = S_0^G[1] \\
&\implies S_1^G[1] = 1, && \text{since we started with } S_1^G[j_1^G] = 1 \\
&\implies S_1^G[i_1^G] = 1, && \text{since we started with } i_1^G = 1.
\end{aligned}
$$

Hence,

$$P(S_1^G[i_1^G] = u, S_1^G[j_1^G] = v) = \begin{cases} \frac{1}{N} & \text{if } u = 1 \text{ and } v = 1; \\ 0 & \text{if } u \neq 1 \text{ and } v = 1; \\ 0 & \text{if } u \neq 1 \text{ and } v = u; \\ \frac{1}{N(N-1)} & \text{otherwise.} \end{cases}$$

The probability distribution of $t_1 = S_1^G[i_1^G] + S_1^G[j_1^G]$ can be computed as follows.

- For odd x,

$$P(t_1 = x) = \sum_{\substack{v=0 \\ v \neq 1}}^{N-1} P(S_1^G[j_1^G] = v, S_1^G[i_1^G] = N - v + x)$$

$$= (N-1) \cdot \frac{1}{N(N-1)} = \frac{1}{N}.$$

- For even $x \neq 2$,

$$P(t_1 = x) = \sum_{\substack{v=0 \\ v \neq 1, \frac{x}{2}, \frac{N+x}{2}}}^{N-1} P(S_1^G[j_1^G] = v, S_1^G[i_1^G] = N - v + x)$$

$$= (N-3) \cdot \frac{1}{N(N-1)} = \frac{1}{N} - \frac{2}{N(N-1)}.$$

- For $x = 2$,

$$P(t_1 = 2) = P(S_1^G[j_1^G] = 1, S_1^G[i_1^G] = 1) +$$

$$\sum_{\substack{v=0 \\ v \neq 1, \frac{N+2}{2}}}^{N-1} P(S_1^G[j_1^G] = v, S_1^G[i_1^G] = N - v + 2)$$

$$= \frac{1}{N} + (N-2) \cdot \frac{1}{N(N-1)} = \frac{2N-3}{N(N-1)} = \frac{2}{N} - \frac{1}{N(N-1)}. \qquad \blacksquare$$

5.4 Characterization of PRGA Evolution

The work [9], for the first time, performed a complete characterization of the RC4 PRGA for one step:

$$i_r^G = i_{r-1}^G + 1; j_r^G = j_{r-1}^G + S_r^G[i_r^G];$$

$$\text{swap } S_{r-1}^G[i_r^G], S_{r-1}^G[j_r^G] \text{ to get } S_r^G;$$

$$z_r = S_r^G[S_r^G[i_r^G] + S_r^G[j_r^G]].$$

Considering all the permutations (keeping in mind the Finney states also), the distribution of z is shown to be non-uniform given i^G, j^G. Further, analysis of two consecutive steps of RC4 PRGA reveals that the index j^G is not produced uniformly at random given the value of j^G two steps ago. The immediate implication of the above results is that information about j^G is leaked in the keystream.

5.4.1 One Step of RC4 PRGA

Let us characterize the input-output distribution of one step RC4. The characterization is based on determination of the number of possible configurations of the internal states S^G and pointers i^G and j^G that can yield certain keystream bytes at a particular time step.

We present the work for the case where N is even, as RC4 in general works for $N = 256$. A similar approach works for odd N, but the detailed calculations are different, which we omit here.

Due to the non-existence of Finney cycles (discussed in Section 5.1) for the initialization in RC4, all the permutations may not occur before each step. In addition to the Finney states, there may exist some other (so far undiscovered) permutations that may not occur at the beginning of each step. In order to make things simple, one may consider that any permutation can occur before any PRGA step. The related result is presented in Theorem 5.4.1. Subsequently, one can count the cases for the Finney cycles in Theorem 5.4.3 which should be subtracted from the cases in Theorem 5.4.1 to get a more complicated exact formula. Further, if some other kinds of non-existing permutations become identified in future, those cases may also be analyzed similar to Theorem 5.4.3 and then can be subtracted from the cases in Theorem 5.4.1.

The Finney cycle occurs when $j^G = i^G + 1$ and $S^G[j^G] = 1$. The number of permutations in the Finney cycles is $(N - 1)!$, which is only $\frac{1}{N}$ of the number $N!$ of all possible RC4 permutations. Hence, excluding the Finney states in the computations does not change the orders of the probabilities in Corollary 5.4.2. This is also confirmed by extensive experimentation.

We only outline the basic idea behind the proofs of Theorems 5.4.1 and 5.4.3 here. For complete proofs, the readers are referred to [9]. There does not seem to exist a shorter proof technique than a case by case analysis of all the possibilities that have been studied in [9]. However, the end results presented in the statements of Theorems 5.4.1 and 5.4.3 are concise and depict the complete view of how RC4 PRGA evolves.

During a one-step update in the RC4 PRGA, i^G is incremented to

$$i'^G = i^G + 1$$

and j^G is updated to

$$j'^G = j^G + S^G[i^G + 1].$$

$$\text{Denote} \quad u = S[i'^G] = S^G[i^G + 1]$$

$$\text{and} \quad v = S^G[j'^G] = S^G[j^G + u].$$

Then, after the swap,

$$S^G[j^G + u] = u \text{ and } S^G[i^G + 1] = v.$$

Further, the output z is given by $S^G[u + v]$.

Let $\psi(i^G, j^G, z)$ denote the number of permutations out of all the $N!$ permutations of $\mathbb{Z}_N$ such that with the given values of i^G, j^G before the execution of one step in RC4 PRGA, the output after the execution is z. $\psi(i^G, j^G, z)$ is computed by conditioning on the values i^G, j^G, z, u and v in a tree-like fashion. At the root level, we condition on i^G. In the next level, given i^G, we condition on j^G. Similarly, we move down the hierarchy as follows: given i^G and j^G, we condition on z; given i^G, j^G and z, we condition on u; and finally at the leaf level, given i^G, j^G, z and u, we condition on v. Given i^G, j^G and z corresponding to any internal node A of the tree, $\psi(i^G, j^G, z)$ is calculated by adding the contributions of each leaf of the subtree rooted at A. For example, if at some leaf, z, u and v are all distinct and can have n_z, n_u and n_v many values respectively, then the contribution of that leaf toward $\psi(i^G, j^G, z)$ is given by

$$n_z \cdot n_u \cdot n_v \cdot (N-3)!.$$

Though we need to condition on i^G, j^G, z to compute $\psi(i^G, j^G, z)$, we involve only z in counting the permutations, since z is generated from one of the permutation bytes.

Theorem 5.4.1. *Let N be even. Then we have the following.*
   (1) If i^G is even, then

$$\psi(i^G, j^G, z) = \begin{cases} (N-1)! - (N-2)! & \text{if } z = j^G, i^G + 1 - j^G; \\ (N-1)! + 2(N-3)! & \text{otherwise.} \end{cases}$$

   (2) If i^G is odd and $j^G = \frac{i^G+1}{2}, \frac{i^G+1+N}{2}$, then

$$\psi(i^G, j^G, z) = \begin{cases} 2(N-1)! - (N-2)! & \text{if } z = j^G; \\ (N-1)! - 2(N-2)! & \text{if } z = j^G + \frac{N}{2}; \\ (N-1)! - (N-2)! + 2(N-3)! & \text{otherwise.} \end{cases}$$

   (3) If i^G is odd and $j^G \neq \frac{i^G+1}{2}, \frac{i^G+1+N}{2}$, then

$$\psi(i^G, j^G, z) = \begin{cases} (N-1)! - (N-2)! + 2(N-3)! & \text{if } z = j^G, i^G + 1 - j^G, \\ & \qquad\qquad \frac{i^G+1}{2}, \frac{i^G+1+N}{2}; \\ (N-1)! + 4(N-3)! & \text{otherwise.} \end{cases}$$

See [9] for the detailed proof.

Corollary 5.4.2. *Considering that the permutations appear uniformly at random before any step of RC4 PRGA, we have the following biased probabilities.*

   1. $P(j^G = z \mid i^G \text{ odd}) \geq \frac{1}{N} + \frac{1}{N^2}$.

   2. $P(j^G = z \mid i^G \text{ even}) \leq \frac{1}{N} - \frac{1}{N^2}$.

   3. $P(j^G = z \mid 2z = i^G + 1) = \frac{2}{N} - \frac{1}{N(N-1)}$.

Proof:

1. Consider the first cases of both the items (2), (3) from Theorem 5.4.1. Thus,

$$
\begin{aligned}
&P(j^G = z \mid i^G \ odd) \\
&= \sum_m P(j^G = m)P(z = m \mid j^G = m, i^G \ odd) \\
&= \frac{1}{N}\left(2 \cdot \frac{2(N-1)! - (N-2)!}{N!}\right. \\
&\qquad \left. + (N-2) \cdot \frac{(N-1)! - (N-2)! + 2(N-3)!}{N!}\right) \\
&= \frac{1}{N} + \frac{2}{N^2} - \frac{1}{N(N-1)} + \frac{2}{N^2(N-1)} \\
&\geq \frac{1}{N} + \frac{1}{N^2}.
\end{aligned}
$$

2. Consider the first case of item (1) from Theorem 5.4.1. Thus,

$$
\begin{aligned}
P(j^G = z \mid i^G \ even) &= \frac{(N-1)! - (N-2)!}{N!} \\
&\leq \frac{1}{N} - \frac{1}{N^2}.
\end{aligned}
$$

3. Consider the first and second cases of item (2) and the first case of item (3) from Theorem 5.4.1. Note that,

$$
P\left(j^G = z \mid z = \frac{i^G + 1}{2}\right) = \frac{P(j^G = z \wedge z = \frac{i^G+1}{2})}{P(z = \frac{i^G+1}{2})}.
$$

Now

$$
P\left(j^G = z \wedge z = \frac{i^G + 1}{2}\right) = \frac{1}{N}\frac{2(N-1)! - (N-2)!}{N!}
$$

and

$$
\begin{aligned}
&P\left(z = \frac{i^G + 1}{2}\right) \\
&= \frac{1}{N}\left(\frac{2(N-1)! - (N-2)!}{N!} + \frac{(N-1)! - 2(N-2)!}{N!}\right. \\
&\qquad \left. + (N-2)\frac{(N-1)! - (N-2)! + 2(N-3)!}{N!}\right).
\end{aligned}
$$

Thus,

$$
P\left(j^G = z \mid z = \frac{i^G + 1}{2}\right) = \frac{2}{N} - \frac{1}{N(N-1)}.
$$

The proof of

$$P\left(j^G = z \mid z = \frac{i^G + 1 + N}{2}\right) = \frac{2}{N} - \frac{1}{N(N-1)}$$

is similar.

Given

$$2z = i^G + 1 \bmod N,$$

the events

$$\left(z = \frac{i^G + 1}{2}\right) \text{ and } \left(z = \frac{i^G + 1 + N}{2}\right)$$

are equally likely. Thus we get the result. ∎

The above results imply that z is either positively or negatively biased to j^G at any stage of PRGA. The bias is approximately $\frac{1}{N^2}$ away from the random association $\frac{1}{N}$ as given in items 1, 2 in Corollary 5.4.2. The bias is quite high when $2z = i^G + 1$ and this bias is of the order of $\frac{1}{N}$ over the random association $\frac{1}{N}$. Experimental results also support the claim, which identify that the assumption of "any permutation can arrive uniformly at random at any step of RC4 PRGA" works with a very good approximation and the absence of the Finney states does not change the order of the results.

The theoretical probabilities appearing on the right-hand side of items 1, 2, and 3 of Corollary 5.4.2 are 0.00392151, 0.00389099 and 0.00779718 respectively. The experimental values are very close to the theoretical ones, but not exact. The reason is as follows. We arrive at the theoretical values from combinatorial results based on 256! different permutations, which is a huge number and cannot be generated in practice in a feasible time. Each run of the experiment is based on samples of one million different keys, i.e., 1 million permutations are considered, which is much less than the number 256!. Also there are some approximations involved in the theoretical results as given in Corollary 5.4.2.

The next result presents an idea on how the cases for the Finney cycle can be tackled. These cases may be subtracted from the cases in Theorem 5.4.1 for exact calculations.

Theorem 5.4.3. *Suppose $j^G = i^G + 1$ and $S^G[j^G] = 1$ (Finney cycle). Then we have the following.*
    (1) If $i^G = 0$, then

$$\psi(i^G, j^G, z) = \begin{cases} 2(N-2)! & \text{if } z = 0; \\ 0 & \text{if } z = 1; \\ (N-2)! - (N-3)! & \text{otherwise.} \end{cases}$$

    (2) If $i^G = 1$, then

$$\psi(i^G, j^G, z) = \begin{cases} (N-2)! & \text{if } z = 1, 2; \\ (N-2)! - (N-3)! & \text{otherwise.} \end{cases}$$

(3) *If* $i^G \neq 0, 1$, *then*

$$\psi(i^G, j^G, z) = \begin{cases} (N-2)! & \text{if } z = 1; \\ 2(N-2)! - (N-3)! & \text{if } z = i^G; \\ (N-2)! - (N-3)! & \text{if } z = i^G + 1; \\ (N-2)! - 2(N-3)! & \text{otherwise.} \end{cases}$$

See [9] for the detailed proof.

5.4.2 Two Steps of RC4 PRGA

The complete characterization of two steps may be attempted as in Theorem 5.4.1, though it will be extremely complicated. Instead of enumerating all such cases, we look into one particular two-step scenario that reveals new weaknesses in RC4.

First let us provide the proof that the hidden index j^G is not produced uniformly at random conditioning on the value of j^G two steps before. Following this, we note another evidence different from Corollary 5.4.2 that the keystream of RC4 leaks information on j^G.

Lemma 5.4.4. *For any $r \geq 0$, the following three are equivalent.*

1. $S_r^G[i_{r+1}^G] = i_r^G - j_r^G + 2.$

2. $j_{r+1}^G = i_r^G + 2.$

3. $j_{r+2}^G = 2j_{r+1}^G - j_r^G.$

Proof: First, we will show that (1) $\Longleftrightarrow$ (2). In RC4 PRGA,

$$j_{r+1}^G = j_r^G + S_r^G[i_{r+1}^G],$$

from which it immediately follows that

$$S_r^G[i_{r+1}^G] = i_r^G - j_r^G + 2$$
$$\Longleftrightarrow \quad j_{r+1}^G = i_r^G + 2.$$

Next, we are going to show that (2) $\Longleftrightarrow$ (3). In the next round,

$$\begin{aligned} j_{r+2}^G &= j_{r+1}^G + S_{r+1}^G[i_{r+2}^G] \\ &= j_{r+1}^G + S_{r+1}^G[i_r^G + 2]. \end{aligned}$$

So,

$$j_{r+1}^G = i_r^G + 2$$

if and only if

$$\begin{aligned} j_{r+2}^G &= j_{r+1}^G + S_{r+1}^G[j_{r+1}^G] \\ &= j_{r+1}^G + S_r^G[i_{r+1}^G] \\ &= j_{r+1}^G + (j_{r+1}^G - j_r^G) \\ &= 2j_{r+1}^G - j_r^G. \end{aligned}$$

■

If $j_r^G = i_r^G + 1$, then the equivalent conditions in Lemma 5.4.4 correspond to a Finney state. In the following results, one should exclude this case to get better accuracy. However, this does not have much effect on the order of probabilities, and further makes the calculations more cumbersome. So the exclusion of Finney states is not considered in Lemma 5.4.4.

The above result leads to the following.

Theorem 5.4.5. $P(j_{r+2}^G = 2i_r^G + 4 - j_r^G) = \frac{2}{N}$, $r \geq 0$.

Proof: The event

$$(j_{r+2}^G = 2i_r^G + 4 - j_r^G)$$

can occur in the following two ways:

Case I: $S_r^G[i_{r+1}^G] = i_r^G - j_r^G + 2$ and hence by Lemma 5.4.4,

$$
\begin{aligned}
j_{r+2}^G &= 2j_{r+1}^G - j_r^G \\
&= 2(i_r^G + 2) - j_r^G \\
&= 2i_r^G + 4 - j_r^G.
\end{aligned}
$$

The contribution of this part is

$$P(S_r^G[i_{r+1}^G] = i_r^G - j_r^G + 2) = \frac{1}{N}.$$

Case II: $S_r^G[i_{r+1}^G] \neq i_r^G - j_r^G + 2$, but j_{r+2}^G still equals $2i_r^G + 4 - j_r^G$ due to random association. From Lemma 5.4.4, we have

$$
\begin{aligned}
&(S_r^G[i_{r+1}^G] \neq i_r^G - j_r^G + 2) \\
\implies\; &(j_{r+2}^G \neq 2j_{r+1}^G - j_r^G).
\end{aligned}
$$

Assuming that j_{r+2}^G takes the other $N - 1$ values (except $2j_{r+1}^G - j_r^G$) uniformly at random, we have the contribution of this part as

$$
P(S_r^G[i_{r+1}^G] \neq i_r^G - j_r^G + 2) \cdot \frac{1}{N-1} = \left(1 - \frac{1}{N}\right) \cdot \frac{1}{N-1}
$$
$$
= \frac{1}{N}.
$$

Adding the above two contributions, we get the result. ■

The above result implies that for any $r \geq 0$, $(j_{r+2}^G | j_r^G)$ is not distributed uniformly at random in $\mathbb{Z}_N$. A closer look reveals that in this process the keystream output during PRGA leaks information about the index j. To prove this, we first need the following result.

Theorem 5.4.6. $P(j_{r+2}^G = i_{r+2}^G + j_{r+1}^G - j_r^G) = \frac{2}{N}$, $r \geq 0$.

Proof: The event $(j_{r+2}^G = i_{r+2}^G + j_{r+1}^G - j_r^G)$ can occur in two ways.

Case I: $j_{r+1}^G = i_r^G + 2$. By Lemma 5.4.4,

$$
\begin{aligned}
j_{r+2}^G &= 2j_{r+1}^G - j_r^G \\
&= j_{r+1}^G + (i_r^G + 2) - j_r^G.
\end{aligned}
$$

Thus, the contribution of this part is

$$P(j_{r+1}^G = i_r^G + 2) = \frac{1}{N}.$$

Case II: $j_{r+1}^G \neq i_r^G + 2$ and still $j_{r+2}^G = i_{r+2}^G + j_{r+1}^G - j_r^G$ due to random association. The probability contribution of this part is

$$\left(1 - \frac{1}{N}\right) \cdot \frac{1}{N-1} = \frac{1}{N}.$$

Adding the above two contributions, we get the result. ∎

Since $S_{r+2}^G[j_{r+2}^G] = S_{r+1}^G[i_{r+2}^G] = j_{r+2}^G - j_{r+1}^G$, we immediately arrive at the following corollary.

Corollary 5.4.7. $P(j_r^G = i_{r+2}^G - S_{r+2}^G[j_{r+2}^G]) = \frac{2}{N}$, $r \geq 0$.

Now we show further leakage (different than Corollary 5.4.2) of information about the pseudo-random index j from z.

Theorem 5.4.8. $P(z_{r+2} = j_r^G) = \frac{1}{N}(1 + \frac{2}{N})$, $r \geq 0$.

Proof: The event $(z_{r+2} = j_r^G)$ can occur in two possible ways.

Case I: Consider the events

$$z_{r+2} = i_{r+2}^G - S_{r+2}^G[j_{r+2}^G]$$

and

$$j_r^G = i_{r+2}^G - S_{r+2}^G[j_{r+2}^G].$$

By Theorem 5.2.1 and Corollary 5.4.7, the probability contribution of this part is

$$\frac{2}{N} \cdot \frac{2}{N} = \frac{4}{N^2}.$$

Case II: Consider that

$$z_{r+2} \neq i_{r+2}^G - S_{r+2}^G[j_{r+2}^G]$$

and still $z_{r+2} = j_r^G$ due to random association. The probability contribution of this part is

$$\left(1 - \frac{2}{N}\right) \cdot \frac{1}{N}.$$

Adding the above two contributions, we get the result.   ■

In the next chapter (in Section 6.3.2), a corollary of this theorem would be used to construct a new distinguisher.

The analysis performed in this section presents some weaknesses in RC4 design. On the positive side, the elegant structure of RC4 allows such an analysis possible by pen and paper and shows that there does not exist any hidden trapdoor in the design. Most of the recent software stream ciphers (e.g. the eSTREAM candidates) are significantly more complicated than RC4. However, the complicated nature of these ciphers may not allow a detailed theoretical analysis of even a single step during the keystream generation.

5.5   Some Biases in First Keystream Byte toward Secret Key

A bias of the first keystream byte z_1 toward $K[2] + 3$, given that the sum of the first two key bytes is zero, was first observed by Roos [146]. This result was subsequently proved in [131], which reported some additional biases in z_1. We present all these biases in this section.

The expression

$$\left(\frac{N-1}{N}\right)^N \left(1 - \frac{1}{N} - \frac{1}{N^2}\right) + \frac{1}{N^2}$$

will be used a number of times in this section. So we denote it by the symbol ϕ_N.

Here we are interested in z_1 that is generated in the first round of PRGA. So let us analyze the first round in detail and introduce some additional notations. Before the PRGA begins, we have

$$i_0^G = j_0^G = 0.$$

Let

$$S_0^G[1] = u \text{ and } S_0^G[u] = v.$$

In the first round, $i_1^G = 1$ and

$$\begin{aligned} j_1^G &= 0 + S_0^G[1] \\ &= u. \end{aligned}$$

The contents u and v of locations 1 and u respectively are interchanged. Thus, after the first round, we have

$$S_1^G[1] = v \text{ and } S_1^G[u] = u.$$

5.5.1 Results for Any Arbitrary Secret Key

Using Theorem 5.3.1 and Proposition 3.1.1, one can prove (Theorem 5.5.2) that for any arbitrary key, the first byte of the keystream output is significantly biased to the first three bytes of the secret key, thus revealing a weakness of the KSA. This result first appeared in [131].

We need the following technical result that will be used in the proof of Theorem 5.5.2.

Proposition 5.5.1. $P(S_1^G[2] = K[0] + K[1] + K[2] + 3) \approx \phi_N.$

Proof: As in the KSA, the index j^G of the PRGA is assumed to take values uniformly at random. During the first round of the PRGA, i_1^G takes the value 1 and j_1^G takes the value $S_0^G[1]$. Thus the index 1 is always involved in the first swap. The other index involved in the first swap depends on the value of $S_0^G[1]$. Let

$$f(K) = K[0] + K[1] + K[2] + 3.$$

Recall from item 2 of Corollary 3.1.2 that

$$P\left(S_0^G[2] = f(K)\right) \approx \left(\frac{N-1}{N}\right)^N.$$

Case I: After the KSA, $S_0^G[2] = f(K)$, and there is no swap involving index 2 in the first round of the PRGA. The contribution of this part is

$$P\left(S_0^G[2] = f(K)\right) \cdot P(S_0^G[1] \neq 2) = \left(\frac{N-1}{N}\right)^N \left(1 - \frac{1}{N}\right).$$

Case II: Following the KSA, $S_0^G[2] \neq f(K)$, and $f(K)$ comes into index 2 from index 1 by the swap in the first round of the PRGA. The contribution of this part is

$$
\begin{aligned}
& P\left(S_0^G[2] \neq f(K)\right) \cdot P\left(S_0^G[1] = f(K), S_0^G[1] = 2\right) \\
= & P\left(S_0^G[2] \neq f(K)\right) \cdot P\left(S_0^G[1] = 2\right) \cdot P(f(K) = 2) \\
\approx & \left(1 - \left(\frac{N-1}{N}\right)^N\right) \cdot \frac{1}{N^2}.
\end{aligned}
$$

Adding the above two contributions, we get

$$
\begin{aligned}
P\left(S_1^G[2] = f(K)\right) &\approx \left(\frac{N-1}{N}\right)^N \left(1 - \frac{1}{N} - \frac{1}{N^2}\right) + \frac{1}{N^2} \\
&= \phi_N.
\end{aligned}
$$

■

Note that $\phi_N \approx 0.37$ for $N = 256$. Instead of approximating $P(S_0^G[1] = 2)$ by $\frac{1}{N}$, one could use Theorem 3.2.3 to substitute the exact expression, i.e.,

$$P(S_0^G[1] = 2) = \frac{1}{N}\left(\left(\frac{N-1}{N}\right)^{N-2} + \left(\frac{N-1}{N}\right)^2\right).$$

However, for $N = 256$, practically there is no change in the value of ϕ_N due to this exact substitution.

Next we present the result that shows the bias of the first keystream output byte toward the first three bytes of the secret key.

Theorem 5.5.2. *For any arbitrary secret key, the correlation between the key bytes and the first byte of the keystream output is given by*

$$P(z_1 = K[0] + K[1] + K[2] + 3) \approx \frac{1}{N}(1 + \phi_N).$$

Proof: For the sake of brevity, once again let

$$f(K) = K[0] + K[1] + K[2] + 3.$$

Then

$$
\begin{aligned}
P\left(z_1 = f(K)\right) &= P\left(S_1^G[t_1] = f(K)\right) \\
&= \sum_{x=0}^{N-1} P(t_1 = x) \cdot P\left(S_1^G[t_1] = f(K) \mid t_1 = x\right) \\
&= \sum_{x=0}^{N-1} P(t_1 = x) \cdot P\left(S_1^G[x] = f(K)\right) \\
&= P(t_1 = 2) \cdot P\left(S_1^G[2] = f(K)\right) \\
&\quad + \sum_{\substack{x=0 \\ even\ x \neq 2}}^{N-1} P(t_1 = x) \cdot P\left(S_1^G[x] = f(K)\right) \\
&\quad + \sum_{\substack{x=0 \\ odd\ x}}^{N-1} P(t_1 = x) \cdot P\left(S_1^G[x] = f(K)\right) \\
&= \left(\frac{2}{N} - \frac{1}{N(N-1)}\right) \cdot P\left(S_1^G[2] = f(K)\right) \\
&\quad + \sum_{\substack{x=0 \\ even\ x \neq 2}}^{N-1} \left(\frac{1}{N} - \frac{2}{N(N-1)}\right) \cdot P\left(S_1^G[x] = f(K)\right) \\
&\quad + \sum_{\substack{x=0 \\ odd\ x}}^{N-1} \frac{1}{N} \cdot P\left(S_1^G[x] = f(K)\right) \qquad \text{(by Theorem 5.3.1)}
\end{aligned}
$$

So

$$P\left(z_1 = f(K)\right) \approx \frac{2}{N} \cdot P\left(S_1^G[2] = f(K)\right)$$

$$+ \sum_{\substack{x=0 \\ even\ x \neq 2}}^{N-1} \frac{1}{N} \cdot P\left(S_1^G[x] = f(K)\right) \quad \left(\text{as } \tfrac{1}{N(N-1)} \ll \tfrac{1}{N}\right)$$

$$+ \sum_{\substack{x=0 \\ odd\ x}}^{N-1} \frac{1}{N} \cdot P\left(S_1^G[x] = f(K)\right)$$

$$= \frac{1}{N} \cdot P\left(S_1^G[2] = f(K)\right) + \frac{1}{N} \cdot P\left(S_1^G[2] = f(K)\right)$$

$$+ \frac{1}{N} \cdot \sum_{\substack{x=0 \\ even\ x \neq 2}}^{N-1} P\left(S_1^G[x] = f(K)\right)$$

$$+ \frac{1}{N} \cdot \sum_{\substack{x=0 \\ odd\ x}}^{N-1} P\left(S_1^G[x] = f(K)\right)$$

$$= \frac{1}{N} \cdot P\left(S_1^G[2] = f(K)\right) + \frac{1}{N} \cdot \sum_{x=0}^{N-1} P\left(S_1^G[x] = f(K)\right)$$

$$= \frac{1}{N} \cdot \phi_N + \frac{1}{N} \cdot 1 \quad \text{(by Proposition 5.5.1)}$$

$$= \frac{1}{N}(1 + \phi_N).$$

∎

Let us detail the interpretation of the probabilities involved in the results that are discussed. The event under consideration is the equality of two random variables X and Y, where X is some byte of the permutation (e.g. $S[y]$ as in Corollary 3.1.2 and Proposition 5.5.1) or some byte of the keystream output (e.g. z_1 as in Theorem 5.5.2), and Y is some function $f(K)$ of the secret key. Each of X and Y takes values from $\mathbb{Z}_N$. Thus, the joint space of (X, Y) consists of N^2 different points (x, y), where $x \in \mathbb{Z}_N$ and $y \in \mathbb{Z}_N$. If X and Y are i.i.d. uniform random variables, then for any (x, y), $P(X = x, Y = y)$ should be $\frac{1}{N^2}$, and

$$P(X = Y) = \sum_{x=0}^{N-1} P(X = x, Y = x)$$

$$= \sum_{x=0}^{N-1} \frac{1}{N^2} = N \cdot \frac{1}{N^2} = \frac{1}{N}.$$

For $N = 256$, this value is approximately 0.0039, whereas the observed values of the probabilities are much higher. In case of Proposition 5.5.1, this is 0.37 and in case of Theorem 5.5.2, this is

$$\frac{1}{N}(1 + \phi_N) \approx \frac{1}{256} \cdot (1 + 0.37)$$
$$\approx 0.0053.$$

It is important to note that the bias in S observed by Roos (Proposition 3.1.1, item 2 of Corollary 3.1.2) and extended in Proposition 5.5.1 do not necessarily imply the bias in z_1. For example, suppose $S_0^G[2]$ (or $S_1^G[2]$) equals some combination of the secret key bytes with probability 1. The output z_1 may still be unbiased, if the index t_1, where z_1 is selected from, is uniformly random. However, it has been proved (Theorem 5.3.1) that t_1 is not uniformly random. In other words, there may exist some bias in $S_0^G[2]$ due to the weakness of the KSA. But it is the bias in t_1 due to the weakness of the PRGA that propagates the bias from $S_0^G[2]$ to z_1. The proof of Theorem 5.5.2 connects the bias in $S_0^G[2]$ and that in t_1 and relates them to the bias in z_1.

5.5.2   Results for Secret Keys Whose First Two Bytes Sum to Zero

Recall that Theorem 5.3.1 demonstrated a significant bias in the index in S, where the first byte of the keystream output is selected from. In this section, Theorem 5.5.3 shows that this bias is increased significantly if the first two key bytes satisfy the condition $K[0] + K[1] = 0$. Further, Theorem 5.5.2 proved a bias in the first byte of the output toward the first three bytes of the secret key. All the three results, i.e., Theorems 5.3.1, 5.5.3 and 5.5.2 were discovered in [131]. On the other hand, Roos [146] experimentally observed that given any RC4 key with the restriction $K[0] + K[1] = 0$, the probability that the first keystream byte generated by RC4 will be $K[2]+3$ ranges between 12% and 16% with an average of 13.8%. The work [131] used Theorem 5.5.3 to derive a conditional variant of Theorem 5.5.2. This variant, presented as Theorem 5.5.4 in this section, provides a theoretical justification of Roos' observation.

Theorem 5.5.3. *Assume that the first two bytes $K[0]$, $K[1]$ of the secret key add to 0 mod N. Then the bias of the output index t_1, that selects the first byte of the keystream output, is given by*

$$P(t_1 = 2 \mid K[0] + K[1] = 0) > \left(\frac{N-1}{N}\right)^N.$$

Proof:

$$P(t_1 = 2 \mid K[0] + K[1] = 0)$$
$$= P(S_1^G[i_1^G] + S_1^G[j_1^G] = 2 \mid K[0] + K[1] = 0)$$
$$= P(S_1^G[i_1^G] = 1, S_1^G[j_1^G] = 1 \mid K[0] + K[1] = 0)$$
$$+ \sum_{\substack{x=0 \\ x \neq 1, \frac{N+2}{2}}}^{N-1} P(S_1^G[j_1^G] = x, S_1^G[i_1^G] = N - x + 2 \mid K[0] + K[1] = 0)$$
$$> P(S_1^G[i_1^G] = 1, S_1^G[j_1^G] = 1 \mid K[0] + K[1] = 0)$$
$$= P(S_0^G[1] = 1 \mid K[0] + K[1] = 0)$$
$$\quad \left(\text{ as } (S_1^G[i_1^G] = 1 \wedge S_1^G[j_1^G] = 1) \iff (S_0^G[1] = 1) \right)$$
$$= P(S_0^G[1] = K[0] + K[1] + 1 \mid K[0] + K[1] = 0)$$
$$\approx P(S_0^G[1] = K[0] + K[1] + 1)$$
$$\approx \left(\frac{N-1}{N} \right)^N \quad \text{(by Corollary 3.1.2, item 1).}$$

Recall that in the derivation of $P(S_0^G[1] = K[0] + K[1] + 1)$ in Corollary 3.1.2, no assumption on $K[0] + K[1]$ was required. Hence, the event $(S_0^G[1] = K[0] + K[1]+1)$ may be considered independent of the event $(K[0]+K[1] = 0)$, which justifies the removal of the conditional in the last line of the proof.[1]   ■

The bias in the output index (Theorem 5.3.1), the conditional bias in the output index (Theorem 5.5.3), and the bias of the first byte of the output toward the first three bytes of the secret key (Theorem 5.5.2) was discovered in [131].

Roos [146] experimentally observed that given any RC4 key with the restriction $K[0] + K[1] = 0$, the probability that the first keystream byte generated by RC4 will be $K[2]+3$ ranges between 12% and 16% with an average of 13.8%. The work [131], for the first time, provided a theoretical justification of this observation that we present in Theorem 5.5.4 below.

Theorem 5.5.4. *Assume that the first two bytes $K[0]$, $K[1]$ of the secret key add to 0 mod N. Then the correlation between the key bytes and the first byte of the keystream output is given by*

$$P(z_1 = K[2] + 3 \mid K[0] + K[1] = 0) > \left(\tfrac{N-1}{N} \right)^N \phi_N.$$

---

[1]In fact, along the same line of proof of Theorem 3.1.5, one can directly show that $P(S_0^G[1] = x + 1 \mid K[0] + K[1] = x) \approx \left(\frac{N-1}{N} \right)^N$, for $0 \leq x \leq N - 1$.

Proof:

$$P(z_1 = K[2] + 3 \mid K[0] + K[1] = 0)$$
$$= P(S_1^G[t_1] = K[2] + 3 \mid K[0] + K[1] = 0)$$
$$= \sum_{x=0}^{N-1} \Big(P(t_1 = x \mid K[0] + K[1] = 0) \cdot$$
$$\qquad\qquad P(S_1^G[t_1] = K[2] + 3 \mid K[0] + K[1] = 0, t_1 = x) \Big)$$
$$= \sum_{x=0}^{N-1} P(t_1 = x \mid K[0] + K[1] = 0) \cdot P(S_1^G[x] = K[2] + 3 \mid K[0] + K[1] = 0)$$
$$> P(t_1 = 2 \mid K[0] + K[1] = 0) \cdot P(S_1^G[2] = K[2] + 3 \mid K[0] + K[1] = 0)$$
$$= P(t_1 = 2 \mid K[0] + K[1] = 0) \cdot$$
$$\qquad P(S_1^G[2] = K[0] + K[1] + K[2] + 3 \mid K[0] + K[1] = 0)$$
$$\approx P(t_1 = 2 \mid K[0] + K[1] = 0) \cdot P(S_1^G[2] = K[0] + K[1] + K[2] + 3)$$
$$\approx \left(\frac{N-1}{N} \right)^N \phi_N \qquad \text{(by Theorem 5.5.3 and Proposition 5.5.1)}.$$

Recall that the derivation of $P(S_1^G[2] = K[0] + K[1] + K[2] + 3)$ in Proposition 5.5.1 used Corollary 3.1.2. In that derivation, no assumption on $K[0] + K[1]$ was required. Hence, the event $(S_1^G[2] = K[0] + K[1] + K[2] + 3)$ is considered independent of the event $(K[0] + K[1] = 0)$ in the last line of the proof.[2]   ∎

For $N = 256$, the value of $(\frac{N-1}{N})^N \phi_N$ is approximately 0.13 that proves the experimental observation of [146].

5.5.3   Cryptanalytic Applications

We present this section to demonstrate applications of Theorem 5.5.2 and Theorem 5.5.4.

Let m_1 and c_1 denote the first bytes of the message and the ciphertext respectively. Then $c_1 = m_1 \oplus z_1$. One can use Theorem 6.1.1 to calculate the number of samples that suffice to mount the cryptanalysis. Let X be the joint distribution of the two variables $(K[0] + K[1] + K[2] + 3)$ and z_1, when $K[0], K[1], K[2], z_1$ are chosen uniformly at random from $\mathbb{Z}_N$, and Y be the joint distribution of the same two variables for RC4 for randomly chosen keys. Let A be the event that these two variables are equal. Using the notations of Theorem 6.1.1 (to be discussed in Chapter 6), we write the probability expression of Theorem 5.5.2 as $p_0(1 + \epsilon)$, where $p_0 = \frac{1}{N}$ and $\epsilon = \phi_N$. Thus

---

[2] In fact, along the same line of proof of Theorem 3.1.5, one can directly show that $P(S_1^G[2] = x + K[2] + 3 \mid K[0] + K[1] = x) \approx (\frac{N-1}{N})^N$, for $0 \le x \le N-1$.

the number of samples required is (see Theorem 6.1.1)

$$\frac{1}{p_0 \epsilon^2} = \frac{N}{\phi_N^2}.$$

For $N = 256$, the value of this expression is 1870.

When the IV Precedes the Secret Key

Consider that the same message is broadcast after being encrypted by RC4 with uniformly distributed keys for multiple recipients. If the first three bytes of each of the keys are known, then one can compute

$$m_1' = c_1 \oplus (K[0] + K[1] + K[2] + 3)$$

for each of the keys. According to Theorem 5.5.2, the most frequent value of m_1' gives the actual value of m_1 with high probability.

Knowing the first three bytes of the key is not always impractical. In Wireless Equivalent Privacy (WEP) protocol [92] a secret key is used with known IV modifiers in RC4 [52, 108]. If the IV bytes precede the secret key bytes then the first three bytes of the key are actually known. This analysis is different from [52, 108], as only the first byte of the ciphertext output is used. In [107], on the other hand, the second byte of the ciphertext is used for cryptanalysis in broadcast RC4.

Further, if one can ensure that the first two bytes of the key add to zero, then following Theorem 5.5.4, one can perform the attack more efficiently. Experimental data establish that only 25 samples suffice to recover m_1 with an average probability of success exceeding 0.80 in this case.

Condition	#Samples n	min	max	avg	sd
Unconditional	1870	0.020000	0.062000	0.037990	0.008738
Unconditional	25000	0.574000	0.667000	0.628230	0.019447
Unconditional	50000	0.914000	0.955000	0.933380	0.007945
$K[0] + K[1] = 0$	25	0.740000	0.922000	0.806840	0.039063
$K[0] + K[1] = 0$	50	0.957000	0.989000	0.975410	0.007297
$K[0] + K[1] = 0$	100	0.998000	1.000000	0.999790	0.000431

TABLE 5.3: Results with 3-byte IV preceding 5-byte secret key.

Table 5.3 shows the success probabilities of recovering m_1 when a 3-byte IV is preceded by the 5-byte secret key. For a given m_1, we choose n (values of n appear in the second column of the table) number of samples, i.e., n number of first bytes c_1 of the ciphertexts corresponding to n runs of RC4, each time with a randomly chosen $< IV(3 \; bytes), Key(5 \; bytes) >$ pair. We calculate

$$m_1' = c_1 \oplus (K[0] + K[1] + K[2] + 3)$$

for each of the n samples and find out the most frequently occurring value m_1'. If this value matches with the given m_1, the recovery is considered to be successful. To estimate the success probability, we repeat the above process with 1000 randomly chosen m_1 values and find out in how many cases, say s, m_1 is recovered successfully. That is, the success probability is $\frac{s}{1000}$.

The above experiment is repeated 100 times to get a set of 100 success probabilities. The minimum, maximum, average and standard deviation of these probabilities are presented in columns min, max, avg and sd respectively of Table 5.3.

When the IV Follows the Secret Key

Suppose we can observe the first byte z_1 of the keystream output. Then with probability 0.0053 we know the value of $K[0] + K[1] + K[2] + 3$. Thus the uncertainty about these 8 bits (of z_1) is reduced by 0.44. If the key is not appended with IV, then the same secret key would give rise to the same keystream (and hence the same z_1) each time. Appending different IVs makes the keystream change and helps in constructing a frequency distribution of z_1. Then the most frequent value of z_1 can be treated as the value of $K[0] + K[1] + K[2] + 3$. We need the same secret key to be appended by different IVs to generate z_1 for recovering the value of $K[0] + K[1] + K[2] + 3$ reliably.

Further, if one can ensure that the first two bytes of the secret key add to zero, then following Theorem 5.5.4, one can perform the attack (e.g. recovering $K[2]$) much more efficiently requiring a very few (practically ≤ 100) samples.

Condition	#Samples n	min	max	avg	sd
Unconditional	1870	0.012000	0.032000	0.022540	0.004318
Unconditional	25000	0.113000	0.164000	0.144340	0.010867
Unconditional	50000	0.125000	0.182000	0.152910	0.011410
$K[0] + K[1] = 0$	25	0.759000	0.821000	0.789010	0.013393
$K[0] + K[1] = 0$	50	0.921000	0.968000	0.943020	0.007708
$K[0] + K[1] = 0$	100	0.952000	0.977000	0.965300	0.005442

TABLE 5.4: Results with 11-byte IV following 5-byte secret key

Table 5.4 shows the success probabilities of recovering the sum of the first three bytes when a 11 byte-IV is followed by the 5-byte secret key. We use the same experimental setup as before. The only difference is that in order to generate a distribution of z_1, we need to assume that a different IV is used with each of the n samples.

Observe that the success probability is less compared to the case where the IV bytes precede the secret key. In the previous case, a different IV and a different key are used for each sample, so that the effective key (i.e., the secret key and IV concatenated) can safely be assumed uniformly random. On the other hand, here we use many different IVs with the same key n number of times. So, the key can not be treated as uniformly randomly distributed across the samples. The lower success probabilities may be attributed to this non-uniformity in the distribution of the effective keys.

5.6 More Biases in Many Keystream Bytes toward Secret Key

Some biases in initial keystream bytes of RC4 have been reported in [86]. Later, additional biases in 256th and 257th keystream bytes, including some biases in the initial keystream bytes were reported in [103]. We present these observations here.

5.6.1 Biases of z_r toward $r - f_r$

In Chapter 3, we discussed different biases of the permutation bytes toward the secret key. On the other hand, Theorem 5.2.1, which is a restatement of Jenkins' correlations, establishes the bias of the keystream bytes toward the permutation bytes. Interestingly, these two biases can be linked to have leakage of secret key information in the keystream in the form of the bias of z_r toward $r - f_r$.

Lemma 5.6.1 shows how the permutation bytes at rounds τ and $r-1$ of the PRGA, for $r \geq \tau+1$, are related. The result first appeared in [103]. However, based on [157], we present a revised version of the result here.

Lemma 5.6.1. *For $\tau + 1 \leq r \leq \tau + N$, we have*

$$P(S_{r-1}^G[r] = X)$$

$$\approx P(S_\tau^G[r] = X) \cdot \left(\frac{N-1}{N}\right)^{r-\tau-1} +$$

$$\sum_{t=\tau+1}^{r-1} \sum_{n=0}^{r-t} \frac{P(S_\tau^G[t] = X)}{n! \cdot N} \left(\frac{r-t-1}{N}\right)^n \left(1 - \frac{1}{N}\right)^{r-\tau-2-n} .$$

Proof: Let us start from the PRGA state S_τ^G, that is, the state that has been updated τ times in the PRGA. Suppose $S_{r-1}^G[r] = X$. There can be two cases by which the event $(S_{r-1}^G[r] = X)$ can happen.

Case I: In the first case, suppose that $(S_\tau^G[r] = X)$ after round τ, and the r-th index is not disturbed for the next $r - \tau - 1$ state updates. Notice that index i^G varies from $\tau + 1$ to $r - 1$ during this period, and hence never touches the r-th index. Thus, the index r will retain its state value r if index j^G does not touch it. The probability of this event is

$$\left(1 - \frac{1}{N}\right)^{r-\tau-1}$$

over all the intermediate rounds. Hence the first part of the probability is

$$P(S_\tau^G[r] = X)\left(1 - \frac{1}{N}\right)^{r-\tau-1}.$$

Case II: In the second case, suppose that $S_\tau^G[r] \neq X$ and $S_\tau^G[t] = X$ for some $t \neq r$. In such a case, only a swap between the positions r and t during rounds $\tau+1$ to $r-1$ of PRGA can make the event $(S_{r-1}^G[r] = X)$ possible. Notice that if t does not fall in the *path of* i^G, that is, if the index i^G does not touch the t-th location, then the value at $S_\tau^G[t]$ can only go to some position behind i^G, and this can never reach $S_{r-1}^G[r]$, as i^G can only go up to $(r-1)$ during this period. Thus, we must have $2 \leq t \leq r-1$ for $S_\tau^G[t]$ to reach $S_{r-1}^G[r]$. Observe that the way $S_\tau^G[t]$ can move to the r-th position may be either a one hop or a multi-hop route.

In the easiest case of single hop, we require j^G not to touch t until i^G touches t, and $j^G = r$ when $i^G = t$, and j^G not to touch r for the next $r - t - 1$ state updates. Total probability comes to be

$$P(S_\tau^G[t] = X)\left(1 - \frac{1}{N}\right)^{t-\tau-1} \cdot \frac{1}{N} \cdot \left(1 - \frac{1}{N}\right)^{r-t-1}$$

$$= P(S_\tau^G[t] = X) \cdot \frac{1}{N}\left(1 - \frac{1}{N}\right)^{r-\tau-2}.$$

Suppose that it requires $(n + 1)$ hops to reach from $S_\tau^G[t]$ to $S_{r-1}^G[r]$. Then the main issue to note is that the transfer will never happen if the position t swaps with any index which does not lie in the future *path of* i^G. Again, this path of i^G starts from probability $\frac{r-t-1}{N}$ for the first hop and decreases approximately to $\frac{r-t-1}{mN}$ at the m-th hop. We would also require j^G not to touch the location of X between the hops, and this happens with probability $\left(1 - \frac{1}{N}\right)^{r-\tau-2-n}$. Combining all, we get the second part of the probability as approximately

$$P(S_\tau^G[t] = X)\left(\prod_{m=1}^{n}\frac{r-t-1}{mN}\right)\left(1 - \frac{1}{N}\right)^{r-\tau-2-n}$$

$$= \frac{P(S_\tau^G[t] = r)}{n! \cdot N}\left(\frac{r-t-1}{N}\right)^n\left(1 - \frac{1}{N}\right)^{r-\tau-2-n}.$$

It is easy to see that for $n = 0$, the multihop case coincides with the single hop case.

Finally, note that the number of hops $(n+1)$ is bounded from below by 1 and from above by $(r-t+1)$, depending on the initial gap between t and r positions. Considering the sum over t and n with this consideration, we get the desired expression for $P(S_{r-1}^G[r] = X)$. ∎

The following is an immediate implication of Lemma 5.6.1.

Corollary 5.6.2. *For $2 \le r \le N - 1$, we have*

$$P(S_{r-1}^G[r] = f_r) \approx p_r \left(1 - \frac{1}{N}\right)^{r-1} +$$

$$\frac{1 - p_r}{N(N-1)} \sum_{t=1}^{r-1} \sum_{n=0}^{r-t} \frac{1}{n!} \left(\frac{r-t-1}{N}\right)^n \left(1 - \frac{1}{N}\right)^{r-2-n},$$

where

$$p_r = \left(\frac{N-r}{N}\right) \cdot \left(\frac{N-1}{N}\right)^{\left[\frac{r(r+1)}{2} + N\right]} + \frac{1}{N}.$$

Proof: Taking $X = f_r$ and $\tau = 0$ in Lemma 5.6.1, one obtains

$$P(S_0^G[r] = f_r)$$
$$= P(S_N[r] = f_r)$$
$$\approx \left(\frac{N-r}{N}\right) \cdot \left(\frac{N-1}{N}\right)^{\left[\frac{r(r+1)}{2} + N\right]} + \frac{1}{N} \quad \text{(by Corollary 3.1.6)}$$
$$= p_r \quad \text{(say)}.$$

Assume for all $t \ne r$, the events $(S_0^G[t] = f_r)$ are all equally likely and hence substitute

$$P(S_0^G[t] = f_r) = \frac{1 - p_r}{N - 1},$$

in Lemma 5.6.1. This gives the desired expression. ∎

Now consider the bias of each keystream output byte to a combination of the secret key bytes. Let w_r denote the probabilities $P(S_{r-1}^G[r] = f_r)$, $r \ge 1$.

Theorem 5.6.3. *For $r \ge 1$, we have*

$$P(z_r = r - f_r) = \frac{1}{N} \cdot (1 + w_r).$$

Proof: Take two separate cases in which the event $(z_r = r - f_r)$ can occur.

Case I: $S_{r-1}^G[r] = f_r$ and $z_r = r - S_{r-1}^G[r]$. By Theorem 5.2.1, the contribution of this case is

$$P(S_{r-1}^G[r] = f_r) \cdot P(z_r = r - S_{r-1}^G[r]) = w_r \cdot \frac{2}{N}$$

Case II: $S_{r-1}^G[r] \neq f_r$, and still $z_r = r - f_r$ due to random association. So the contribution of this case is

$$P(S_{r-1}^G[r] \neq f_r) \cdot \frac{1}{N} = (1 - w_r) \cdot \frac{1}{N}.$$

Adding the above two contributions, the total probability is

$$w_r \cdot \frac{2}{N} + (1 - w_r) \cdot \frac{1}{N} = \frac{1}{N} \cdot (1 + w_r).$$

∎

Klein Biases

In the first round, i.e., when $r = 1$, we have

$$
\begin{aligned}
w_1 &= P(S_0^G[1] = f_1) \\
&= P(S_N[1] = f_1) \\
&\approx \left(\frac{N-1}{N}\right) \cdot \left(\frac{N-1}{N}\right)^{[\frac{1(1+1)}{2}+N]} + \frac{1}{N} \\
&= \left(\frac{N-1}{N}\right)^{N+2} + \frac{1}{N} \qquad \text{(by Corollary 3.1.6).}
\end{aligned}
$$

Now, using Theorem 5.6.3, we get

$$
\begin{aligned}
P(z_1 = 1 - f_1) &= \frac{1}{N} \cdot (1 + w_1) \\
&\approx \frac{1}{N} \cdot \left(1 + \left(\frac{N-1}{N}\right)^{N+2} + \frac{1}{N}\right).
\end{aligned}
$$

Theorem 5.6.3 quantifies the same bias for the other rounds as well, i.e., the bias of z_r toward $r - f_r$ for $2 \leq r \leq 255$. These biases were first discovered by A. Klein in [86].

To have a clear understanding of the quantity of these biases, Table 5.5 lists the numerical values of the probabilities according to the formula of Theorem 5.6.3. Close to the round 48, the biases tend to disappear. This is indicated by the convergence of the values to the probability $\frac{1}{256} = 0.00390625 \approx 0.0039$ of random association.

Observe that

$$P(z_1 = 1 - f_1) \approx \frac{1}{N}(1 + 0.36)$$

and this bias decreases to

$$P(z_{32} = 32 - f_{32}) \approx \frac{1}{N}(1 + 0.05),$$

but still it is 5% more than the random association.

r	$P(z_r = r - f_r)$
1-8	0.0053 0.0053 0.0053 0.0053 0.0052 0.0052 0.0051 0.0051
9-16	0.0050 0.0050 0.0049 0.0049 0.0048 0.0048 0.0047 0.0047
17-24	0.0046 0.0046 0.0045 0.0045 0.0044 0.0044 0.0043 0.0043
25-32	0.0043 0.0042 0.0042 0.0042 0.0041 0.0041 0.0041 0.0041
33-40	0.0040 0.0040 0.0040 0.0040 0.0040 0.0040 0.0040 0.0040
41-48	0.0040 0.0040 0.0039 0.0039 0.0039 0.0039 0.0039 0.0039

TABLE 5.5: Klein bias probabilities computed following Theorem 5.6.3

Extension to 256-th and 257th Keystream Bytes

Interestingly, the Klein type biases again reappear after rounds 256 and 257 [103].

During the N-th round of the PRGA, $i_N^G = N \bmod N = 0$. Taking $X = f_0$, $\tau = 0$ and $r = N$ in Lemma 5.6.1, we can compute the initial probability as

$$P(S_0^G[0] = f_0) = P(S_N[0] = f_0)$$
$$\approx \left(\frac{N-1}{N}\right)^N + \frac{1}{N} \quad \text{(by Corollary 3.1.6).}$$

and hence the final probability

$$\omega_N = P(S_{N-1}^G[0] = f_0)$$

can be easily computed. Now, using Theorem 5.6.3, the bias in 256th round is given by

$$P(z_N = N - f_0)$$
$$= \frac{1}{N} \cdot (1 + \omega_N).$$

For $N = 256$, $\omega_N = \omega_{256} \approx 0.1383$ and the bias turns out to be $\frac{1}{256}(1 + 0.1383) \approx 0.0044$. Experimental results also confirm this bias.

The bias in the 257-th keystream output byte follows from Theorem 3.1.14, i.e.,

$$P(S_N[S_N[1]] = K[0] + K[1] + 1) \approx \left(\frac{N-1}{N}\right)^{2(N-1)}.$$

During the $(N+1)$-th round, we have, $i_{N+1}^G = (N+1) \bmod N = 1$. Taking $X = f_1$, $\tau = 1$ and $r = N+1$ in Lemma 5.6.1, we have the initial probability as

$$P(S_1^G[1] = f_1) = P(S_N[S_N[1]] = f_1)$$
$$= \left(\frac{N-1}{N}\right)^{2(N-1)}.$$

Once the final probability

$$\omega_{N+1} \quad = \quad P(S_N^G[1] = f_1)$$

is computed, one can use Theorem 5.6.3 to get

$$P(z_{N+1} = N + 1 - f_1)$$
$$= \quad \frac{1}{N} \cdot (1 + \omega_{N+1}).$$

For $N = 256$, $\omega_{N+1} = \omega_{257} \approx 0.0522$ and

$$P(z_{257} \quad = \quad 257 - f_1)$$
$$\approx \quad \frac{1}{N} \cdot (1 + 0.0522)$$
$$\approx \quad 0.0041.$$

This also conforms to experimental observation.

5.6.2   Biases of z_r toward r for Initial Keystream Bytes

The biases of z_r with $r - f_r$ for the initial keystream output bytes have already been pointed out. Interestingly, experimental observation reveals bias of z_r with f_{r-1} too. The results are presented in Table 5.6 which is obtained from over a hundred million (10^8) trials with randomly chosen keys of 16 bytes. For proper random association, $P(z_r = f_{r-1})$ should have been $\frac{1}{256}$, i.e., approximately 0.0039.

r	$P(z_r = f_{r-1})$							
1-8	0.0043	0.0039	0.0044	0.0044	0.0044	0.0044	0.0043	0.0043
9-16	0.0043	0.0043	0.0043	0.0042	0.0042	0.0042	0.0042	0.0042
17-24	0.0041	0.0041	0.0041	0.0041	0.0041	0.0040	0.0040	0.0040
25-32	0.0040	0.0040	0.0040	0.0040	0.0040	0.0040	0.0040	0.0040
33-40	0.0039	0.0039	0.0039	0.0039	0.0039	0.0039	0.0039	0.0039
41-48	0.0039	0.0039	0.0039	0.0039	0.0039	0.0039	0.0039	0.0039

TABLE 5.6: Additional bias of the keystream bytes toward the secret key.

A detailed theoretical analysis of the event $z_r = f_{r-1}$ in general and explicit derivation of a formula for $P(z_r = f_{r-1})$ appeared in [103]. Before giving the general proof, let us first explain the case corresponding to $r = 3$, i.e.,

$$P(z_3 = f_2) > \frac{1}{256}.$$

Assume that after the third round of the KSA, $S_3[2]$ takes the value f_2, and is hit by j later in the KSA. Then f_2 is swapped with $S_\kappa[\kappa]$ and consider that

$S_\kappa[\kappa]$ has remained κ so far. Further, suppose that $S_N[3] = 0$ holds. Hence $S_N[2] = \kappa$, $S_N[k] = f_2$ and $S_N[3] = 0$ at the end of the KSA. In the second round of the PRGA, $S_1^G[2] = \kappa$ is swapped with a more or less random location $S_1^G[l]$. Therefore, $S_2^G[l] = \kappa$ and $j_2^G = l$. In the next round, $i_3^G = 3$ and points to $S_2^G[3] = 0$. Hence j^G does not change from its value in the previous round and so $j_3^G = l = j_2^G$. Thus, $S_2^G[l] = \kappa$ is swapped with $S_2^G[3] = 0$, and one has $S_3^G[l] = 0$ and $S_3^G[3] = \kappa$. The output z_3 is now

$$
\begin{aligned}
S_3^G[S_3^G[i_3^G] + S_3^G[j_3^G]] &= S_3^G[\kappa + 0] \\
&= S_3^G[\kappa] \\
&= f_2.
\end{aligned}
$$

Along the same line of argument given above, the general proof for any r depends on $P(S_N[r] = 0)$. Recall that Item (2) of Theorem 3.2.11 presents an explicit formula for the probabilities $P(S_N[u] = v)$, $v \le u$. Plugging in r for u and 0 for v, we can state the following corollary.

Corollary 5.6.4. *For $0 \le r \le N - 1$, $P(S_N[r] = 0) = \gamma_r$, where*

$$
\gamma_r = \frac{1}{N} \cdot \left(\frac{N-1}{N}\right)^{N-1-r} + \frac{1}{N} \cdot \left(\frac{N-1}{N}\right) - \frac{1}{N} \cdot \left(\frac{N-1}{N}\right)^{N-r}.
$$

Theorem 5.6.5. $P(z_1 = f_0)$

$$
= \left(\frac{N-1}{N}\right)^2 \cdot \left(\frac{N-2}{N}\right)^{N-1} \cdot \gamma_1 + \frac{1}{N},
$$

where

$$
\gamma_1 = \frac{1}{N} \cdot \left(\frac{N-1}{N}\right)^{N-2} + \frac{1}{N} \cdot \left(\frac{N-1}{N}\right) - \frac{1}{N} \cdot \left(\frac{N-1}{N}\right)^{N-1}.
$$

Proof: Substituting $r = 1, y = 0$ in Proposition 3.1.5, we have

$$
P(S_1[0] = f_0) = 1.
$$

After the first round, suppose that the index 0 is touched for the first time by $j_{\tau+1}$ in round $\tau + 1$ of the KSA and due to the swap the value f_0 is moved to the index τ, $1 \le \tau \le N - 1$ and also prior to this swap the value at the index τ was τ itself, which now comes to the index 0. This means that from round 1 to round τ (both inclusive), the pseudo-random index j has not taken the values 0 and τ. So, after round $\tau + 1$,

$$
\begin{aligned}
&P\left((S_{\tau+1}[0] = \tau) \wedge (S_{\tau+1}[\tau] = f_0)\right) \\
= \ &P\left((S_\tau[0] = f_0) \wedge (S_\tau[\tau] = \tau) \wedge (j_{\tau+1} = 0)\right) \\
= \ &\left(\frac{N-2}{N}\right)^\tau \cdot \frac{1}{N}.
\end{aligned}
$$

From the end of round $\tau + 1$ to the end of the KSA, f_0 remains in index τ and the value τ remains in index 0 with probability $(\frac{N-2}{N})^{N-\tau-1}$. Thus,

$$
\begin{aligned}
P\left((S_N[0] = \tau) \wedge (S_N[\tau] = f_0)\right) &= \left(\frac{N-2}{N}\right)^{\tau} \cdot \frac{1}{N} \cdot \left(\frac{N-2}{N}\right)^{N-\tau-1} \\
&= \left(\frac{N-2}{N}\right)^{N-1} \cdot \frac{1}{N} \\
&= \beta_1 \quad \text{(say)}.
\end{aligned}
$$

In the first round of the PRGA,

$$
\begin{aligned}
j_1^G &= 0 + S_0^G[1] \\
&= S_N[1].
\end{aligned}
$$

From Corollary 5.6.4, we have

$$
\begin{aligned}
P(S_N[1] = 0) &= \gamma_1 \\
&= \frac{1}{N} \cdot \left(\frac{N-1}{N}\right)^{N-2} + \frac{1}{N} \cdot \left(\frac{N-1}{N}\right) - \frac{1}{N} \cdot \left(\frac{N-1}{N}\right)^{N-1}.
\end{aligned}
$$

If $S_N[1] = 0$, then $j_1^G = 0$ and

$$
\begin{aligned}
z_1 &= S_1^G\left[S_1^G[1] + S_1^G[j_1^G]\right] \\
&= S_1^G\left[S_0^G[1] + S_0^G[j_1^G]\right] \\
&= S_1^G[0 + \tau] \\
&= S_1^G[\tau].
\end{aligned}
$$

Since $\tau \neq 0$, the swap of the values at the indices 0 and 1 does not move the value at the index τ. Thereby

$$
S_1^G[\tau] = S_0^G[\tau] = S_N[\tau] = f_0
$$

and $z_1 = f_0$ with probability

$$
\beta_1 \cdot \gamma_1 = \delta_1 \quad \text{(say)}.
$$

Since, τ can take values $1, 2, 3, \ldots, N-1$, the total probability is $\delta_1 \cdot (N-1)$. Substituting the values of $\beta_1, \gamma_1, \delta_1$, we get the probability that the event $(z_1 = f_0)$ occurs in the above path is

$$
p = \left(\frac{N-1}{N}\right)^1 \cdot \left(\frac{N-2}{N}\right)^{N-1} \cdot \gamma_1.
$$

If the above path is not followed, still there is $(1 - p) \cdot \frac{1}{N}$ probability of occurrence of the event due to random association. Adding these two probabilities, we get the result. ∎

Theorem 5.6.6. *For* $3 \le r \le N$, $P(z_r = f_{r-1})$

$$
\approx \left(\frac{N-1}{N}\right) \cdot \left(\frac{N-r}{N}\right) \cdot \left(\left(\frac{N-r+1}{N}\right) \cdot \left(\frac{N-1}{N}\right)^{\left[\frac{r(r-1)}{2}+r\right]} + \frac{1}{N}\right) \cdot
$$

$$
\left(\frac{N-2}{N}\right)^{N-r} \cdot \left(\frac{N-3}{N}\right)^{r-2} \cdot \gamma_r + \frac{1}{N},
$$

where

$$
\gamma_r = \frac{1}{N} \cdot \left(\frac{N-1}{N}\right)^{N-1-r} + \frac{1}{N} \cdot \left(\frac{N-1}{N}\right) - \frac{1}{N} \cdot \left(\frac{N-1}{N}\right)^{N-r}.
$$

Proof: Substituting $y = r - 1$ in Theorem 3.1.5, we have

$$
P(S_r[r-1] = f_{r-1}) = \alpha_r,
$$

where for $1 \le r \le N$,

$$
\alpha_r \approx \left(\frac{N-r+1}{N}\right) \cdot \left(\frac{N-1}{N}\right)^{\left[\frac{r(r-1)}{2}+r\right]} + \frac{1}{N}.
$$

After round r, suppose that the index $r - 1$ is touched for the first time by $j_{\tau+1}$ in round $\tau + 1$ of the KSA and due to the swap the value f_{r-1} is moved to the index τ, $r \le \tau \le N - 1$ and also prior to this swap the value at the index τ was τ itself, which now comes to the index $r - 1$. This means that from round $r + 1$ to round τ (both inclusive), the pseudo-random index j has not taken the values $r - 1$ and τ. So, after round $\tau + 1$,

$$
\begin{aligned}
&P\left((S_{\tau+1}[r-1] = \tau) \wedge (S_{\tau+1}[\tau] = f_{r-1})\right) \\
&= P\left((S_\tau[r-1] = f_{r-1}) \wedge (S_\tau[\tau] = \tau) \wedge (j_{\tau+1} = r-1)\right) \\
&= \alpha_r \cdot \left(\frac{N-2}{N}\right)^{\tau-r} \cdot \frac{1}{N}.
\end{aligned}
$$

From the end of round $\tau + 1$ until the end of the KSA, f_{r-1} remains in index τ and the value τ remains in index $r - 1$ with probability $\left(\frac{N-2}{N}\right)^{N-\tau-1}$. Thus,

$$
\begin{aligned}
&P\left((S_N[r-1] = \tau) \wedge (S_N[\tau] = f_{r-1})\right) \\
&= \alpha_r \cdot \left(\frac{N-2}{N}\right)^{\tau-r} \cdot \frac{1}{N} \cdot \left(\frac{N-2}{N}\right)^{N-\tau-1} \\
&= \alpha_r \cdot \left(\frac{N-2}{N}\right)^{N-r-1} \cdot \frac{1}{N} \\
&= \beta_r \quad \text{(say)}.
\end{aligned}
$$

Also, from Corollary 5.6.4, we have

$$
\begin{aligned}
P(S_N[r] = 0) &= \gamma_r \\
&= \frac{1}{N} \cdot \left(\frac{N-1}{N}\right)^{N-1-r} + \frac{1}{N} \cdot \left(\frac{N-1}{N}\right) - \frac{1}{N} \cdot \left(\frac{N-1}{N}\right)^{N-r}.
\end{aligned}
$$

Suppose the indices $r - 1$, τ and r are not touched by the pseudo-random index j^G in the first $r - 2$ rounds of the PRGA. This happens with probability $(\frac{N-3}{N})^{r-2}$. In round $r - 1$ of the PRGA, due to the swap, the value τ at index $r - 1$ moves to the index j^G_{r-1} with probability 1, and $j^G_{r-1} \notin \{\tau, r\}$ with probability $(\frac{N-2}{N})$. Further, if $S^G_{r-1}[r]$ remains 0, then in round r of the PRGA, $j^G_r = j^G_{r-1}$ and

$$
\begin{aligned}
z_r &= S^G_r \left[S^G_r[r] + S^G_r[j^G_r] \right] \\
&= S^G_r \left[S^G_{r-1}[r] + S^G_{r-1}[j^G_{r-1}] \right] \\
&= S^G_r[0 + \tau] \\
&= S^G_r[\tau] \\
&= f_{r-1}
\end{aligned}
$$

with probability

$$
\beta_r \cdot \gamma_r \cdot \left(\frac{N-3}{N} \right)^{r-2} \cdot \left(\frac{N-2}{N} \right) = \delta_r \quad \text{(say)}.
$$

Since, τ can values $r, r+1, r+2, \ldots, N-1$, the total probability is $\delta_r \cdot (N - r)$. Substituting the values of $\alpha_r, \beta_r, \gamma_r, \delta_r$, we get the probability that the event $(z_r = f_{r-1})$ occurs in the above path is

$$
\begin{aligned}
p \approx \; & \left(\frac{N-r}{N} \right) \cdot \left(\left(\frac{N-r+1}{N} \right) \cdot \left(\frac{N-1}{N} \right)^{\lceil \frac{r(r-1)}{2} + r \rceil} + \frac{1}{N} \right) \cdot \\
& \left(\frac{N-2}{N} \right)^{N-r} \cdot \left(\frac{N-3}{N} \right)^{r-2} \cdot \gamma_r.
\end{aligned}
$$

If the above path is not followed, still there is $(1 - p) \cdot \frac{1}{N}$ probability of occurrence of the event due to random association. Adding these two probabilities, we get the result. ∎

The theoretically computed values of the probabilities according to the expressions derived in the above two theorems match with the experimental values provided in Table 5.6. Theorem 5.6.6 does not cover the cases $r = 1$ and $r = 2$. Case $r = 1$ has already separately been presented in Theorem 5.6.5. It is interesting to justify the absence of bias in case $r = 2$ as observed experimentally in Table 5.6.

5.6.3   Cryptanalytic Applications

Consider the first keystream output byte z_1 of the PRGA. We find the results that $P(z_1 = 1 - f_1) = 0.0053$ (from Theorem 5.6.3) and that $P(z_1 = f_0) = 0.0043$ (from Theorem 5.6.5 and Table 5.6). Further, from Theorem 5.5.2, we have the result that $P(z_1 = f_2) = 0.0053$. Taking them together, one can verify that

$$P(z_1 = f_0 \vee z_1 = 1 - f_1 \vee z_1 = f_2)$$
$$= 1 - (1 - 0.0043) \cdot (1 - 0.0053) \cdot (1 - 0.0053)$$
$$= 0.0148.$$

The independence assumption in calculating the probability is supported by experimental results. The above result implies that out of randomly chosen 10000 secret keys, in 148 cases on an average, z_1 reveals f_0 or $1 - f_1$ or f_2, i.e., $K[0]$ or $1 - (K[0] + K[1] + 1)$ or $(K[0] + K[1] + K[2] + 3)$. If, however, one considers only random association, the probability that z_1 will be among three randomly chosen values v_1, v_2, v_3 from $\{0, \ldots, 255\}$, is given by

$$P(z_1 = v_1 \vee z_1 = v_2 \vee z_1 = v_3) = 1 - \left(1 - \frac{1}{256}\right)^3$$
$$= 0.0117.$$

Thus, one can guess z_1 with an additional advantage of

$$\frac{0.0148 - 0.0117}{0.0117} \cdot 100\% = 27\%$$

over the random guess.

Looking at z_2, from Theorem 5.6.3 and Table 5.5, we have

$$P(z_2 = 2 - f_2) = 0.0053$$

which provides an advantage of

$$\frac{0.0053 - 0.0039}{0.0039} \cdot 100\% = 36\%.$$

Similarly, referring to Theorem 5.6.3 and Theorem 5.6.6 (and also Table 5.5 and Table 5.6), significant biases can be observed in the events $(z_r = f_{r-1})$ given $z_r = r - f_r$, for $r = 3$ to 32, over random association.

Next, consider the following scenario with the events $A_1, \ldots, A_{32}$, where

$$A_1 : (z_1 = f_0 \vee z_1 = 1 - f_1 \vee z_1 = f_2),$$
$$A_2 : (z_2 = 2 - f_2), \quad \text{and}$$
$$A_r : (z_r = f_{r-1} \vee z_r = r - f_r) \quad \text{for } 3 \leq r \leq 32.$$

Observing the first 32 keystream output bytes $z_1, \ldots, z_{32}$, one may attempt at guessing the secret key, assuming that 3 or more of the events $A_1, \ldots, A_{32}$ occur. Experimenting with 10 million randomly chosen secret keys of length 16 bytes, it is found that 3 or more of the events occur in 0.0028 proportion of cases, which is true for 0.0020 proportion of cases for random association. This demonstrates a substantial advantage (40%) over random guess.

5.6.4   Further Biases when Pseudo-Random Index Is Known

In all the biases discussed in this chapter so far, it is assumed that the value of the pseudo-random index j^G is unknown. In this section, we are going to show that the biases in the permutation at some stage of the PRGA propagates to the keystream at a later stage, if the value of the index j^G at the earlier stage is known.

Suppose that we know the value j_τ^G of j^G after the round τ in the PRGA. With high probability, the value V at the index j_τ^G would remain there, until j_τ^G is touched by the deterministic index i^G for the first time after a few more rounds depending on what was the position ($\tau \bmod N$) of i^G at the τ-th stage. This immediately leaks V in keystream. More importantly, if the value V is biased to the secret key bytes, then that information will be leaked too.

Formally, let

$$P(S_\tau^G[j_\tau^G] = V) = \eta_\tau$$

for some V. j_τ^G would be touched by i^G in round r, where

$$r = \tau + (j_\tau^G - \tau \bmod N) \text{ or } \tau + (N - \tau \bmod N) + j_\tau^G$$

depending on whether

$$j_\tau^G > \tau \bmod N \, \text{or} \, j_\tau^G \le \tau \bmod N$$

respectively. By Lemma 5.6.1, we have

$$P(S_{r-1}^G[j_\tau^G] = V) = \eta_\tau \cdot \left[\left(\frac{N-1}{N}\right)^{r-\tau-1} - \frac{1}{N}\right] + \frac{1}{N}.$$

Now, Lemma 5.6.3 immediately gives

$$P(z_r = r - V) = \frac{1}{N} \cdot \left(1 + \eta_\tau \cdot \left[\left(\frac{N-1}{N}\right)^{r-\tau-1} - \frac{1}{N}\right] + \frac{1}{N}\right).$$

For some special V's, the form of η_τ may be known. In that case, it would be advantageous to probe the values of j^G at particular rounds. For instance, according to Corollary 5.6.2, after the ($\tau - 1$)-th round of the PRGA, $S_{\tau-1}^G[\tau]$ is biased to the linear combination f_τ of the secret key bytes with probability

$$\eta_\tau \approx \left[\left(\frac{N-\tau}{N}\right) \cdot \left(\frac{N-1}{N}\right)^{\left[\frac{\tau(\tau+1)}{2} + N\right]} + \frac{1}{N}\right] \cdot \left[\left(\frac{N-1}{N}\right)^{\tau-1} - \frac{1}{N}\right] + \frac{1}{N}.$$

At round τ, f_τ would move to the index j_τ^G due to the swap, and hence $S_\tau^G[j_\tau^G]$ would be biased to f_τ with the same probability. So, the knowledge of j_τ^G would leak information about f_τ in round

$$r = \tau + (j_\tau^G - \tau \bmod N) \text{ or } \tau + (N - \tau \bmod N) + j_\tau^G$$

depending on whether

$$j_\tau^G > \tau \bmod N \text{ or } j_\tau^G \leq \tau \bmod N$$

respectively. If we know the values of j^G at multiple stages of the PRGA (it may be possible to read some values of j^G through side-channel attacks), then the biases propagate further down the keystream. The following example illustrates how the biases propagate down the keystream output bytes when single as well as multiple j^G values are known.

Example 5.6.7. *Suppose we know the value of j_5^G as 18. With probability η_5, $S_4^G[5]$ would have remained f_5 which would move to index 18 due to the swap in round 5, i.e., $S_5^G[18] = f_5$. With approximately*

$$\eta_5 \cdot \left[\left(\frac{N-1}{N} \right)^{18-5-1} - \frac{1}{N} \right] + \frac{1}{N}$$

probability, f_5 would remain in index 18 till the end of the round 18-1 = 17. Thus, we immediately get a bias of z_{18} with $18 - f_5$.

Further, in round 18, f_5 would move from index 18 to j_{18}^G. If the value of j_{18}^G is also known, say $j_{18}^G = 3$, then we have $S_{18}^G[3] = f_5$. We can apply the same line of arguments for round $256 + 3 = 259$ to get a bias of z_{259} with $259 - f_5$. Experiments with 1 billion random keys demonstrate that in this case the bias of z_{18} toward $18 - f_5$ is 0.0052 and the bias of z_{259} toward $259 - f_5$ is 0.0044. These conform to the theoretical values and show that the knowledge of j^G during the PRGA helps in finding non-random association (away from $\frac{1}{256} = 0.0039$) between the keystream bytes and the secret key.

5.7 Exhaustive Enumeration of All Biases

Recently, experimental results of an exhaustive search for biases in all possible linear combinations of the state variables and the keystream bytes of RC4 were published in [158]. In this work, the linear equations are modeled as

$$a_0 i_r^G + a_1 j_r^G + a_2 S_r^G[i_r^G] + a_3 S_r^G[j_r^G] + a_4 z_r = b, \tag{5.1}$$

where the a_i's and b belong to $\mathbb{Z}_N$ and the additions are modulo N. This gives 2^{48} linear equations corresponding to each round r of RC4 PRGA. Now, $b - a_0 i_r^G$ can be regarded as a single variable C, reducing Equation (5.1) to

$$c_0 j_r^G + c_1 S_r^G[i_r^G] + c_2 S_r^G[j_r^G] + c_3 z_r = C. \tag{5.2}$$

The number of equations 5.2 are further reduced as follows. The equations are grouped into 256 classes, each corresponding to a specific value of $C \in \mathbb{Z}_N$.

Further, though c_i's are elements of $\mathbb{Z}_N$, the coefficients set of the c_i's are restricted to $\{-1, 0, 1\}$. This gives 256 groups of $3^4 = 81$ linear equations for each round. Experimental verification of these equations with 10^9 randomly chosen secret keys of 16 bytes reveals the already known biases as well as some new biases. In Table 5.7, we present some of the new biases reported in [158].

c_0	c_1	c_2	c_3	C	Probability
0	1	1	-1	1	$\frac{0.89}{N}$
⋮	⋮	⋮	⋮	⋮	⋮
0	1	1	-1	255	$\frac{1.25}{N}$
0	1	1	1	0	$\frac{0.95}{N}$
⋮	⋮	⋮	⋮	⋮	⋮
0	1	1	1	255	$\frac{0.95}{N}$
1	1	0	0	0	$\frac{0.95}{N}$
⋮	⋮	⋮	⋮	⋮	⋮
1	1	0	0	255	$\frac{0.95}{N}$
1	1	-1	0	i	$\frac{2}{N}$
1	-1	1	0	i	$\frac{2}{N}$
1	-1	0	0	1	$\frac{0.9}{N}$
⋮	⋮	⋮	⋮	⋮	⋮
1	-1	0	0	255	$\frac{1.25}{N}$

TABLE 5.7: Some keystream-state correlations as observed in [158].

To achieve faster than exhaustive search, the work [158] proposed a spectral approach that revealed some more new biases. Additionally, some new biases involving the secret key bytes were also found by considering the following linear model.

$$a_0 K[0] + \cdots + a_{l-1} K[l-1] + a_l z_1 + \cdots + a_{2l-1} z_l = b. \qquad (5.3)$$

Some of the new secret key biases are presented in Table 5.8.

Relation	Probability
$z_1 + K[0] + K[1] - K[2] = 3$	$\frac{1.14116}{N}$
$z_2 + K[1] + K[2] = -3$	$\frac{1.36897}{N}$
$z_4 - K[1] + K[4] = 4$	$\frac{1.04463}{N}$

TABLE 5.8: Some keystream-key correlations as observed in [158].

Proofs of these biases have recently been published in [156].

5.8 State Recovery from Keystream

RC4 can be completely broken if one can reconstruct the permutation S^G by observing the keystream output bytes. Such attacks are called *state recovery attacks*.

The RC4 state consists of two 8-bit indices i and j and a permutation of 256 possible 8-bit elements. Thus, the size of the state space is $2^8! \times (2^8)^2 \approx 2^{1700}$, making the exhaustive search completely infeasible.

5.8.1 Partial Recovery Using Predictive States

The notion of *fortuitous state* was introduced in [51, Section 5]. The RC4 state in which only a consecutive permutation elements are known and only those a elements participate in producing the next a successive keystream bytes, is called a fortuitous state of length a. The authors analyzed these special states to demonstrate that sometimes the attacker may be able to determine portions of the internal state by observing the keystream with non-trivial probability.

In [110, Chapter 2], Mantin generalizes the notion of fortuitous states into three wider classes as follows.

1. *b-Predictive a-States*: A state $A = (S^G, i^G, j^G)$ is called an *a-state*, if only a elements of S^G are known. An a-state A is called *b-predictive*, if it can predict the outputs of b consecutive rounds.

2. *Profitable States*: An a-state A is called *weak a-profitable*, if all the states compatible with A have the same sequence of j^G's during the next a rounds. Further, if these a values of j^G lie in the interval $[i^G + 1, \ldots, i^G + a]$, then A is called *a-profitable*.

3. *Useful States*: An RC4 state (S^G, i^G, j^G) is called *useful* if $S^G[i^G] + S^G[j^G] = i^G$.

Note that every fortuitous state of length a is an a-predictive a-state, but the converse is not true. Also, it is pointed out in [110, Chapter 2] that though the profitable states occur more frequently and they can be stored more efficiently than the fortuitous states of the same order, an attack based on a single profitable state takes significantly more time than the time of a similar attack based on a single fortuitous state. Interestingly, the useful states are the origin of the glimpse bias discovered by Jenkins [78] (see Theorem 5.2). A sequence of a useful states provides $2a$ indications to entries in S^G.

As already mentioned, an a-predictive a-state may or may not be a fortuitous state of length a. In [138], analysis of *non-fortuitous predictive states* of length a is performed in great detail. This analysis reveals that if ϕ_a and

ν_a denote the numbers of fortuitous and non-fortuitous states of length a respectively, then the knowledge of non-fortuitous states reduces the length of the keystream output segment required for the occurrence of any a-predictive a-state by $N \cdot N^{a+1}(\frac{1}{\phi_a} - \frac{1}{\phi_a + \nu_a})$.

The concept of *recyclable states* is introduced in [109, Section 4]. Suppose $A = (S^G, i^G, j^G)$ is an a-state and let I be the minimal interval containing the a permutation entries of A. Suppose that in every a-compliant state, the permutation S'^G after i^G leaves I satisfies again the permutation constraints of A. Then A is said to be recyclable. Based on the analysis of these states, one can mount what is called the *recycling attack* (described in [109, Section 4]) which can predict a single byte from 2^{50} samples of keystream bytes with a success probability of 82%.

The above methods of partial state recovery do not have much cryptanalytic significance and hence we avoid detailed discussion on them. Rather, we focus on the important milestones of complete state recovery in the subsequent sections.

5.8.2   Knudsen et al.'s Attack and Its Improvements

Knudsen et al. [87] discovered for the first time that a branch and bound strategy reduces the complexity for recovering the internal state much below that of exhaustive search.

The output of RC4 at time (PRGA round) r is

$$z_r = S_r^G[S_r^G[i_r^G] + S_r^G[j_r^G]] \tag{5.4}$$

The attack of [87] first considers a simplified version of RC4 in which there is no swap and then proceeds to incorporate the feature of swap.

The following result is immediate.

Theorem 5.8.1. *If the swap operation of RC4 is omitted, the keystream becomes cyclic with a period of 2^{n+1}, where each index and each permutation entry is of n bits.*

Proof: Since there is no swap operation, the permutation is constant and so we denote it by S^G, without any subscript. From Equation (5.4),

$$z_{r+2^{n+1}} = S^G[S^G[i_{r+2^{n+1}}^G] + S^G[j_{r+2^{n+1}}^G]].$$

Now, $i_{r+2^n}^G = i_r^G$. Applying $j_r^G = j_{r-1}^G + S^G[i_r^G]$ repeatedly, we have

$$\begin{aligned}
j_{r+2^n}^G &= j_r^G + \sum_{u=0}^{2^n-1} S^G[u] \\
&= j_r^G + 2^{n-1} \pmod{2^n}.
\end{aligned}$$

After 2^n more steps, $i^G_{r+2^{n+1}} = i^G_r$ and

$$
\begin{aligned}
j^G_{r+2^{n+1}} &= j^G_r + 2 \cdot 2^{n-1} \\
&= j^G_r \pmod{2^n}.
\end{aligned}
$$

Hence $z_{r+2^{n+1}} = z_r$. ∎

At any point of time r, i^G_r is always known. In the known plaintext attack model, z_r is also assumed to be known. Thus, there are four unknowns, namely, $j^G_r, S^G[i^G_r], S^G[j^G_r], (S^G)^{-1}[z_r]$. For any three of these, the fourth one can be calculated as follows.

$$j^G_r = (S^G)^{-1}[(S^G)^{-1}[z_r] - S^G[i^G_r]] \tag{5.5}$$

$$S^G[i^G_r] = (S^G)^{-1}[z_r] - S^G[j^G_r] \tag{5.6}$$

$$S^G[j^G_r] = (S^G)^{-1}[z_r] - S^G[i^G_r] \tag{5.7}$$

$$(S^G)^{-1}[z_r] = S^G[i^G_r] + S^G[j^G_r] \tag{5.8}$$

The algorithm to recover S^G works as follows. Initially, we guess a small subset of the values of S^G. As r progresses, Equations (5.5) to (5.8) are used to derive additional entries of S^G. If a contradiction arises, then the initial guess was wrong. We repeat this process for all possible guesses.

At the beginning, i^G_0 and j^G_0 are both known (0). If we guess the first x values of S^G, then j^G_0 through j^G_{x-1} becomes known. For each of these, Equations (5.7) and (5.8) can be used to determine more values of S^G. Once i^G_r goes past x, we lose the knowledge of j^G_r, if $S^G[x+1]$ is not known. We discard the next values of z_r until we can use Equation (5.5) to discover j^G_r again using a known $S^G[i^G_r]$. Once j^G_r is recovered, we can then work backward using Equation (5.6) to recover more entries of S^G. Once the backward movement is finished, we continue as we started, using Equations (5.7) and (5.8) to discover values of S^G until we lose the value of j^G_r.

A modification of the above strategy allows one to attack the real RC4. In the modified version, no values of S^G are guessed initially. A permutation entry is guessed only when it is needed. For times $r = 1, 2, \ldots$, if $S^G_{r1}[i^G_r]$ or $S^G_{r-1}[j^G_r]$ have not already been assigned values in a previous time $t < r$, we need to choose a value v for $S^G_{r-1}[i^G_r]$, $0 \le v < 2^n$, derive j^G_r and then choose $S^G_{r-1}[j^G_r]$. The next update of the RC4 PRGA is then followed. We proceed in such a way that at each time r, an output word is generated with the property that it matches with the actual z_r that is observed. This imposes the following three restrictions on the possible choices for $S^G_{r1}[i^G_r]$ and $S^G_{r-1}[j^G_r]$.

1. Since S^G is a permutation, every new value assigned to $S^G_{r1}[i^G_r]$ or $S^G_{r-1}[j^G_r]$ must be different from a value already assigned to some other entry in S^G.

2. If the known z_r differs from all entries previously fixed in S^G, then $t_r = S^G_r[i^G_r] + S^G_r[j^G_r]$ must also differ from all index positions which have already been assigned values. If this condition is met, then we set $S^G_r[i^G_r] = z_r$, otherwise we have a contradiction in the search.

3. If z_r is equal to the value v previously assigned to index u of S^G, then $t_r = S_r^G[i_r^G] + S_r^G[j_r^G]$ becomes equal to u. This either uniquely determines $S_r^G[j_r^G]$ or leads to a contradiction.

The above algorithm can be efficiently implemented using recursion on the time step r. The recursion steps backward if a contradiction is reached, due to the previously wrong guesses. If M (out of N) many permutation entries are *a priori* known, the complexity can be reduced further. For $N = 256$, the complete attack complexity turns out to be around 2^{779}. The time complexity of the attack for various values of N and M are enumerated in Tables D.1 and D.2 of [110, Appendix D.4].

In [123], the cycle structures in RC4 are investigated in detail and a "tracking" attack is proposed. This strategy recovers the RC4 state, if a significant fraction of the full cycle of keystream bits is available. For example, the state of a 5-bit RC4-like cipher can be derived from a portion of the keystream using 2^{42} steps, while the nominal key-space of the system is 2^{160}.

The work [165] showed that Knudsen's attack [87] requires 2^{220} search complexity if 112 entries of the permutation are known *a priori*. It also presents an improvement whereby state recovery with the same complexity can be achieved with prior knowledge of only 73 permutation entries in certain cases.

In [181], an improvement over [87] is proposed using a tree representation of RC4. At time-step r, the nodes are distributed at $r + 1$ levels. Nodes at level h, $0 < h \leq r$, correspond to the set of all possible positions in S_{r-h}^G where z_r can be found. The nodes are linked by the branches which represent the conditions to pass from one node to another. In order to find the internal state, such a tree of general conditions is searched by a hill-climbing strategy. This approach reduces the time complexity of the full RC4 state recovery from 2^{779} to 2^{731}.

5.8.3   Maximov et al.'s Attack

The best known result for state recovery has been reported in [116]. It shows that the permutation can be reconstructed from the keystream in around 2^{241} complexity. This renders RC4 insecure when the key length is more than 30 bytes (240 bits).

The basic idea of cryptanalysis in [116] is as follows. Corresponding to a window of $w + 1$ keystream output bytes from some round r, it is assumed that all the j^G's are known, i.e., $j_r^G, j_{r+1}^G, \ldots, j_{r+w}^G$ are known. Thus w many state variable values $S_{r+1}^G[i_{r+1}^G], \ldots, S_{r+w}^G[i_{r+w}^G]$ can immediately be computed as

$$S_{t+1}^G[i_{t+1}^G] = j_{t+1}^G - j_t^G, \quad t = r, r+1, \ldots, r+w-1.$$

Then w many equations of type

$$(S_t^G)^{-1}[z_t] = S_t^G[i_t^G] + S_t^G[j_t^G], \quad t = r, r+1, \ldots, r+w. \quad (5.9)$$

can be formed, where each equation contains only two unknowns,

$$(S_t^G)^{-1}[z_t], S_t^G[j_t^G], \qquad (5.10)$$

instead of four unknowns $j^G, S^G[i^G], S^G[j^G], (S^G)^{-1}[z]$ as in Knudsen's attack. The set of w equations of the above form is called *a window of length* w.

The algorithm for recovery is a recursive one. It begins with a window of w equations of the form 5.9 and tries solving them. An equation is called *solved* or *processed* if the corresponding unknowns (5.10) have been either explicitly derived or guessed, and an equation is termed as *active* if it is not yet processed. The window dynamically grows and shrinks. When window length w is sufficiently long (say, $w = 2N$), all equations are likely to be solved and the initial state S_r^G is likely to be fully recovered. The algorithm has four major components.

1. **Iterative Recovering (IR) Block.** It receives a number of *active* equations in the window as input and attempts to derive as many unknowns $(S_t^G)^{-1}[z_t], S_t^G[j_t^G]$ as possible. Upon contradiction, the recursion steps backward.

2. **Find and Guess Maximum Clique (MC) Block.** If no more active equation can be solved, $(S_t^G)^{-1}[z_t]$ for one t needed to be guessed. Construct a graph using the active equations as vertices v_t. Two vertices v_{t_1} and v_{t_2} are connected by an edge, if either $z_{t_1} = z_{t_2}$ or $S_{t_1}^G[j_{t_1}^G] = S_{t_2}^G[j_{t_2}^G]$. Guessing any unknown variable in such a subgraph solves all the equations involved and hence these subgraphs are cliques. The role of the MC block is to search for a maximum clique and guess one $(S_t^G)^{-1}[z_t]$ for one of the equations belonging to the clique.

3. **Window Expansion (WE) Block.** A new equation is added to the system as soon as the missing value $S_t^G[i_t^G]$ in the beginning or in the end of the window is derived. The WE block checks for this event and dynamically extends the window.

4. **Guessing One $S[i]$ (GSi) Block.** If no active equations are left and the initial state is not yet fully recovered, the window is then expanded by a direct guess of $S_t^G[i_t^G]$, at one end (in front or in back) of the window.

Algorithm 5.8.1 gives the sequence of steps that need to be executed in order to perform the state recovery.

Some precomputation may be performed to identify a certain position in the keystream where the internal state is compliant to a specific pattern. In [116], many different kinds of patterns and their effects on the key recovery algorithm are discussed. For the sake of simplicity, we only define the primary patterns and sketch the technique of utilizing these patterns. Interested readers may look into [116, Sections 3, 4] for more details.

Input: A window of w equations of the form 5.9, starting from round r.

Output: The recovered initial state S_r^G or FAIL.

1 Execute IR;

2 **if** *a contradiction is reached* **then**

3 | return (*recurse backward*);

end

4 **else if** *new equations are available* **then**

5 | Execute WE;

6 | Go to Step 1;

end

7 **else if** *all equations in the window are solved* **then**

8 | Execute GSi;

9 | Go to Step 5 (*recurse forward*);

end

10 **else**

11 | Execute MC;

12 | Go to Step 1 (*recurse forward*);

end

Algorithm 5.8.1: RecoverState

Definition 5.8.2. (d-order Pattern) *A d-order pattern is a tuple $A = \{i, j, U, V\}$, where U and V are two vectors from $\mathbb{Z}_N^d$ with pairwise distinct elements.*

Definition 5.8.3. (Compliant State) *At time step r, the internal state is compliant with A if $i_r^G = i$, $j_r^G = j$, and d cells of S_r^G with indices from U have corresponding values from V.*

Definition 5.8.4. (w-generative Pattern) *A pattern A is called w-generative if for any internal state compliant with A, the next w clockings allow to derive w equations of the form $S_r^{G^{-1}}[z_r] = S_r^G[i_r^G] + S_r^G[j_r^G]$, i.e., if consecutive w values of j^G are known.*

The basic strategy is to look for d-order w-generative patterns with small d and large w. Whenever the observed keystream indicates such a pattern of the internal state, iterative recovery of the unknowns is performed using the *RecoverState* algorithm and the window w is dynamically expanded. A general time complexity estimate is performed in [117, Appendix C]. The success probability of the full attack turns out to be at least 98%.

A very recent work [61] revisits Maximov's attack and presents an iterative probabilistic state reconstruction method. It also discusses how one can practically approach the complexity of [116].

Research Problems

Problem 5.1 Does there exist any cycle in real RC4 states?

Problem 5.2 Apart from Finney cycle, does there exist any other cycle in the state space of RC4 corresponding to different initialization conditions?

Problem 5.3 Recall the conditional bias of z_1 toward $K[2]+3$ (given $K[0]+K[1] = 0$) in Theorem 5.5.4. Can similar strong conditional biases be present in subsequent keystream bytes?

Chapter 6

Distinguishing Attacks

The keystream of a stream cipher should ideally be random, i.e., just by observing the keystream, one should not be able to differentiate it from any random bitstream.

The best method of generating a random bitstream is to toss an unbiased coin or to record thermal noise in an environment or to probe any such stochastic process in nature. These random number generators are called True Random Number Generators (TRNG).

However, when the sender and receiver are far apart, establishing a common keystream between them by synchronizing TRNGs at both ends is not feasible in practice. A pragmatic method would be to design a Finite State Machine (FSM) that generates the same pseudo-random sequence when fed with a common seed shared between the sender and the receiver. Such an FSM is called a Pseudo-Random Number Generator (PRNG).

A stream cipher is nothing but a PRNG with the secret key as the seed. If by analyzing the cipher, one can establish a bias in the probability of occurrence of some keystream-based event, then we have a *distinguishing attack* or a *distinguisher* on the cipher.

6.1 A Theoretical Framework of Distinguishing Attacks

Before describing the distinguishing attacks on RC4, let us discuss the necessary theoretical framework.

The effectiveness of a distinguishing attack is measured by the number of keystream bits that needs to be inspected for the frequencies of the desired event to be substantially different in the two streams, the keystream of the cipher and the random stream. The less this number, the more efficient is the attack. The following technical result on the number of samples required to mount a successful attack appears as Theorem 2 in [107, Section 3.3].

Theorem 6.1.1. *Suppose the event A happens in distribution X with probability p_0 and in distribution Y with probability $p_0(1+\epsilon)$. Then for small p_0 and ϵ, $O(\frac{1}{p_0\epsilon^2})$ samples suffice to distinguish X from Y with a constant probability of success.*

The above result involves some approximation and does not associate a desired success probability with the required number of samples. We formulate this connection in a general setting of statistical hypothesis testing.

Consider an event A with $P(A) = p$, while observing samples of keystream words of a stream cipher. Let $X_r = 1$, if the event A occurs in the r-th sample; $X_r = 0$, otherwise. In other words, $P(X_r = 1) = p$ for all r. Thus,

$$X_r \sim \mathcal{B}er(p).$$

If we observe n many samples, then

$$\sum_{r=1}^{n} X_r \sim \mathcal{B}(n, p).$$

When X_r's are independent and identically distributed (i.i.d.) random variables and n is large enough,

$$\sum_{r=1}^{n} X_r \sim \mathcal{N}\left(np, np(1-p)\right).$$

We are interested in testing the null hypothesis

$$H_0 : p = p_0(1 + \epsilon), \qquad \epsilon > 0,$$

against the alternative hypothesis

$$H_1 : p = p_0.$$

The objective is to find a threshold c in $[np_0, np_0(1 + \epsilon)]$ such that

$$P\left(\sum_{r=1}^{n} X_r \leq c \mid H_0\right) \leq \alpha$$

and

$$P\left(\sum_{r=1}^{n} X_r > c \mid H_1\right) \leq \beta.$$

For such a c to exist, we need

$$np_0(1 + \epsilon) - np_0 > \kappa_1 \sigma_1 + \kappa_2 \sigma_2,$$

where

$$\sigma_1^2 = np_0(1 + \epsilon)\left(1 - p_0(1 + \epsilon)\right),$$
$$\sigma_2^2 = np_0(1 - p_0),$$
$$\Phi(-\kappa_1) = \alpha$$
$$\text{and} \quad \Phi(\kappa_2) = 1 - \beta.$$

This gives,

$$n > \frac{\left(\kappa_1\sqrt{1-p_0} + \kappa_2\sqrt{(1+\epsilon)\,(1-p_0(1+\epsilon))}\right)^2}{p_0\epsilon^2}.$$

When both $p_0, \epsilon \ll 1$, the numerator is approximately equal to $(\kappa_1 + \kappa_2)^2$, and one needs at least $\frac{(\kappa_1+\kappa_2)^2}{p_0\epsilon^2}$ many samples to perform the test.

Now $\kappa_1 = \kappa_2 = 1$ gives $\alpha = \beta = 1 - 0.8413$, giving a success probability 0.8413. Similarly, $\kappa_1 = \kappa_2 = 2$ gives $\alpha = \beta = 1 - 0.9772$, i.e., a success probability 0.9772. Observe that $\kappa_1 = \kappa_2 = 0.5$ gives $\alpha = \beta = 1 - 0.6915$ and corresponds to the special case when the least number of samples needed is $\frac{1}{p_0\epsilon^2}$ as mentioned in Theorem 6.1.1. The success probability associated with this number of samples is 0.6915.

Since $0.6915 > 0.5$ is a reasonably good success probability, $\Sigma(\frac{1}{p_0\epsilon^2})$ many samples are enough to mount a distinguisher and this threshold is indeed used as a benchmark to compare the data complexities of different distinguishing attacks in practice.

6.2   Distinguishers Based on Initial Keystream Bytes

One of the early results in RC4 distinguishers is due to Golic [58]. This work used linear approximations to non-linear functions to establish that the second binary derivative of the least significant bit of the keystream bytes (i.e., the LSB of $z_r \oplus z_{r+2}$, for $r \geq 1$) is correlated to 1 with the correlation coefficient $15 \cdot 2^{-3n}$ (with $n = 8$ for typical RC4). According to [58], this leads to a distinguisher for RC4 requiring about $2^{40.2}$ keystream bytes for a false positive (i.e., detection of bias when there is none) and false negative (i.e., no detection of bias when there is a bias) rate of 10%.

Subsequently, the work [51] used information theoretic bounds to show that Golic's estimates were optimistic and in fact $2^{44.7}$ keystream bytes are needed to achieve the false positive and false negative rate of 10%. Analyzing the probabilities of each successive pair of n-bit outputs, called the *digraph probabilities*, the work [51, Section 3] presented a more efficient distinguisher that requires only $2^{30.6}$ keystream bytes for the same success probability. Interestingly, the digraph distribution with positive biases can be attributed to the classes of 2-predictive 3-states and 2-predictive 2-states (see Section 5.8.1).

The distinguisher of [52, Section 5.1] is based on the fact that for a significant fraction of keys, many initial keystream bytes contain an easily recognizable pattern and this pattern is not completely flattened even when the keys are chosen from a uniform distribution. For 64-bit keys, it is claimed in [52] that this distinguisher requires 2^{21} keystream bytes for a 10% false positive and false negative rate.

In [110, Section 7.2], Glimpse Theorem and Finney cycles are considered together to prove that $P(z_r = i_r^G - 1) \approx \frac{1}{N}(1 - \frac{1}{N^2})$. Applying Theorem 6.1.1, the number of keystream bytes required for a distinguisher based on this bias turns out to be 2^{40}.

In [122, Section 6], a very small non-uniformity in the distribution of the first byte of the keystream output is observed (without any proof). This bias is attributed to the non-uniformity in the permutation after the KSA. The first detailed theoretical analysis of the distribution of the first keystream byte z_1 is performed in [157].

Another work [139] reported a negative bias in the equality of the first two keystream output bytes. It claims that $P(z_1 = z_2) = \frac{1}{N}(1 - \frac{1}{N})$. The distinguisher based on this bias requires 2^{24} pairs of keystream bytes as per Theorem 6.1.1.

6.2.1   Negative Bias in the First Byte toward Zero

Mironov [122] observed that the first keystream byte of RC4 is slightly negatively biased toward zero. After almost a decade, it is recently proved in [157]. We present this result here.

Note that PRGA begins with $i_0^G = j_0^G = 0$. In the first round, $i_1^G = 1$ and $j_1^G = S_0^G[i_1^G] = S_0^G[1]$. First, we show that there is a special path in which z_1 can never be equal to zero.

Lemma 6.2.1. *If $S_0^G[j_1^G] = 0$, z_1 cannot be 0.*

Proof: Suppose $S_0^G[j_1^G] = 0$. After the swap in the first round, $S_1^G[1] = S_0^G[j_1^G] = 0$ and $S_1^G[j_1^G] = S_0^G[1] = j_1^G$. Now,

$$
\begin{aligned}
z_1 &= S_1^G[S_1^G[1] + S_1^G[j_1^G]] \\
&= S_1^G[0 + j_1^G] \\
&= S_1^G[j_1^G] \\
&= j_1^G \\
&= S_0^G[1].
\end{aligned}
$$

Now, if z_1 were 0, one must have $S_0^G[1] = 0 = S_0^G[j_1^G]$, implying

$$
1 = j_1^G = S_0^G[1] = 0,
$$

which is a contradiction. Hence, z_1 cannot be 0.   ∎

Next, let us present the main result.

Theorem 6.2.2. *Assume that the initial permutation S_0^G is randomly chosen from the set of all possible permutations of $\{0, \ldots, N-1\}$. Then*

$$
P(z_1 = 0) = \frac{1}{N} - \frac{1}{N^2}.
$$

Proof: Let us consider two possible cases as follows.

Case I: Suppose that $S_0^G[j_1^G] = 0$. We already proved in Lemma 6.2.1 that

$$P(z_1 = 0 \mid S_0^G[j_1^G] = 0) = 0.$$

Case II: Suppose that $S_0^G[j_1^G] \neq 0$. Assuming z_1 is uniformly random in this case, we have

$$P(z_1 = 0 \mid S_0^G[j_1^G] \neq 0) = \frac{1}{N}.$$

Combining these two cases, we get

$$
\begin{aligned}
P(z_1 = 0) &= P(S_0^G[j_1^G] = 0) \cdot P(z_1 = 0 \mid S_0^G[j_1^G] = 0) \\
&\quad + P(S_0^G[j_1^G] \neq 0) \cdot P(z_1 = 0 \mid S_0^G[j_1^G] \neq 0) \\
&= \frac{1}{N} \cdot 0 + \left(\frac{N-1}{N}\right) \cdot \frac{1}{N} \\
&= \frac{1}{N} - \frac{1}{N^2}.
\end{aligned}
$$

Thus, z_1 has a negative bias of $\frac{1}{N^2}$ toward 0. $\blacksquare$

Theorem 6.2.2 immediately gives a distinguisher. Let X and Y be the distributions corresponding to random stream and RC4 keystream respectively and define A as the event "$z_1 = 0$." The bias can be written as $p_0(1+\epsilon)$, where $p_0 = \frac{1}{N}$ and $\epsilon = -\frac{1}{N}$. According to Theorem 6.1.1, the number of samples required to distinguish RC4 from random stream with a constant probability of success is $O(\frac{1}{p_0\epsilon^2})$, i.e., $O(N^3)$.

6.2.2 Strong Positive Bias in the Second Byte toward Zero

Currently the best distinguisher for RC4 keystream is due to Mantin and Shamir [107], that requires around 2^8 samples. This is based on the observation that $P(z_2 = 0) \approx \frac{2}{N}$, which is two times the probability of random association.

Theorem 6.2.3. *Assume that the initial permutation S_0^G is randomly chosen from the set of all possible permutations of $\{0, \ldots, N-1\}$. Then*

$$P(z_2 = 0) \approx \frac{2}{N}.$$

Proof: PRGA starts with $i_0^G = j_0^G = 0$. In round 1, $i_1^G = 1$ and $j_1^G = j_0^G + S_0^G[i_1^G] = S_0^G[i_1^G] = S_0^G[1]$. In round 2, $i_2^G = 2$ and $j_2^G = j_1^G + S_1^G[i_2^G]$. Let us consider two possible cases as follows.

Case I: Suppose that $S_0^G[2] = 0$. Now, if $j_1^G = S_0^G[i_1^G] \neq 2$, then this 0 is not swapped out and we have $S_1^G[i_2^G] = S_1^G[2] = 0$. Hence, $j_2^G = S_0^G[i_1^G] = j_1^G$ and

$$
\begin{aligned}
z_2 &= S_2^G[S_2^G[i_2^G] + S_2^G[j_2^G]] \\
&= S_2^G[S_1^G[j_2^G] + S_1^G[i_2^G]] && \text{(reverting the swap in round 2)} \\
&= S_2^G[S_1^G[j_2^G]] && \text{(since } S_1^G[i_2^G] = 0) \\
&= S_2^G[S_1^G[j_1^G]] && \text{(since } j_2^G = j_1^G) \\
&= S_2^G[S_0^G[i_1^G]] && \text{(reverting the swap in round 1)} \\
&= S_2^G[j_2^G] && \text{(since } j_2^G = S_0^G[i_1^G]) \\
&= S_1^G[i_2^G] && \text{(reverting the swap in round 2)} \\
&= 0.
\end{aligned}
$$

$$
\begin{aligned}
Thus, P(z_2 = 0 \mid S_0^G[2] = 0) &= P(j_1^G \neq 2 \mid S_0^G[2] = 0) \\
&= P(S_0^G[1] \neq 2 \mid S_0^G[2] = 0) \\
&= \frac{N-2}{N-1},
\end{aligned}
$$

assuming j_1^G to be uniformly random over the set $\mathbb{Z}_N \setminus \{0\}$ of its possible values. Note that since the permutation S_0^G has the entry 0 at index 2, it cannot have the entry 0 at index 1.

Case II: Suppose that $S_0^G[2] \neq 0$. Assuming z_2 is uniformly random in this case, we have $P(z_2 = 0 \mid S_0^G[2] \neq 0) = \frac{1}{N}$.

Combining these two cases, we get

$$
\begin{aligned}
P(z_2 = 0) &= P(S_0^G[2] = 0) \cdot P(z_2 = 0 \mid S_0^G[2] = 0) \\
&\quad + P(S_0^G[2] \neq 0) \cdot P(z_2 = 0 \mid S_0^G[2] \neq 0) \\
&= \frac{1}{N} \cdot \left(\frac{N-2}{N-1}\right) + \left(\frac{N-1}{N}\right) \cdot \frac{1}{N} \\
&= \frac{2}{N} - \frac{1}{N^2} - \frac{1}{N(N-1)} \approx \frac{2}{N}.
\end{aligned}
$$

Hence the result.   ∎

According to Theorem 3.2.3, the underlying assumption of [107] regarding the randomness of the permutation is violated in practice. However, it does not affect the final result of Theorem 6.2.3 much.

Theorem 6.2.3 gives a distinguisher which is currently the best known distinguisher for RC4. Let X and Y be the distributions corresponding to random stream and RC4 keystream respectively and define A as the event "$z_2 = 0$." The bias $\frac{2}{N}$ can be written as $p_0(1 + \epsilon)$, where $p_0 = \frac{1}{N}$ and $\epsilon = 1$. According to Theorem 6.1.1, the number of samples required to distinguish

RC4 from random stream with a constant probability of success is $O(\frac{1}{p_0\epsilon^2})$, i.e., $O(N)$.

Moreover, the bias of z_2 to 0 can be used to mount a linear-time ciphertext-only attack on broadcast RC4 in which the same plaintext is sent to multiple recipients under different keys. As mentioned in [107], a classical problem in distributed computing is to allow N Byzantine generals to coordinate their actions, when at most one third of them can be traitors. This problem is solved by a multi-round protocol in which each general broadcasts the same plaintext message (initially consisting of either "Attack" or "Retreat") to all the other generals. Each copy is encrypted under a different key agreed to in advance between any two generals. In [107], the authors propose a practical attack against an RC4 implementation of the broadcast scheme, based on the bias observed in the second keystream byte.

Theorem 6.2.4. *Let M be a plaintext, and let $C_1, C_2, \ldots, C_k$ be the RC4 encryptions of M under k uniformly distributed keys. Then if $k = \Omega(N)$, the second byte z_2 can be reliably extracted from $C_1, C_2, \ldots, C_k$.*

Proof: Recall from Theorem 6.2.3 that $P(z_2 = 0) \approx \frac{2}{N}$. Thus, for each encryption key chosen during the broadcast scheme, M has probability $\frac{2}{N}$ to be XOR-ed with 0.

Due to this bias of z_2 toward zero, $\frac{2}{N}$ fraction of the second ciphertext bytes would have the same value as the second plaintext byte, with a higher probability. When $k = \Omega(N)$, the attacker can identify the most frequent character in $C_1[2], C_2[2], \ldots, C_k[2]$ as $M[2]$ with constant probability of success. ∎

Many users send the same email message to multiple recipients, where the messages are encrypted under different keys. Many groupware applications also enable multiple users to synchronize their documents by broadcasting encrypted modification lists to all the other group members. All such applications are vulnerable to this attack.

6.2.3   Positive Biases in Bytes 3 to 255 toward Zero

Contrary to the claim of Mantin and Shamir [107], it was demonstrated in [100] that all the initial 253 bytes of RC4 keystream from round 3 to 255 are biased to zero. The approximations in the results of [100] has recently been improved in [157]. Here we present the latter results that have better approximations.

The main result is a positive bias for each of the events $(z_r = 0)$, $3 \le r \le 255$. We would decompose the event $(z_r = 0)$ into two mutually exclusive and exhaustive paths, one in conjunction with the event $(S_{r-1}^G[r] = r)$ and the other in conjunction with the complimentary event $(S_{r-1}^G[r] \ne r)$. Hence, we first need to estimate the probability of the event $(S_{r-1}^G[r] = r)$ for all rounds $r \ge 3$.

Let $p_{t,r}$ denote the probability $P(S_t^G[r] = r)$, after t rounds of PRGA, where $t \geq 0$, $0 \leq r \leq N - 1$. Note that $p_{0,r} = P(S_0^G[r] = r)$ can be directly found from Theorem 3.2.3.

A natural method of calculating $p_{r-1,r} = P(S_{r-1}^G[r] = r)$ would be to use Lemma 5.6.1 with $X = r$. But unlike Corollary 5.6.2, one cannot take the initial round to be $\tau = 0$ here. The reason is as follows. We know that after the swap in the first round of the PRGA,

$$
\begin{aligned}
S_1^G[S_0^G[1]] &= S_1^G[j_1^G] \\
&= S_0^G[i_1^G] \\
&= S_0^G[1]
\end{aligned}
$$

So, if $S_0^G[1] = r$, then the event $(S_1^G[r] = r)$ occurs with probability 1. If $S_0^G[1] \neq r$, then $S_1^G[r]$ may contain r due to random association. Thus, for any given r, there is a sharp jump between the probabilities $p_{0,r}$ and $p_{1,r}$. This means, one should consider $\tau = 1$ as the initial round and then apply Lemma 5.6.1 to calculate $p_{r-1,r}$. This gives the following result.

Lemma 6.2.5. *For $r \geq 3$, the probability that $S_{r-1}^G[r] = r$ is*

$$
p_{r-1,r} \approx p_{1,r} \left(1 - \frac{1}{N}\right)^{r-2} +
$$
$$
\sum_{t=2}^{r-1} \sum_{n=0}^{r-t} \frac{P(S_1^G[t] = r)}{n! \cdot N} \left(\frac{r-t-1}{N}\right)^n \left(1 - \frac{1}{N}\right)^{r-3-n}.
$$

Observe that for $t = r$, $P(S_1^G[t] = r)$ coincides with $p_{1,r}$. In order to have the explicit values of $p_{r-1,r}$, we need the probability distribution $P(S_1^G[t] = r)$, $0 \leq t \leq N - 1$. This is given in the following result.

Lemma 6.2.6. *After the completion of the first round of RC4 PRGA, the probability that $S_1^G[t] = r$, $0 \leq r \leq N - 1$, is given by*
$$P(S_1^G[t] = r)$$

$$
= \begin{cases}
\displaystyle\sum_{X=0}^{N-1} P(S_0^G[1] = X) \cdot P(S_0^G[X] = r), & t = 1; \\[2ex]
P(S_0^G[1] = r) + (1 - P(S_0^G[1] = r)) \cdot P(S_0^G[r] = r), & t = r; \\[2ex]
(1 - P(S_0^G[1] = t)) \cdot P(S_0^G[t] = r), & otherwise.
\end{cases}
$$

Proof: After the first round of RC4 PRGA, we obtain the state S_1^G from the initial state S_0^G through a single swap operation between the positions $i_1^G = 1$ and $j_1^G = S_0^G[i_1^G] = S_0^G[1]$. Thus, all other positions of S_0^G remain the same apart from these two. This gives us the value of $S_1^G[t]$ as follows.

$$
S_1^G[t] = \begin{cases}
S_0^G[S_0^G[1]], & t = 1; \\
S_0^G[1], & t = S_0^G[1]; \\
S_0^G[t], & otherwise.
\end{cases}
$$

Now, we can compute the probabilities $P(S_1^G[t] = r)$ based on the probabilities for S_0^G, which are in turn derived from Theorem 3.2.3. We have three cases:

Case $t = 1$: In this case, using the recurrence relation $S_1^G[1] = S_0^G[S_0^G[1]]$, we can write

$$P(S_1^G[1] = r) = \sum_{X=0}^{N-1} P(S_0^G[1] = X) \cdot P(S_0^G[X] = r).$$

Case $t = r$: In this situation, if $S_0^G[1] = r$, we will surely have $S_1^G[r] = r$ as these are the positions swapped in the first round, and if $S_0^G[1] \neq r$, the position $t = r$ remains untouched and $S_1^G[r] = r$ is only possible if $S_0^G[r] = r$. Thus, we have

$$P(S_1^G[r] = r) = P(S_0^G[1] = r) + (1 - P(S_0^G[1] = r)) \cdot P(S_0^G[r] = r).$$

Case $t \neq 1, r$: In all other cases where $t \neq 1, r$, it can either take the value $S_0^G[1]$ with probability $P(S_0^G[1] = t)$, or not. If $t = S_0^G[1]$, the value $S_0^G[t]$ will get swapped with $S_0^G[1] = t$ itself, i.e., we will get $S_1^G[t] = t \neq r$ for sure. Otherwise, the value $S_1^G[t]$ remains the same as $S_0^G[t]$. Hence,

$$P(S_1^G[t] = r) = (1 - P(S_0^G[1] = t)) \cdot P(S_0^G[t] = r).$$

Combining all the above cases together, we obtain the desired result. ∎

Now, let us analyze the keystream z_r. RC4 PRGA begins with $j_0^G = 0$. In the first round, i.e., for $r = 1$, we have $j_1^G = j_0^G + S_0^G[1] = S_0^G[1]$ which, due to Theorem 3.2.3, is not uniformly distributed. In the second round, i.e., for $r = 2$, we have $j_2^G = j_1^G + S_1^G[2] = S_0^G[1] + S_1^G[2]$, which is closer to uniformly random distribution than j_1^G. In round 3, another pseudo-random byte $S_2^G[3]$ would be added to form j_3^G. From round 3 onwards, j^G can safely be assumed to be uniform over $\mathbb{Z}_N$. Experimental observations also confirm this.

Before we discuss the next two lemma, note that

$$
\begin{aligned}
z_r &= S_r^G[S_r^G[i_r^G] + S_r^G[j_r^G]] \\
&= S_r^G[S_r^G[r] + S_{r-1}^G[i_r^G]] \\
&= S_r^G[S_r^G[r] + S_{r-1}^G[r]].
\end{aligned}
$$

We would make use of the above identity in proving Lemma 6.2.7 and 6.2.8.

Lemma 6.2.7. *For $3 \leq r \leq 255$, $P\left(z_r = 0 \ \& \ S_{r-1}^G[r] = r\right) = p_{r-1,r} \cdot \frac{2}{N}$.*

Proof: First, let us calculate the following conditional probability.

$$P(z_r = 0 \mid S^G_{r-1}[r] = r)$$
$$= P\left(S^G_r[S^G_r[r] + S^G_{r-1}[r]] = 0 \mid S^G_{r-1}[r] = r\right)$$
$$= P\left(S^G_r[S^G_r[r] + r] = 0 \mid S^G_{r-1}[r] = r\right)$$
$$= \sum_{x=0}^{N-1} P\left(S^G_r[x + r] = 0 \ \& \ S^G_r[r] = x \mid S^G_{r-1}[r] = r\right)$$
$$= \sum_{x=0}^{N-1} P\left(S^G_r[x + r] = 0 \ \& \ S^G_r[r] = x\right). \qquad (6.1)$$

Observe that $S^G_{r-1}[r] = r$ gets swapped to produce the new state S^G_r and so we can claim the independence of $S^G_r[r]$ and $S^G_{r-1}[r]$.

Now, let us compute $P(S^G_r[x + r] = 0 \ \& \ S^G_r[r] = x) = P(S^G_r[x + r] = 0) \cdot P(S^G_r[r] = x \mid S^G_r[x + r] = 0)$. If there exists any bias in the event $(S^G_r[x+r] = 0)$, it must have originated from a similar bias in $(S^G_0[x+r] = 0)$ (similar to the case of $(S^G_{r-1}[r] = r)$ in Lemma 6.2.5). However, $P(S^G_0[x+r] = 0) = \frac{1}{N}$ by Theorem 3.2.3, and thus we can safely assume $S^G_r[x + r]$ to be random as well. This provides us with $P(S^G_r[x + r] = 0) = \frac{1}{N}$.

For $P(S^G_r[r] = x \mid S^G_r[x+r] = 0)$, observe that when $x = 0$, the indices $x+r$ and r in the state S^G_r point to the same location, and the events $(S^G_r[x + r] = S^G_r[r] = 0)$ and $(S^G_r[r] = x = 0)$ denote identical events. So in this case, $P(S^G_r[r] = x \mid S^G_r[x + r] = 0) = 1$. On the other hand, when $x \neq 0$, the indices $x + r$ and $[r]$ refer to two distinct locations in the permutation S^G_r, containing two different values. In this case,

$$P(S^G_r[r] = x \mid S^G_r[x + r] = 0) = P(S^G_r[r] = x \mid x \neq 0) = \frac{1}{N - 1}.$$

The randomness assumption of $S^G_r[r]$ for $x \neq 0$ falls back upon the randomness assumption of j^G_r, the value at which index comes to index r due to the swap.

According to the discussion above, we obtain

$$P\left(S^G_r[x + r] = 0 \ \& \ S^G_r[r] = x\right) = \begin{cases} \frac{1}{N} \cdot 1 = \frac{1}{N} & \text{if } x = 0, \\ \frac{1}{N} \cdot \frac{1}{N-1} = \frac{1}{N(N-1)} & \text{if } x \neq 0. \end{cases}$$
$$(6.2)$$

Substituting these probability values in Equation (6.1), we get

$$P\left(z_r = 0 \mid S^G_{r-1}[r] = r\right)$$
$$= \sum_{x=0}^{N-1} P\left(S^G_r[x + r] = 0 \ \& \ S^G_r[r] = x\right)$$
$$= \frac{1}{N} + \sum_{x=1}^{N-1} \frac{1}{N(N - 1)}$$
$$= \frac{1}{N} + (N - 1) \cdot \frac{1}{N(N - 1)} = \frac{2}{N}.$$

Multiplying the above conditional probability with $P(S_{r-1}^G[r] = r)$, we get the result. ∎

Lemma 6.2.8. *For $3 \leq r \leq 255$, $P\left(z_r = 0 \,\&\, S_{r-1}^G[r] \neq r\right) = \frac{N-2}{N(N-1)} \cdot (1 - p_{r-1,r})$.*

Proof: Here, instead of splitting the joint probability into the product of one marginal probability and one conditional probability, we proceed to calculate directly as follows.

$$
\begin{aligned}
&P\left(z_r = 0 \,\&\, S_{r-1}^G[r] \neq r\right) \\
=\; &P\left(S_r^G[S_r^G[r] + S_{r-1}^G[r]] = 0 \,\&\, S_{r-1}^G[r] \neq r\right) \\
=\; &\sum_{y \neq r} P(S_r^G[S_r^G[r] + y] = 0 \,\&\, S_{r-1}^G[r] = y) \\
=\; &\sum_{y \neq r} \sum_{x=0}^{N-1} P\left(S_r^G[x + y] = 0 \,\&\, S_r^G[r] = x \,\&\, S_{r-1}^G[r] = y\right)
\end{aligned}
$$

Consider a special case, namely, $x = r - y$. In this case, $S_r^G[x+y] = S_r^G[r] = 0$ for the first event and $S_r^G[r] = x = r - y \neq 0$ for the second event (note that $y \neq r$). This poses a contradiction (an event with probability of occurrence 0), and hence one can write

$$
\begin{aligned}
&P\left(z_r = 0 \,\&\, S_{r-1}^G[r] \neq r\right) \\
=\; &\sum_{y \neq r} \sum_{x \neq r-y} P\left(S_r^G[x + y] = 0 \,\&\, S_r^G[r] = x \,\&\, S_{r-1}^G[r] = y\right) \\
=\; &\sum_{y \neq r} \sum_{x \neq r-y} P\left(S_r^G[x + y] = 0 \,\&\, S_r^G[r] = x\right) \cdot P\left(S_{r-1}^G[r] = y\right), \quad (6.3)
\end{aligned}
$$

where the last expression results from the fact that the events $(S_r^G[x+y] = 0)$ and $(S_r^G[r] = x)$ are both independent of $(S_{r-1}^G[r] = y)$, since a state update has occurred in the process, with a swap involving the index r.

Similar to the derivation of Equation (6.2), we obtain

$$
P\left(S_r^G[x + y] = 0 \,\&\, S_r^G[r] = x\right) = \begin{cases} 0 & \text{if } x = 0, \\ \frac{1}{N(N-1)} & \text{if } x \neq 0. \end{cases} \quad (6.4)
$$

Note that in the case $x = 0$, simultaneous occurrence of the events $(S_r^G[x+y] = S_r^G[y] = 0)$ and $(S_r^G[r] = x = 0)$ pose a contradiction, as the two locations y and r of S_r^G are distinct (since $y \neq r$), and they can not have the same value 0.

Substituting the values above in Equation (6.3), we get

$$P\left(z_r = 0 \ \& \ S^G_{r-1}[r] \neq r\right)$$

$$= \sum_{y \neq r} P\left(S^G_{r-1}[r] = y\right) \left[\sum_{x \neq r-y} P\left(S^G_r[x+y] = 0 \ \& \ S^G_r[r] = x\right)\right]$$

$$= \sum_{y \neq r} P\left(S^G_{r-1}[r] = y\right) \left[0 + \sum_{\substack{x \neq r-y \\ x \neq 0}} \frac{1}{N(N-1)}\right]$$

$$= \sum_{y \neq r} P\left(S^G_{r-1}[r] = y\right) \left[(N-2) \cdot \frac{1}{N(N-1)}\right]$$

$$= \frac{N-2}{N(N-1)} \sum_{y \neq r} P\left(S^G_{r-1}[r] = y\right)$$

$$= \frac{N-2}{N(N-1)} \cdot \left(1 - P\left(S^G_{r-1}[r] = r\right)\right) = \frac{N-2}{N(N-1)} \cdot \left(1 - p_{r-1,r}\right)$$

∎

Now, let us state the main theorem on the bias of RC4 initial bytes from rounds 3 to 255.

Theorem 6.2.9. *For $3 \leq r \leq 255$, the probability that the r-th RC4 keystream byte is equal to 0 is*

$$P(z_r = 0) = \frac{1}{N} + \frac{c_r}{N^2},$$

where the value of c_r is $\frac{N^2}{N-1}\left(p_{r-1,r} - \frac{1}{N}\right)$.

Proof: Adding the expressions of Lemma 6.2.7 and 6.2.8, one obtains

$$P(z_r = 0) = p_{r-1,r} \cdot \frac{2}{N} + \frac{N-2}{N(N-1)} \cdot \left(1 - p_{r-1,r}\right)$$

$$= \frac{p_{r-1,r}}{N-1} + \frac{N-2}{N(N-1)}$$

$$= \frac{1}{N} + \frac{1}{N-1} \cdot \left(p_{r-1,r} - \frac{1}{N}\right). \qquad (6.5)$$

Hence, $P(z_r = 0) = \frac{1}{N} + \frac{c_r}{N^2}$, with the value of c_r as $\frac{N^2}{N-1}\left(p_{r-1,r} - \frac{1}{N}\right)$. Note that the values $p_{r-1,r}$ can be explicitly calculated using Lemma 6.2.5. ∎

In Theorem 6.2.9, the parameter c_r that quantifies the bias is a function of r. The next result is a corollary of Theorem 6.2.9 that provides exact numeric bounds on $P(z_r = 0)$ within the interval $3 \leq r \leq 255$, depending on the corresponding bounds of c_r within the same interval.

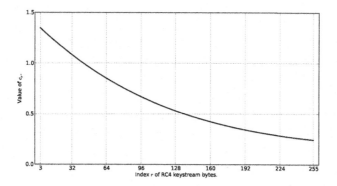

FIGURE 6.1: Value of c_r versus r during RC4 PRGA $(3 \leq r \leq 255)$.

Corollary 6.2.10. *For* $3 \leq r \leq 255$, *the probability that the r-th RC4 keystream byte is equal to 0 is*

$$\frac{1}{N} + \frac{1.3471679056}{N^2} \geq P(z_r = 0) \geq \frac{1}{N} + \frac{0.2428109804}{N^2}.$$

Proof: We calculate all the values of c_r (as in Theorem 6.2.9) for the range $3 \leq r \leq 255$ and find that c_r is a decreasing function in r where $3 \leq r \leq 255$ (one may refer to the plot in Fig. 6.1). Therefore, we get

$$\max_{3 \leq r \leq 255} c_r = c_3 = 1.3471679056 \quad \text{and} \quad \min_{3 \leq r \leq 255} c_r = c_{255} = 0.2428109804.$$

Hence the result. ∎

Fig. 6.2 depicts a comparison between the theoretically derived vs. experimentally obtained values of $P(z_r = 0)$ versus r, where $3 \leq r \leq 255$. The experimentation has been carried out with 1 billion trials, each trial with a randomly generated 16-byte key.

Denote X and Y to be the distributions corresponding to random stream and RC4 keystream respectively and define A_r as the event "$z_r = 0$" for $r = 3$ to 255. From the formulation of Theorem 6.1.1, we can write $p_0 = \frac{1}{N}$ and $\epsilon = \frac{c_r}{N}$. Thus, to distinguish RC4 keystream from random stream based on the event "$z_r = 0$," one would need samples of the order of

$$\left(\frac{1}{N}\right)^{-1} \left(\frac{c_r}{N}\right)^{-2} \approx O(N^3).$$

We can combine the effect of all these distinguishers by counting the number of zeros in the initial keystream of RC4, according to Theorem 6.2.11, as follows.

Theorem 6.2.11. *The expected number of 0's in RC4 keystream rounds 3 to 255 is approximately 0.9906516923.*

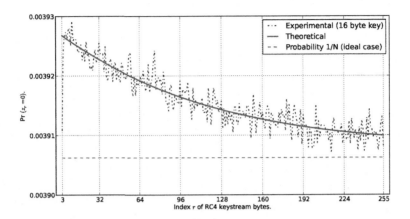

FIGURE 6.2: $P(z_r = 0)$ versus r during RC4 PRGA $(3 \leq r \leq 255)$.

Proof: Let X_r be a random variable taking values $X_r = 1$ if $z_r = 0$, and $X_r = 0$ otherwise. Hence, the total number of 0's in rounds 1 to 255 excluding round 2 is given by

$$C = \sum_{r=3}^{255} X_r.$$

We have $E(X_r) = P(X_r = 1) = P(z_r = 0)$ from Theorem 6.2.9. By linearity of expectation,

$$E(C) = \sum_{r=3}^{255} E(X_r) = \sum_{r=3}^{255} P(z_r = 0).$$

Substituting the numeric values of the probabilities $P(z_r = 0)$ from Theorem 6.2.9, we get $E(C) \approx 0.9906516923$. Hence the result. ∎

One may note that for a random stream of bytes, this expectation turns out to be $E(C) = \frac{254}{256} = 0.98828125$. Thus, the expectation for RC4 is approximately 0.24% higher than that for the random case. The inequality of this expectation $E(C)$ in a RC4 keystream compared to that in a random stream of bytes may be used to design a distinguisher.

In a similar line of action of Mantin's broadcast attack based on the second byte, we may exploit the bias observed in bytes 3 to 255 of the RC4 keystream to mount a similar attack on RC4 broadcast scheme. Notice that we obtain a bias of the order of $\frac{1}{N^2}$ in each of the bytes z_r where $3 \leq r \leq 255$. Thus, roughly speaking, if the attacker obtains about N^3 ciphertexts corresponding to the same plaintext M (from the broadcast scheme), then he can check the frequency of occurrence of bytes to deduce the r-th $(3 \leq r \leq 255)$ byte of M.

The most important point to note is that this technique will work for each r where $3 \leq r \leq 255$, and hence will reveal *all the 253 initial bytes* (number

3 to 255 to be specific) of the plaintext M. We can formally state the result (analogous to Theorem 6.2.4) as follows.

Theorem 6.2.12. *Let M be a plaintext, and let $C_1, C_2, \ldots, C_k$ be the RC4 encryptions of M under k uniformly distributed keys. Then if $k = \Omega(N^3)$, the bytes 3 to 255 of M can be reliably extracted from $C_1, C_2, \ldots, C_k$.*

Proof: Recall from Theorem 6.2.9 that $P(z_r = 0) = \frac{1}{N} + \frac{c_r}{N^2}$ for all $3 \leq r \leq 255$ in the RC4 keystream. Thus, for each encryption key chosen during the broadcast scheme, $M[r]$ has probability $\frac{1}{N} + \frac{c_r}{N^2}$ to be XOR-ed with 0.

Due to this bias of z_r toward zero, $\frac{1}{N} + \frac{c_r}{N^2}$ fraction of the r-th ciphertext bytes will have the same value as the r-th plaintext byte, with a higher probability. When $k = \Omega(N^3)$, the attacker can identify the most frequent character in $C_1[r], C_2[r], \ldots, C_k[r]$ as $M[r]$ with constant probability of success. $\blacksquare$

6.2.4 Distinguishers Based on Combination of Biases in RC4 Permutation

In Section 3.2, we discussed the bias of the permutation elements after the KSA toward many values in $\mathbb{Z}_N$. Here we are going to show how combinations of these biases propagate to the keystream and thus help distinguish the RC4 keystream from random streams. The distinguishers presented here first appeared in [135].

Lemma 6.2.13. *Consider $B \subset \mathbb{Z}_N$ with $|B| = b$. Let*

$$P(S_N[r] \in B) = \frac{b}{N} + \epsilon,$$

where ϵ can be positive or negative. Then for $r \geq 1$,

$$P(S^G_{r-1}[r] \in B) = \frac{b}{N} + \delta,$$

where

$$
\delta = \left(\frac{b}{N} + \epsilon\right) \cdot \left(\left(\frac{N-1}{N}\right)^{r-1} + \left(1 - \left(\frac{N-1}{N}\right)^{r-1}\right) \cdot \left(\frac{b-1}{N-1} - \frac{b}{N}\right)\right)
$$
$$
- \frac{b}{N} \cdot \left(\frac{N-1}{N}\right)^{r-1}.
$$

Proof: The event $(S^G_{r-1}[r] \in B)$ can occur in three ways.

1. $S_N[r] \in B$ and the index r is not touched by any of the $r-1$ many j^G values during the first $r-1$ rounds of the PRGA. The contribution of this part is $(\frac{b}{N} + \epsilon) \cdot (\frac{N-1}{N})^{r-1}$.

2. $S_N[r] \in B$ and index r is touched by at least one of the $r-1$ many j^G values during the first $r-1$ rounds of the PRGA. Further, after the swap(s), the value $S_N[r]$ remains in the set B. This will happen with probability $\left(\frac{b}{N} + \epsilon\right) \cdot \left(1 - \left(\frac{N-1}{N}\right)^{r-1}\right) \cdot \frac{b-1}{N-1}$.

3. $S_N[r] \notin B$ and index r is touched by at least one of the $r-1$ many j^G values during the first $r-1$ rounds of the PRGA. Due to the swap(s), the value $S_N[r]$ comes to the set B. This will happen with probability $\left(1 - \frac{b}{N} - \epsilon\right) \cdot \left(1 - \left(\frac{N-1}{N}\right)^{r-1}\right) \cdot \frac{b}{N}$.

Adding these contributions, we get the total probability as given by δ. ∎

Lemma 6.2.14. *If* $P(S^G_{r-1}[r] \in B) = \frac{b}{N} + \delta$, *then* $P(z_r \in C) = \frac{b}{N} + \frac{2\delta}{N}$, *where* $C = \{c' \mid c' = r - b' \text{ where } b' \in B\}$, $r \geq 1$.

Proof: The event $(z_r \in C)$ can happen in two ways.

1. $S^G_{r-1}[r] \in B$ and $z_r = r - S^G_{r-1}[r]$. From Corollary 5.2.2, we have $P(z_r = r - S^G_{r-1}[r]) = \frac{2}{N}$ for $r \geq 1$. Thus, the contribution of this part is $\frac{2}{N}\left(\frac{b}{N} + \delta\right)$.

2. $S^G_{r-1}[r] \notin B$ and still $z_r \in C$ due to random association. The contribution of this part is $\left(1 - \frac{2}{N}\right)\frac{b}{N}$.

Adding these two contributions, we get the result. ∎

Theorem 6.2.15. *If* $P(S_N[r] \in B) = \frac{b}{N} + \epsilon$, *then* $P(z_r \in C)$

$$= \frac{b}{N} + \frac{2}{N} \cdot \left[\left(\frac{b}{N} + \epsilon\right) \cdot \left(\left(\frac{N-1}{N}\right)^{r-1} + \left(1 - \left(\frac{N-1}{N}\right)^{r-1}\right) \cdot \right. \right.$$
$$\left. \left. \left(\frac{b-1}{N-1} - \frac{b}{N}\right)\right) - \frac{b}{N} \cdot \left(\frac{N-1}{N}\right)^{r-1}\right],$$

where $C = \{c' \mid c' = r - b' \text{ where } b' \in B\}$, $r \geq 1$.

Proof: The proof immediately follows by combining Lemma 6.2.13 and Lemma 6.2.14. ∎

The above results imply that for a single value v, if

$$P(S_N[r] = v) = \frac{1}{N} + \epsilon,$$

then

$$P(z_r = r - v) = \frac{1}{N} + \frac{2\delta}{N},$$

where the value of δ can be computed by substituting $b = 1$ in Lemma 6.2.14. This reveals a non-uniform distribution of the initial keystream output bytes z_r for small r.

When the bias of $S_N[r]$ toward a single value $v \in \mathbb{Z}_N$ is propagated to z_r, the final bias at z_r is quite small and difficult to observe experimentally. Rather, if we start with the bias of $S_N[r]$ toward many values in some suitably chosen set $B \subset \mathbb{Z}_N$, then according to Theorem 6.2.15, a sum of $b = |B|$ many probabilities is propagated to z_r, making the bias of z_r empirically observable too. For instance, given $1 \le r \le 127$, consider the set $B = \{r+1, \ldots, r+128\}$, i.e., $b = |B| = 128$. The formulae of Theorem 3.2.11 gives

$$P(S_N[r] \in B) > 0.5,$$

and in turn we get

$$P(z_r \in C) > 0.5,$$

which is clearly observable at the r-th keystream output byte of RC4. It is important to note that the non-uniform distribution can be observed even at the 256-th output byte z_{256}, since the deterministic index i at round 256 becomes 0 and $S_N[0]$ has a non-uniform distribution according to Theorem 3.2.11. For random association, $P(z_r \in C)$ should be $\frac{b}{N}$, which is not the case here, and thus all these results provide several distinguishers for RC4.

In Section 3.2.3, it has already been pointed out that for short key lengths, there exist many anomaly pairs. One may exploit these to construct some additional distinguishers by including in the set B those values which are far away from being random. We illustrate this in the following two examples. For 5-byte secret keys, the experimentally observed value of $P(S_N[9] \in B)$ is 0.137564, calculated from 100 million runs, where B is the set of all even integers greater than or equal to 128 and less than 256, i.e., $b = |B| = 64$ and $\frac{b}{N} = 0.25$. The value 0.137564 is much less than the theoretical value 0.214785. Applying Theorem 6.2.15 we get

$$P(z_9 \in C) = 0.249530 < 0.25,$$

where

$$C = \{c' \mid c' = 9 - b' \text{ where } b' \in B\}.$$

Again, for 8-byte secret keys, we experimentally find that $P(S_N[15] \in B) = 0.160751$, which is much less than the theoretical value 0.216581, where B is the set of all odd integers greater than or equal to 129 and less than 256, i.e., $b = |B| = 64$ once again. Theorem 6.2.15 gives

$$P(z_{15} \in C) = 0.249340 < 0.25,$$

where

$$C = \{c' \mid c' = 15 - b' \text{ where } b' \in B\}.$$

Direct experimental data on the keystream also confirm these biases of z_9 and z_{15}. Further, given the values of δ approximately -0.1 in the above two examples, one can form new linear distinguishers for RC4 with 5-byte and 8-byte keys.

It is interesting to note that since the anomaly pairs are different for different key lengths, one can also distinguish among RC4 of different key lengths, by suitably choosing the anomaly pairs in the set B.

6.3   Distinguishers Based on Any Stage of PRGA

So far, we have discussed distinguishing attacks based on biases in the initial keystream bytes. If the initial keystream output of RC4 (say, the first 1024 bytes) are thrown away, which is the standard practice for real-life applications, the above attack fails. However, there are a few biased events that hold at any stage during the PRGA. This section is devoted to these special distinguishers.

6.3.1   Digraph Repetition Bias

The work in [109] discovered the first distinguisher for RC4 when any amount of initial keystream bytes are thrown away. This distinguisher is based on the digraph distribution of RC4. The term *digraph* means a pair of consecutive keystream words. In [109, Section 3], it has been shown that getting strings of the pattern $ABTAB$, where A, B are bytes and T is a string of bytes of small length Δ (typically $\Delta \leq 16$) are more probable in RC4 keystream than in random stream. This is called the *digraph repetition bias*.

In order to prove this bias, the following two lemma are required.

Lemma 6.3.1. *Let R be a set of m permutation locations. Suppose that RC4 is in a state where the index i^G in the next t rounds does not visit R. Then the probability of the permutation t rounds later to have the same values in R is approximately $e^{-\frac{mt}{N}}$.*

Proof: The index i^G does not reach any of the indices in R. The index j^G does not touch a particular position in each of the t rounds with probability $1 - \frac{1}{N}$. The set R is untouched, if j^G does not touch any of the m positions in any of the t rounds. The probability of this event is $(1 - \frac{1}{N})^{mt} \approx e^{-\frac{mt}{N}}$. ∎

Lemma 6.3.2. *Suppose that $S_{r-1}^G[i_r^G] = 1$ and let $g = j_{r-1}^G - i_{r-1}^G$. If $j_{r+g-1}^G = i_{r-1}^G$, then the probability of the digraph that is outputted in rounds $[r + g - 1, r + g]$ to be identical to the digraph outputted in rounds $[r - 1, r]$ is at least $e^{\frac{8-8g}{N}}$.*

Proof: By Lemma 6.3.1, the four locations i_{r-1}^G, i_r^G, i_{r+g-1}^G and i_{r+g}^G remain unchanged during the $g - 2$ intermediate rounds with probability $e^{-\frac{4(g-2)}{N}}$.

The locations of the initial digraph are

$$
\begin{aligned}
t_{r-1} &= S_{r-1}^G[i_{r-1}^G] + S_{r-1}^G[i_{r+g-1}^G] \\
\text{and } t_r &= S_r^G[i_r^G] + S_r^G[i_{r+g}^G].
\end{aligned}
$$

Again by Lemma 6.3.1, the two locations t_{r-1} and t_r remain unchanged for next g rounds (t_{r-1} unchanged in rounds $r, \ldots, r + g - 1$ and t_r unchanged in rounds $r + 1, \ldots, r + g$) with probability $e^{-\frac{2g}{N}}$, given that the index i^G in these g rounds does not touch them. But each of t_{r-1} and t_r does not fall in the i^G-affected area with probability $1 - \frac{g}{N}$ and hence the probability of the values at t_{r-1} and t_r to remain in place is

$$\left(1 - \frac{g}{N}\right)^2 \cdot e^{-\frac{2g}{N}} \approx e^{-\frac{4g}{N}} \quad \text{(for small } g\text{)}.$$

Thus the overall probability of all indices to be in place is at least

$$e^{-\frac{4(g-2)}{N}} \cdot e^{-\frac{4g}{N}} = e^{\frac{8-8g}{N}}.$$

∎

Now, using the above two results, we are going to prove the main result on the digraph repetition bias.

Theorem 6.3.3. *For small values of* Δ*, the probability of the pattern* $ABTAB$ *in RC4 keystream, where* T *is a* Δ*-word string is* $\frac{1}{N^2}(1 + \frac{e^{\frac{-8-8\Delta}{N}}}{N})$*.*

Proof: If $g = j^G_{r-1} - i^G_{r-1}$ as defined in Lemma 6.3.2, the real gap between the digraphs is $\Delta = g - 2$. The digraph repetition can occur in two ways.

1. When the conditions of Lemma 6.3.2 are satisfied, the digraph repetition occurs with probability $e^{\frac{8-8g}{N}} = e^{\frac{-8-8\Delta}{N}}$. The probability of the conditions being satisfied is

   $$P(S^G_{r-1}[i^G_r] = 1) \cdot P(j^G_{r-1} = i^G_{r+\Delta+1}) \cdot P(j^G_{r+\Delta+1} = i^G_{r-1}) = \frac{1}{N^3}.$$

   Thus, the contribution of this part is $\frac{e^{\frac{-8-8\Delta}{N}}}{N^3}$.

2. When the conditions of Lemma 6.3.2 are not satisfied (the probability of which is $1 - \frac{1}{N^3}$), it may be assumed that the digraph repetition occurs due to random association (the probability of which is $\frac{1}{N^2}$). Hence the contribution of this part is $\frac{1 - \frac{1}{N^3}}{N^2}$.

Adding the above two components, we get the total probability to be approximately $\frac{1}{N^2}(1 + \frac{e^{\frac{-8-8\Delta}{N}}}{N})$.

∎

Note that in uniformly random stream, the probability of digraph repetition is $\frac{1}{N^2}$, irrespective of the length Δ of T. Whereas in RC4, the probability of such an event is more than $\frac{1}{N^2}$ and little less than $\frac{1}{N^2} + \frac{1}{N^3}$.

Using information theoretic arguments [21], it is claimed in [109] that the number of samples required to distinguish RC4 from uniform random stream are $2^{29}, 2^{28}, 2^{26}$ for success probabilities $0.9, 0.8$ and $\frac{2}{3}$ respectively.

6.3.2   A Conditional Bias in Equality of Any Two Consecutive Bytes

The following technical result is a consequence of the leakage of j^G in keystream as quantified in Theorem 5.4.8.

Theorem 6.3.4. $P(z_{r+1} = z_r \mid 2z_r = i_r^G) = \frac{1}{N}\left(1 + \frac{2}{N^2}\right),\ r \geq 1.$

Proof: Item 3 of Corollary 5.4.2 gives, for $r \geq 0$,

$$P(j_r^G = z_{r+1} \mid 2z_{r+1} = i_{r+1}^G) = \frac{2}{N} - \frac{1}{N(N-1)}.$$

From Theorem 5.4.8, we get

$$P(z_{r+2} = j_r^G) = \frac{1}{N}\left(1 + \frac{2}{N}\right).$$

Now,

$$
\begin{aligned}
&P(z_{r+2} = z_{r+1} \mid 2z_{r+1} = i_{r+1}^G) \\
={}& P(j_r^G = z_{r+1} \mid 2z_{r+1} = i_{r+1}^G) \cdot \\
&\quad P(z_{r+2} = z_{r+1} \mid j_r^G = z_{r+1}, 2z_{r+1} = i_{r+1}^G) + \\
&\quad P(j_r^G \neq z_{r+1} \mid 2z_{r+1} = i_{r+1}^G) \cdot \\
&\quad P(z_{r+2} = z_{r+1} \mid j_r^G \neq z_{r+1}, 2z_{r+1} = i_{r+1}^G) \\
={}& P(j_r^G = z_{r+1} \mid 2z_{r+1} = i_{r+1}^G) \cdot \\
&\quad P(z_{r+2} = j_r^G \mid j_r^G = z_{r+1}, 2z_{r+1} = i_{r+1}^G) + \\
&\quad P(j_r^G \neq z_{r+1} \mid 2z_{r+1} = i_{r+1}^G) \cdot \\
&\quad P(z_{r+2} = z_{r+1} \mid j_r^G \neq z_{r+1}, 2z_{r+1} = i_{r+1}^G) \\
={}& P(j_r^G = z_{r+1} \mid 2z_{r+1} = i_{r+1}^G) \cdot P(z_{r+2} = j_r^G) + \\
&\quad P(j_r^G \neq z_{r+1} \mid 2z_{r+1} = i_{r+1}^G) \cdot \\
&\quad P(z_{r+2} = z_{r+1} \mid j_r^G \neq z_{r+1}, 2z_{r+1} = i_{r+1}^G) \\
={}& \left(\frac{2}{N} - \frac{1}{N(N-1)}\right) \cdot \left(\frac{1}{N} + \frac{2}{N^2}\right) \\
&\quad + \left(1 - \frac{2}{N} + \frac{1}{N(N-1)}\right) \cdot \left(\frac{1}{N-1} \cdot \left(1 - \frac{1}{N} - \frac{2}{N^2}\right)\right) \\
={}& \frac{1}{N} + \frac{2}{N^3},
\end{aligned}
$$

neglecting the higher order terms. Note that the event $(z_{r+2} = j_r^G)$ is considered to be independent of the event $(j_r^G = z_{r+1}\ \&\ 2z_{r+1} = i_{r+1}^G)$. Further, the term $\frac{1}{N-1}$ comes due to the uniformity assumption for the events

$$(j_r^G = z_{r+1} + \gamma \mid 2z_{r+1} = i_{r+1}^G)$$

for $\gamma \neq 0$.

Thus, for $r \geq 0$,

$$P(z_{r+2} = z_{r+1} \mid 2z_{r+1} = i_{r+1}^G) = \frac{1}{N}\left(1 + \frac{2}{N^2}\right).$$

In other words, for $r \geq 1$,

$$P(z_{r+1} = z_r \mid 2z_r = i_r^G) = \frac{1}{N}\left(1 + \frac{2}{N^2}\right).$$

∎

The result of Theorem 6.3.4 provides a new distinguisher for RC4. Using the formulation of distinguishing attacks from Section 6.1, let $p_0 = \frac{1}{N}$ and $\epsilon = \frac{2}{N^2}$. Given $N = 2^8$, we require $2^{32}, 2^{38}, 2^{40}$ and 2^{42} keystream bytes with success probabilities 0.5249, 0.6915, 0.8413 and 0.9772 respectively to distinguish RC4 keystream from a uniformly random stream by observing the event

$$(z_{r+1} = z_r) \text{ given } 2z_r = i_r^G.$$

The event $2z_r = i_r^G \bmod N$ is expected to occur once in a sequence of N consecutive keystream output bytes. Thus the number of keystream bytes needed to mount the distinguisher would be multiplied by N. Hence the distinguisher actually requires $2^{40}, 2^{46}, 2^{48}$ and 2^{50} keystream bytes to get success probabilities 0.5249, 0.6915, 0.8413 and 0.9772 respectively.

Following Item 1 of Corollary 5.4.2 and Theorem 5.4.8, one may note that

$$P(z_{r+1} = z_r \mid i_r^G \text{ even})$$

is positively biased. However, this bias is weaker than that presented in Theorem 6.3.4.

6.3.3 Extension of "$z_2 = 0$" Bias: Best Long-Term Bias in Keystream

Since the work of Mantin [109] in 2005, it has been an open question whether there is any other significant *long-term* bias of RC4 in the keystream. The bias described in Section 6.3.2 is much weaker than that of [109]. Recently, the best long-term bias in RC4 keystream has been discovered [157]. Below we present this result.

RC4 PRGA rounds are typically seen in multiples of $N = 256$:

$$\{1, 2, \ldots, 255\} \bigcup \{256, 258, \ldots, 511\} \bigcup \cdots,$$

where the first period was 1 short of 256 rounds. The analysis in this section views the PRGA rounds in the following structure

$$\{1, 2, \ldots, 255\} \bigcup \{256\} \bigcup \{257, 258, \ldots, 511\} \bigcup \{512\} \bigcup \cdots,$$

where each long period is exactly 255 rounds. Thus,

- Each RC4 period starts with $i_1^G = 1$ now.

- First period of RC4 has $j_0^G = 0$ to start with.

- This will be repeated if $j_{0 \bmod N}^G = 0$ in any period. And if so, the new period will be identical to the initial period of RC4, in terms of the statistical behavior of the permutation and the keystream.

Recall that the biases in $z_1 = 0$ and $z_2 = 0$ did not depend on KSA. Rather, they both depended on $i_0^G = 0$ and $j_0^G = 0$. So a natural question is whether $z_{\rho N+1} = 0$ and $z_{\rho N+2} = 0$ are biased as well, for integer $\rho \geq 1$.

Note that the event $(j_{\rho N}^G = 0)$, for any integer $\rho \geq 1$, can be safely assumed to be random and hence happens with probability $\frac{1}{N}$. Theorem 6.2.2 directly implies

$$P(z_{\rho N+1} = 0 \mid j_{\rho N}^G = 0) = \frac{1}{N} - \frac{1}{N^2}.$$

When $j_{\rho N}^G \neq 0$, it is assumed that the event $z_{\rho N+1} = 0$ happens due to random association. Thus, one gets

$$
\begin{aligned}
&P(z_{\rho N+1} = 0) \\
={} & P(j_{\rho N}^G = 0) \cdot P(z_{\rho N+1} = 0 \mid j_{\rho N}^G = 0) \\
& + P(j_{\rho N}^G \neq 0) \cdot P(z_{\rho N+1} = 0 \mid j_{\rho N}^G \neq 0) \\
\approx{} & \frac{1}{N} \left(\frac{1}{N} - \frac{1}{N^2} \right) + \left(1 - \frac{1}{N} \right) \frac{1}{N} \\
={} & \frac{1}{N} - \frac{1}{N^3}.
\end{aligned}
$$

This bias gives rise to a very weak distinguisher and hence is not so interesting. On the other hand, Theorem 6.2.3 implies

$$P(z_{\rho N+2} = 0 \mid j_{\rho N}^G = 0) \approx \frac{2}{N}. \tag{6.6}$$

Lemma 6.3.5 establishes a similar result for $z_{\rho N}$.

Lemma 6.3.5. *For any integer $\rho \geq 1$, assume that the permutation $S_{\rho N}^G$ is randomly chosen from the set of all possible permutations of $\{0, \ldots, N-1\}$. Then*

$$P(z_{\rho N} = 0 \mid j_{\rho N}^G = 0) \approx \frac{2}{N}.$$

Proof: We have $i_{\rho N}^G = 0$. When $j_{\rho N}^G = 0$, we have no swap in round $0 \bmod N$ and the output is

$$z_{\rho N} = S_{\rho N}^G[2 S_{\rho N}^G[0]].$$

Let us consider two possible cases.

Case I: $S_{\rho N}^G[0] = 0$. Then $z_{\rho N} = S_{\rho N}^G[0] = 0$.

Case II: $S_{\rho N}^G[0] \neq 0$. In this case, it is assumed that $z_{\rho N}$ takes the value 0 due to random association only.

Combining the above two cases, one gets

$$
\begin{aligned}
& P(z_{\rho N} = 0 \mid j_{\rho N}^G = 0) \\
= \quad & P(S_{\rho N}^G[0] = 0) \cdot P(z_{\rho N} = 0 \mid S_{\rho N}^G[0] = 0, j_{\rho N}^G = 0) \\
& + P(S_{\rho N}^G[0] \neq 0) \cdot P(z_{\rho N} = 0 \mid S_{\rho N}^G[0] \neq 0, j_{\rho N}^G = 0) \\
= \quad & \frac{1}{N} \cdot 1 + \left(1 - \frac{1}{N}\right) \cdot \frac{1}{N} \\
= \quad & \frac{2}{N} - \frac{1}{N^2} \\
\approx \quad & \frac{2}{N}.
\end{aligned}
$$

∎

Now, consider the scenario when $j_{\rho N}^G \neq 0$. We claim that if $S_{\rho N}^G[2] = 0$, then $z_{\rho N+2}$ can never become zero. The reason is as follows. Suppose $S_{\rho N}^G[1] = X$, which cannot be zero, $S_{\rho N}^G$ being a permutation. Further, suppose $S_{\rho N}^G[j_{\rho N}^G + X] = Y$ (which is also not zero). In round $\rho N + 1$, $i_{\rho N+1}^G$ becomes 1 and

$$
\begin{aligned}
j_{\rho N+1}^G &= j_{\rho N}^G + S_{\rho N}^G[1] \\
&= j_{\rho N}^G + X.
\end{aligned}
$$

After the swap, Y comes to index 1 and X goes to index $j_{\rho N}^G + X$. In round $\rho N + 2$, $i_{\rho N+2}^G$ takes the value 2 and

$$
\begin{aligned}
j_{\rho N+2}^G &= j_{\rho N+1}^G + S_{\rho N}^G[2] \\
&= j_{\rho N}^G + X.
\end{aligned}
$$

Thus, after the swap, X comes to index 2 and 0 goes to index $j_{\rho N}^G + X$. Hence the output is

$$
\begin{aligned}
z_{\rho N+2} &= S_{\rho N+2}^G[X + 0] \\
&= S_{\rho N+2}^G[X] \\
&\neq 0.
\end{aligned}
$$

Thus, given $j_{\rho N}^G \neq 0$, we must have $S_{\rho N}^G[2] \neq 0$ for $z_{\rho N+2}$ to be zero. We assume that given $j_{\rho N}^G \neq 0$ and $S_{\rho N}^G[2] \neq 0$, $z_{\rho N+2}$ is uniformly distributed.

Hence,

$$P(z_{\rho N+2} = 0 \ \& \ j_{\rho N}^G \neq 0) = \sum_{\substack{u \neq 0 \\ S_{\rho N}[2] \neq 0}} P(z_{\rho N+2} = 0 \ \& \ j_{\rho N}^G = u)$$

$$= \left(\frac{N-1}{N}\right) \sum_{u \neq 0} P(z_{\rho N+2} = 0 \mid j_{\rho N}^G = u) \cdot P(j_{\rho N}^G = u)$$

$$\approx \left(\frac{N-1}{N}\right) \cdot (N-1) \cdot \frac{1}{N} \cdot \frac{1}{N}$$

$$= \frac{1}{N} - \frac{2}{N^2} + \frac{1}{N^3}.$$

By similar calculations, it can be shown that the same result holds for the event $z_{\rho N} = 0$ as well. Thus, one can write

$$P(z_{\rho N} = 0 \ \& \ j_{\rho N}^G \neq 0) = P(z_{\rho N+2} = 0 \ \& \ j_{\rho N}^G \neq 0)$$

$$= \frac{1}{N} - \frac{2}{N^2} + \frac{1}{N^3}. \tag{6.7}$$

By combining Equation (6.6), Lemma 6.3.5 and Equation (6.7), one can easily see that none of $z_{\rho N}$ and $z_{\rho N+2}$ has bias toward 0 (of order $\frac{1}{N^2}$ or higher). Thus, for $\rho \geq 1$,

$$P(z_{\rho N} = 0) = P(z_{\rho N+2} = 0) = \frac{1}{N}. \tag{6.8}$$

The main idea behind the search for a long-term distinguisher is to combine Equation (6.6) and Lemma 6.3.5 to eliminate $j_{\rho N}^G$.

Theorem 6.3.6. *For any integer $\rho \geq 1$, assume that the permutation $S_{\rho N}^G$ is randomly chosen from the set of all possible permutations of $\{0, \ldots, N-1\}$. Then*

$$P(z_{\rho N+2} = 0 \mid z_{\rho N} = 0) \approx \frac{1}{N} + \frac{1}{N^2}.$$

Proof: We have

$$P(z_{\rho N+2} = 0 \mid z_{\rho N} = 0)$$

$$= \frac{P(z_{\rho N+2} = 0 \ \& \ z_{\rho N} = 0)}{P(z_{\rho N} = 0)}$$

$$= N \cdot P(z_{\rho N+2} = 0 \ \& \ z_{\rho N} = 0), \quad \text{(using Equation (6.8))}$$

$$= N \cdot P(j_{\rho N}^G = 0) \cdot P(z_{\rho N+2} = 0 \ \& \ z_{\rho N} = 0 \mid j_{\rho N}^G = 0)$$

$$+ N \cdot P(j_{\rho N}^G \neq 0) \cdot P(z_{\rho N+2} = 0 \ \& \ z_{\rho N} = 0 \mid j_{\rho N}^G \neq 0).$$

$$= P(z_{\rho N+2} = 0 \ \& \ z_{\rho N} = 0 \mid j_{\rho N}^G = 0)$$

$$+ (N-1) \cdot P(z_{\rho N+2} = 0 \ \& \ z_{\rho N} = 0 \mid j_{\rho N}^G \neq 0).$$

One may assume that given $j_{\rho N}^G$, the random variables $z_{\rho N+2}$ and $z_{\rho N}$ are independent. Thus, from the first part one gets

$$P(z_{\rho N+2} = 0 \ \& \ z_{\rho N} = 0 \mid j_{\rho N}^G = 0)$$
$$= \ P(z_{\rho N+2} = 0 \mid j_{\rho N}^G = 0) \cdot P(z_{\rho N} = 0 \mid j_{\rho N}^G = 0)$$
$$\approx \ \frac{2}{N} \cdot \frac{2}{N} \qquad \text{(from Equation (6.6) and Lemma 6.3.5)}$$
$$= \ \frac{4}{N^2}$$

and from the second part one has

$$(N-1) \cdot P(z_{\rho N+2} = 0 \ \& \ z_{\rho N} = 0 \mid j_{\rho N}^G \neq 0)$$
$$= \ (N-1) \cdot P(z_{\rho N+2} = 0 \mid j_{\rho N}^G \neq 0) \cdot P(z_{\rho N} = 0 \mid j_{\rho N}^G \neq 0)$$
$$= \ (N-1) \cdot \frac{P(z_{\rho N+2} = 0 \ \& \ j_{\rho N}^G \neq 0) \cdot P(z_{\rho N} = 0 \ \& \ j_{\rho N}^G \neq 0)}{\left(P(j_{\rho N}^G \neq 0) \right)^2}$$
$$= \ (N-1) \frac{\left(\frac{1}{N} - \frac{2}{N^2} + \frac{1}{N^3} \right)^2}{\left(\frac{N-1}{N} \right)^2} \qquad \text{(by Equation (6.7))}$$
$$= \ \frac{1}{N} - \frac{3}{N^2} + \frac{3}{N^3} - \frac{1}{N^4}.$$

Adding the two expressions from the two parts, we have

$$P(z_{\rho N+2} = 0 \mid z_{\rho N} = 0) \approx \frac{1}{N} + \frac{1}{N^2}.$$

∎

Thus, the distinguisher works as follows.

- Acquire $z_{\rho N+2}$ when $z_{\rho N} = 0$.

- Find the probability $P(z_{\rho N+2} = 0 \mid z_{\rho N} = 0)$.

According to Theorem 6.1.1, the number of samples required to mount a distinguishing attack on RC4 using the above bias is $O(N^3)$, i.e., around 2^{24} for $N = 256$.

Along the same line of analysis, it can be shown [157] that

$$P(z_{\rho N+1} = 0 \mid z_{\rho N} = 0) \approx \frac{1}{N} + \frac{1}{N^2}, \qquad \text{for } \rho \geq 1$$

and

$$P(z_{\rho N+2} = 0 \mid z_{\rho N+1} = 0) \approx \frac{1}{N} + \frac{2}{N^2}, \qquad \text{for } \rho \geq 0.$$

The above trio of best long-term biases of RC4 gives a trio of distinguishers of approximately the same complexity.

Research Problems

Problem 6.1 Does there exist any other distinguisher of complexity $O(N)$ involving the initial keystream bytes than the event $(z_2 = 0)$?

Problem 6.2 Investigate the existence of a long-term keystream bias of order $\frac{1}{N}$, where the random association is also of order $\frac{1}{N}$.

Problem 6.3 Can the existing biases be combined to create a bias greater than the maximum of the biases?

Chapter 7

WEP and WPA Attacks

In the last decade of twentieth century, wireless LANs (WLANs) became prevalent, where devices are connected without wire to an IP network. Because wireless data is accessible publicly, extra security measures are required to protect such networks.

IEEE LAN/MAN Standards Committee has specified a set of standards, collectively called IEEE 802.11, for carrying out WLAN communication. The first version of the standard was released in 1997 and allows a maximum speed of 2 Mbits/s at a frequency band of 2.4 GHz. It has three alternative physical layer technologies: diffuse infrared operating at 1 Mbit/s, frequency-hopping spread spectrum operating at 1 or 2 Mbit(s)/s and direct-sequence spread spectrum operating at 1 or 2 Mbit(s)/s. This standard also specifies a simple security protocol called *Wired Equivalent Privacy* or WEP [92] with the idea that it should provide the same level of privacy to the legitimate users of an IEEE 802.11 network, as they would have with a wired network. WEP incorporates RC4 for encrypting the network traffic.

In 1999, two enhancements of the original standard, namely, IEEE 802.11a and 802.11b, were released. They did not introduce any new security feature, but offered wider bandwidth and/or higher speed. The IEEE standard 802.11a operates in the 5 GHz band with a maximum data rate of 54 Mbits/s. It uses the same data link layer protocol and frame format as the original standard, but with an orthogonal frequency-division multiplexing (OFDM) based air interface. IEEE standard 802.11b has a maximum raw data rate of 11 Mbits/s and uses the same media access method defined in the original standard. This, with direct sequence spread spectrum, is popularly known as *Wi-Fi* (comes from the word *Wireless Fidelity* and is a trademark of the Wi-Fi Alliance).

IEEE 802.11g, proposed in 2003, works in the 2.4 GHz band like 802.11b, but uses the same OFDM based transmission scheme as 802.11a and operates at a maximum physical layer bit rate of 54 Mbits/s. In 2009, an amendment was released with the name of 802.11n which improves upon the previous 802.11 standards by adding multiple-input multiple-output (MIMO) antennas. It operates on both 2.4GHz and 5GHz bands.

In 2004, IEEE 802.11i standard was released which defines the successor protocol for WEP, called *Wi-Fi Protected Access* or WPA [93]. Later, WPA was replaced by WPA2 [94] which uses *Counter Mode with Cipher Block Chaining Message Authentication Code Protocol* (CCMP, an AES-based [2] encryption mode with strong security) as a mandatory requirement (was optional in WPA).

In 2001, Fluhrer, Mantin and Shamir discovered [52] for the first time that RC4 can be broken if the secret key is appended or prepended with known IVs. Shortly after the publication of [52], a full attack on WEP based on the analysis in [52, Sections 6,7] was implemented by Stubblefield et al. [175]. This attack is known as the FMS attack.

WEP is still widely used and some vendors still produce devices that can only connect to unsecured or WEP networks. Perhaps due to this reason, researchers have continued to explore WEP mode attacks [86, 108, 180, 184]. In this chapter we outline the working principles behind the important WEP attacks on RC4. In the end, we also mention some of the recent attacks on WPA. For a detailed survey on WEP and WPA, see [177, 178].

7.1   RC4 in WEP and the Attack Principle

WEP uses a 40-bit secret key, called the *root key*, shared between all the users and the network access point. For every packet, the sender chooses a new 24-bit IV and appends to it the 40-bit secret key to form a 64-bit RC4 session key, called Per Packet Key (PPK). An extended 128-bit WEP protocol is also available that uses a 104-bit key (called WEP-104). Since the IV is prepended with the secret key, the attacker knows the first 3 bytes of the key.

In general, if size of the IV is x bytes and if $\kappa[0], \ldots, \kappa[l-1-x]$ denote the root key, then the session key is given by

$$K[0 \ldots (l-1)] = IV[0 \ldots (x-1)] \| \kappa[0 \ldots (l-1-x)]$$

For WEP, x is taken to be 3.

In principle, all the WEP attacks are based on some event A that relates one unknown key byte (or modulo $N = 256$ sum of some unknown key bytes, which is again an unknown byte) u with the known IV bytes and the keystream bytes. If the probability p of such an event is greater than the perfectly random association, i.e., if $p > \frac{1}{N}$, then we have a successful attack.

Without any prior information, the uncertainty [31] about u is $H_1 = \log_2 N$. The uncertainty about u when it takes one value with probability p and each of the remaining $N-1$ values with probability $q = \frac{1-p}{N-1}$ is given by

$$H_2 = -p \log_2 p - (N-1)q \log_2 q.$$

Thus, the information gained due to the event A is given by

$$\Delta_p = H_1 - H_2 = \log_2 N + p \log_2 p + (N-1)q \log_2 q.$$

Since in each session we observe one sample, we need at least $\frac{\log_2 N}{\Delta_p}$ sessions to reliably estimate u (here "reliably" means with probability $> 50\%$). The experimental observations related to various WEP attacks support this information-theoretic lower bound.

7.2 FMS Attack

The main idea of the FMS attack [52] is as follows. For $x > 1$, consider that the first x bytes of the secret key, i.e., $K[0], \ldots, K[x-1]$ are known. Hence, all the permutations $S_1, \ldots, S_x$ and the indices $j_0, \ldots, j_x$ are also known. In round $x+1$, we have $i_{x+1} = x$ and $j_{x+1} = j_x + S_x[x] + K[x]$. After the swap in this round, $S_{x+1}[x]$ gets the value of $S_x[j_{x+1}]$. Suppose we have the following.

$$S_x[1] < x \tag{7.1}$$

and

$$S_x[1] + S_x[S_x[1]] = x. \tag{7.2}$$

Conditions 7.1 and 7.2 together are called the *resolved condition*. If $j_{x+1} \notin \{1, S_x[1]\}$ (the probability of which is $\frac{N-2}{N}$) then $S_{x+1}[1] = S_x[1]$ and $S_{x+1}[S_{x+1}[1]] = S_x[S_x[1]]$. Thus,

$$S_{x+1}[1] + S_{x+1}[S_{x+1}[1]] \overset{\frac{N-2}{N}}{=} x. \tag{7.3}$$

Recall that $\overset{p}{=}$ means that the equality holds with probability p. In the next $N - x$ rounds, i.e., in

- the next $(N - x - 1)$ rounds of the KSA (corresponding to $r = x+2, x+3, \ldots, N$), and

- the first round of the PRGA,

index i_r takes values in $\{x+1, x+2, \ldots, N-1, 0\}$ and hence does not visit the indices 1, $S_{x+1}[1] = S_x[1] < x$ and $x = S_{x+1}[1] + S_{x+1}[S_{x+1}[1]]$. If j_r also does not visit these three indices in those rounds, the probability of which is $(\frac{N-3}{N})^{N-x}$, then the values at indices 1, $S_{x+1}[1]$ and $S_{x+1}[1] + S_{x+1}[S_{x+1}[1]]$ of the permutation S_{x+1} are not swapped throughout the KSA and also during the first round of the PRGA. Hence, the first keystream output byte is given by

$$
\begin{aligned}
z_1 &= S_1^G[S_1^G[i_1^G] + S_1^G[j_1^G]] \\
&= S_1^G[S_1^G[1] + S_1^G[S_N[1]]] \quad \text{(since } i_1^G = 1 \text{ and } j_1^G = S_N[1]) \\
&= S_1^G[S_N[1] + S_N[S_N[1]]] \quad \text{(reverting the swap)} \\
&\overset{(\frac{N-3}{N})^{N-x}}{=} S_{x+1}[S_{x+1}[1] + S_{x+1}[S_{x+1}[1]]] \\
&\overset{\frac{N-2}{N}}{=} S_{x+1}[x] \quad \text{(by Equation (7.3))} \\
&= S_x[j_{x+1}] \quad \text{(due to the swap in round } x+1) \\
&= S_x[j_x + S_x[x] + K[x]] \quad \text{(expanding } j_{x+1}).
\end{aligned}
$$

Thus, when the resolved condition is satisfied, the event

$$
K[x] = S_x^{-1}[z_1] - j_x - S_x[x] \tag{7.4}
$$

holds with probability $\frac{N-2}{N}(\frac{N-3}{N})^{N-x}$. For $x \geq 3$ and $N = 256$, this probability is always greater than 0.05. Hence if one can find around 60 IVs for which S_x satisfies the resolved condition, then $K[x]$ will be identified properly in approximately 3 cases.

An attacker starts with some IVs and their corresponding first keystream bytes. The attacker chooses a subset of these keys that satisfies the resolved condition after the first x steps of the RC4 KSA. For each such key, (s)he finds a candidate for $K[x]$ using Equation (7.4). The value having the maximum frequency is taken as the estimate of $K[x]$. Thus, the attacker knows the first $(x + 1)$ key bytes $K[0], \ldots, K[x]$. One can then repeat the above strategy to derive $K[x + 1], K[x + 2], \ldots, K[l - 1]$ iteratively for all the $l - 1 - x$ unknown key bytes. When all the key bytes are estimated, the resulting key combined with a few IVs can be used to generate RC4 keystream and thus can be verified for correctness. If the generated keystream byte does not match with the observed keystream, at least one of the key bytes must have been guessed wrongly. The attacker may assume that $K[y]$ is wrong for some y (for example, (s)he may consider the key byte whose most frequent value and the second most frequent value differs by the smallest amount) and proceed with the second most frequent value among the candidates for $K[y]$ as its estimate. The process continues until the correct key is found or a predetermined number of attempts have failed.

The IVs that satisfy the resolved condition are termed as *weak IVs*. In the WEP scenario, the attacker cannot choose the IV of a packet that a sender is going to use. He has to collect enough samples of packets until he is lucky to encounter some weak IVs. In practice, the attack requires 4 to 6 million packets for a success probability of at least 50% depending on the environment and implementation.

7.3 Mantin's Attack

In [52, Section 9], the authors suggested to throw away the first 256 keystream output bytes of RC4 PRGA to resist the FMS attack. However, in 2005, Mantin proposed a WEP mode attack on RC4 [108, Section 4.2] along the same line of the the original FMS attack [52, Sections 6,7] using Glimpse Theorem (see Theorem 5.2.1), by exploiting the the 257th keystream byte.

As in the FMS attack, for $x > 1$, the key bytes $K[0], \ldots, K[x-1]$ are assumed to be known (hence S_x and j_x are also known) and the target is to derive the next key byte $K[x]$. In round $x + 1$, we have $i_{x+1} = x$ and $j_{x+1} = j_x + S_x[x] + K[x]$. After the swap in this round, $S_x[j_{x+1}]$ moves to $S_{x+1}[x]$. Suppose IVs are selected in such a way that the following condition holds.

$$S_x[1] = x. \tag{7.5}$$

If $j_{x+1} \neq 1$, the probability of which is $\frac{N-1}{N}$, then $S_{x+1}[1] = S_x[1]$. Thus,

$$S_{x+1}[1] \overset{\frac{N-1}{N}}{=} x. \tag{7.6}$$

In the next $N - x - 1$ rounds till the end of the KSA, i.e., in rounds $r = x+2, x+3, \ldots, N$, index i_r takes values in $\{x+1, x+2, \ldots, N-1\}$ and hence does not visit the indices 1 and $x = S_{x+1}[1]$. If j_r also does not visit these two indices in those rounds, the probability of which is $(\frac{N-2}{N})^{N-x-1}$, then the values at indices 1 and $S_{x+1}[1]$ are not swapped. Thus, we get the following probabilities involved.

$$
\begin{array}{lll}
N + 1 - z_{N+1} & \overset{\frac{2}{N}}{=} & S_N^G[1] \quad \text{(according to Corollary 5.2.2)} \\[6pt]
& \overset{(\frac{N-1}{N})^{N-1}}{=} & S_1^G[1] \quad \text{(when } j_r^G \neq 1, r = 2 \text{ to } N) \\[6pt]
& = & S_N[j_1^G] \quad \text{(swap in PRGA round 1)} \\[6pt]
& = & S_N[S_N[1]] \quad \text{(since } j_1^G = S_N[1]) \\[6pt]
& \overset{(\frac{N-2}{N})^{N-x-1}}{=} & S_{x+1}[S_{x+1}[1]] \\[6pt]
& \overset{\frac{N-1}{N}}{=} & S_{x+1}[x] \quad \text{(by Equation (7.6))} \\[6pt]
& = & S_x[j_{x+1}] \quad \text{(swap in KSA round } x+1) \\[6pt]
& = & S_x[j_x + S_x[x] + K[x]].
\end{array}
$$

For $x \geq 3$ and $N = 256$, the above chain of events occurs approximately with probability

$$\frac{2}{N} \cdot \left(\frac{N-1}{N}\right)^{N-1} \cdot \left(\frac{N-2}{N}\right)^{N-3-1} \cdot \left(\frac{N-1}{N}\right) \approx \frac{2}{N} \cdot \left(\frac{N-1}{N}\right)^{3N}$$

$$\approx \frac{2}{e^3 N}$$

and given the above chain, the relation

$$K[x] = S_x^{-1}[N + 1 - z_{N+1}] - j_x - S_x[x] \qquad (7.7)$$

holds with probability 1. When the above chain does not occur, Equation (7.7) holds with probability $\frac{1}{N}$ due to random association. Thus,

$$
\begin{aligned}
P(K[x] = S_x^{-1}[N + 1 - z_{N+1}] - j_x - S_x[x]) &\approx \frac{2}{e^3 N} \cdot 1 + \left(1 - \frac{2}{e^3 N}\right) \cdot \frac{1}{N} \\
&\approx \frac{1}{N}\left(1 + \frac{2}{e^3}\right) \\
&\approx \frac{1.1}{N}.
\end{aligned}
$$

As reported in [108], the empirical estimate of the above probability is $\frac{1.075}{N}$ and this attack recovers a 16-byte key with 80% success probability from 2^{25} chosen IVs. If choice of IVs is not possible, then the attacker needs to test approximately 2^{48} different keys for correctness.

7.4   Klein's Attack

Both the FMS and Mantin's attacks require certain weak IVs. Klein improved these attacks in [86] by showing how to attack RC4 in the WEP mode, without requiring any weak IV condition.

As before, suppose for $x > 1$, $K[0], \dots, K[x-1]$ (and hence S_x and j_x) are all known and the goal is to to find $K[x]$. In round $x+1$, we have $i_{x+1} = x$ and $j_{x+1} = j_x + S_x[x] + K[x]$ and the swap moves $S_x[j_{x+1}] = S_x[j_x + S_x[x] + K[x]]$ into $S_{x+1}[x]$.

Now, we consider two possible cases in which the event $(x - z_x = S_{x+1}[x])$ can occur.

Case I: $S_{x-1}^G[x] = S_{x+1}[x]$.

After round x of the KSA, we need to wait for $N - x - 1$ remaining steps of the KSA and the first x-1 steps of the PRGA for index i (i.e., the index i_x^G) to visit the index x again. Thus, $S_{x-1}^G[x]$ would remain the same as $S_{x+1}[x]$ if the index x is also not visited by j in those $N - x - 1 + x - 1 = N - 2$ rounds (not visited by j_r in rounds $r = x+2$ to N of the KSA and not visited j_r^G in rounds $r = 1$ to $x - 1$ of the PRGA), the probability of which is $\left(\frac{N-1}{N}\right)^{N-2}$. We also have

$$
\begin{aligned}
&P(x - z_x = S_{x+1}[x] \mid S_{x-1}^G[x] = S_{x+1}[x]) \\
=~ &P(x - z_x = S_{x-1}^G[x]) \\
=~ &\frac{2}{N} \text{ (according to Corollary 5.2.2)}.
\end{aligned}
$$

Case II: $S^G_{x-1}[x] \neq S_{x+1}[x]$.

This happens with probability $1 - (\frac{N-1}{N})^{N-2}$. Given this, we assume that $x - z_x$ can be equal to $S_{x+1}[x]$ having any one of $N-1$ remaining values other than $S^G_{x-1}[x]$ with equal probabilities. Thus,

$$P(x - z_x = S_{x+1}[x] \mid S^G_{x-1}[x] \neq S_{x+1}[x]) = \frac{1 - \frac{2}{N}}{N-1} = \frac{N-2}{N(N-1)}.$$

Combining these two cases, we have

$$
\begin{aligned}
& P(x - z_x = S_x[j_x + S_x[x] + K[x]) \\
= \ & P(x - z_x = S_{x+1}[x]) \\
= \ & P(S^G_{x-1}[x] = S_{x+1}[x]) \cdot P(x - z_x = S_{x+1}[x] \mid S^G_{x-1}[x] = S_{x+1}[x]) \\
& + P(S^G_{x-1}[x] \neq S_{x+1}[x]) \cdot P(x - z_x = S_{x+1}[x] \mid S^G_{x-1}[x] \neq S_{x+1}[x]) \\
= \ & \left(\frac{N-1}{N}\right)^{N-2} \cdot \frac{2}{N} + \left(1 - \left(\frac{N-1}{N}\right)^{N-2}\right) \cdot \frac{N-2}{N(N-1)} \\
\approx \ & \frac{1.36}{N} \quad \text{(for } N = 256\text{)}.
\end{aligned}
$$

Thus, the relation

$$K[x] = S_x^{-1}[x - z_x] - j_x + S_x[x] \tag{7.8}$$

holds with probability $\frac{1.36}{N}$. To achieve a success probability greater than 50% for a 16-byte key with 3-byte IV, Klein's attack requires approximately 43000 IVs.

7.5 PTW and VX Attacks

In [180], Tews, Weinmann and Pyshkin extended Klein's Attack to guess sums of key bytes instead of individual key bytes. Let $\sigma_m = \sum_{r=0}^{m} \kappa[r]$. Note that $\kappa[0] = \sigma_0$. Once $\sigma_0, \sigma_1, \sigma_2, \ldots$ are known, one can derive $\kappa[r]$ as $\sigma_r - \sigma_{r-1}$.

As usual, assume that for $x > 1$, $K[0], \ldots, K[x-1]$ (and hence S_x and j_x) are known. After the next m steps of the KSA, we have $i_{x+m} = x + m - 1$ and

$$j_{x+m} = j_x + \sum_{r=x}^{x+m-1} (S_r[r] + K[r]). \tag{7.9}$$

Then the swap in round $x + m$ moves $S_{x+m-1}[j_{x+m}]$ into $S_{x+m}[x + m - 1]$.

Let us now calculate the lower bound of the probabilities of the following two events.

Event 1: $\displaystyle\sum_{r=x}^{x+m-1} S_r[r] = \sum_{r=x}^{x+m-1} S_x[r].$

The equality holds necessarily when $S_r[r] = S_x[r]$, for $r = x, \ldots, x + m - 1$. This means that for $d = 1, \ldots, m - 1$, j_{x+d} cannot take values from $\{x + d, \ldots, x + m - 1\}$, i.e., cannot take exactly $m - d$ different values. Hence, Event 1 happens with a minimum probability of

$$\prod_{d=1}^{m-1} \left(\frac{N - (m - d)}{N} \right) = \prod_{d=1}^{m-1} \left(\frac{N - d}{N} \right).$$

Event 2: $S_{x+m-1}[j_{x+m}] = S_x[j_{x+m}].$

If neither i_r nor j_r visits j_{x+m}, the value at the index j_{x+m} remains unchanged during the rounds $r = x + 1, \ldots, x + m - 1$. During these $m - 1$ rounds, i_r takes $m - 1$ distinct values, none of which equals j_{x+m} with probability $\frac{N-m+1}{N}$. Also, None of j_r's equal j_{x+m} with probability $\left(\frac{N-1}{N}\right)^{m-1}$. Thus, Event 2 occurs with a minimum probability of

$$\left(\frac{N - m + 1}{N} \right) \left(\frac{N - 1}{N} \right)^{m-1}.$$

When both the above event occurs, we have

$$
\begin{aligned}
S_{x+m}[x + m - 1] &= S_{x+m-1}[j_{x+m}] \quad &&\text{(due to swap in round } x + m) \\
&= S_x[j_{x+m}] \quad &&\text{(due to Event 2)} \\
&= S_x\left[j_x + \sum_{r=x}^{x+m-1} (S_r[r] + K[r]) \right] \quad &&\text{(by Equation (7.9))} \\
&= S_x\left[j_x + \sum_{r=x}^{x+m-1} (S_x[r] + K[r]) \right] \quad &&\text{(due to Event 2)}
\end{aligned}
$$

or in other words,

$$\sum_{r=x}^{x+m-1} K[r] = S_x^{-1}[S_{x+m}[x + m - 1]] - j_x - \sum_{r=x}^{x+m-1} S_x[r]. \qquad (7.10)$$

A lower bound of the probability of Equation (7.10) is given by

$$\alpha_m = \left(\frac{N - m + 1}{N} \right) \left(\frac{N - 1}{N} \right)^{m-1} \cdot \prod_{d=1}^{m-1} \left(\frac{N - d}{N} \right). \qquad (7.11)$$

Similar to Case I in Section 7.4, $S_{x+m-2}^G[x + m - 1] = S_{x+m}[x + m - 1]$ with probability $\left(\frac{N-1}{N}\right)^{N-2}$. When this holds along with Equation (7.10), then due to the Glimpse bias (Corollary 5.2.2), $x+m-1-z_{x+m-1}$ equals $S_{x+m}[x+m-1]$

with probability $\frac{2}{N}$. Otherwise, when one of these conditions is not satisfied, the probability of which is $1 - \alpha_m(\frac{N-1}{N})^{N-2}$, $x + m - 1 - z_{x+m-1}$ equals $S_{x+m}[x + m - 1]$ with probability $\frac{N-2}{N(N-1)}$ as argued in Case II of Section 7.4. Hence, the event

$$\sum_{r=x}^{x+m-1} K[r] = S_x^{-1}[x + m - 1 - z_{x+m-1}] - j_x - \sum_{r=x}^{x+m-1} S_x[r] \qquad (7.12)$$

holds with probability

$$\alpha_m \cdot \left(\frac{N-1}{N}\right)^{N-2} \cdot \frac{2}{N} + \left(1 - \alpha_m \left(\frac{N-1}{N}\right)^{N-2}\right) \cdot \frac{N-2}{N(N-1)}.$$

where α_m is given by Equation (7.11). When $m = 1$, we have $\alpha_m = 1$ and this case corresponds to Klein's attack. The basic attack strategy is to form a frequency table for each σ_m. Then the attacker can employ different key ranking strategies to improve the search.

Around the same time of the publication of [180], Vaudenay and Vuagnoux [184] independently extended Mantin's and Klein's basic attacks to guess the sum of the key bytes with an aim to reduce key byte dependency. The work [184] additionally exploited the repetition of the key and IV bytes to mount the attack. The *active* attack in [180] requires 2^{20} RC4 key setups, around 40,000 frames of 104-bit WEP to give a success probability of 0.5. On the other hand, the *passive* attack in [184] requires 2^{15} data frames to achieve the same success rate.

7.6   RC4 in WPA and Related Attacks

WPA was designed as a wrapper for the WEP to prevent the FMS attack. The major improvement in WPA over WEP is the *Temporal Key Integrity Protocol* (TKIP), a key management scheme [95] to avoid key reuse. TKIP consists of a key hash function [75] to defend against the FMS attack, and a message integrity code (MIC) [48].

A 16-byte Temporal Key (TK) is derived from a Pre-Shared Key (PSC) during the authentication. TK, in addition to the 6-byte Transmitter Address (TA) and a 6-byte IV (the IV is also called the TKIP Sequence Counter or TSC), goes into the key hash function h as inputs. The output, i.e., $h(TK, TA, IV)$, becomes a 16-byte RC4 key where the first three bytes are derived from the IV. A $\langle TK, IV \rangle$ pair is used only once by a sender and hence none of the WEP attacks are applicable.

MIC ensures the integrity of the message. It takes as inputs a MIC key, TA, receiver address and the message, and outputs the message concatenated with a MIC-tag. If required, this output is split into multiple fragments before it enters the WEP protocol.

The derivation $RC4KEY = h(TK, TA, IV)$ in the TKIP is done in two phases as follows.

$$
\begin{aligned}
P1K &= phase1(TK, TA, IV32) \quad \text{and} \\
RC4KEY &= phase2(P1K, TK, IV16),
\end{aligned}
$$

where P1K is an 80-bit intermediate key, called the TKIP-mixed Transmit Address and Key (TTAK), and $IV32$ is the 32 most significant bits of the IV and $IV16$ is the 16 least significant bits of the IV. The first phase does not depend on IV16 and is performed once every 2^{16} packets. The second phase first generates a 96-bit Per Packet Key (PPK), which is turned into $RC4KEY$. The first three bytes of $RC4KEY$ depends only on $IV16$ and hence can be assumed to be known, as in the WEP. The remaining thirteen bytes of $RC4KEY$ depends on the PPK and TK.

The first attack on WPA [124] came in 2004. It can find the TK and the MIC Key, given only two RC4 packet keys. Compared to the brute force attack with complexity 2^{128}, this attack has a complexity of 2^{105}. In 2009, Tews et al. [178] reported a practical attack against WPA. It shows that if an attacker has a few minutes access to the network, he or she is able to decrypt an ARP request or response and send 7 packets with custom content to the network. Recently, in [159], some weak bits of TK are recovered based on combination of known biases of RC4 and this is turned into a distinguisher of WPA with the complexity 2^{43} and advantage 0.5, that uses 2^{40} packets. This work further recovers the full 128-bit TK by using 2^{38} packets in a search complexity of 2^{96}, using a similar strategy as in [124].

Research Problems

Problem 7.1 Can the success probability of WEP attacks be improved beyond the PTW and the VX attacks?

Problem 7.2 Can the success probability of the WPA attack in [159] be improved further?

Problem 7.3 Identify if RC4 is involved in any other standardized protocol and find out if there exists any weakness.

Chapter 8

Fault Attacks

Several types of faults may occur in a cryptographic device, leading to the vulnerability of the cipher being used. In the *data fault* model, the attacker flips some bits in RAM or internal registers. In the *flow fault* model, the attacker makes small changes in the flow of execution, such as skipping an instruction, changing a memory address etc.

The fault models usually rely on optimistic assumptions and consider a weaker version of the cipher than the original one. It has been commented in [71, Section 1.3] that the attacker should have partial control in terms of number, location, and timing of fault injections. It is also assumed that the attacker can reset the system with the same key as many times as he wants. In other words, he can cancel the effects of the previously made faults, go back to the initial configuration and re-execute the cipher with the same key. Though the assumptions are optimistic, they are not unrealistic. In addition, the fault models serve as platforms to evaluate the strength and weaknesses of the cipher.

So far, only a few fault attacks on RC4 are known. In this chapter, we give an overview of each of them.

8.1 Hoch and Shamir's Attack

This is the first reported fault attack [71, Section 3.3] on RC4. It is assumed that the attacker can introduce a fault on a single byte of S after the KSA is over but before the PRGA begins. Further, the attacker can reset the system and inject a new fault as many times as desired. Under such a model, the attacker can analyze the resulting keystreams to reconstruct the internal state.

Note that the attacker can identify the value that is faulted, but not the location of that value. In other words, if $S_N[y] = v$ and the fault changes v to v', then it is possible to recognize both v and v', but not y. The strategy is to observe the frequency of each value in the keystream. v would never appear and v' would appear with twice the expected frequency. A keystream of length 10,000 bytes can reliably find out v and v'.

The next step is to identify faults in $S_N[1]$. In the first step of the PRGA, $i_1^G = 1$ and $j_1^G = S_N[1]$. After the swap, $S_1^G[1] = S_N[S_N[1]]$ and $S_1^G[S_N[1]] =$

$S_N[1]$ and the first keystream output byte is given by

$$
\begin{aligned}
z_1 &= S_1^G[S_1^G[1] + S_1^G[S_N[1]] \\
&= S_1^G[S_N[S_N[1]] + S_N[1]] \\
&\overset{w.h.p.}{=} S_N[S_N[S_N[1]] + S_N[1]],
\end{aligned}
\tag{8.1}
$$

where w.h.p. stands for "with high probability." Thus, if z_1 is changed as a result of the fault, then one of the following three cases must be true.

1. $S_N[1]$ was at fault.

2. $S_N[S_N[1]]$ was at fault.

3. $S_N[S_N[1] + S_N[S_N[1]]]$ was at fault.

Case 3 is the easiest to detect, since we know the original value at the index $S_N[1] + S_N[S_N[1]]$. For distinguishing between Cases 1 and 2, the attacker needs to inspect the second byte z_2. In the second round of PRGA, $i_2^G = 2$ and $j_2^G = j_1^G + S_1^G[2] = S_N[1] + S_1^G[2]$. Thus, after the swap, the second keystream byte is given by

$$
\begin{aligned}
z_2 &= S_2^G[S_2^G[2] + S_2^G[S_N[1] + S_1^G[2]]] \\
&= S_2^G[S_1^G[S_N[1] + S_1^G[2]] + S_1^G[2]].
\end{aligned}
\tag{8.2}
$$

Thus, if Case 1 occurs, then both z_1 and z_2 would be at fault. If Case 2 occurs, then only z_1 is at fault and with high probability z_2 would not be faulted.

For faults which affected $S_N[1]$ and thereby changed its value from v to v', the attacker can utilize Equation (8.1) to write

$$
S_N[v' + S_N[v']] = z_1'
\tag{8.3}
$$

where z_1' is the first byte of the faulty keystream. For each fault in $S_N[1]$, an equation similar to Equation (8.3) can be found. After collecting many such equations, the attacker starts deducing values at other indices of S_N, using the knowledge of $S_N[1]$.

Example 8.1.1. *Suppose the value at $S_N[1]$ was $v = 26$. Assume that the first fault gives $v' = 1$ and $z_1' = 41$. From Equation (8.3), we get $S_N[1+26] = S_N[27]$ to be 41. Suppose another fault gives $v' = 27$ and $z_1' = 14$. Then $S_N[27 + 41] = S_N[68] = 14$.*

The attackers continue deriving the entries of S_N as above in a chain-like fashion until the equations no longer reveal any new entry of S_N. If in the end, the value of $S_N[S_N[1]]$ is not yet recovered, its value can be guessed. From the knowledge of $S_N[S_N[1]]$, the attacker may continue a similar analysis with the second output byte. Assuming $S_N[2] = t$ is known and noting the fact that S_1^G and S_2^G of Equation (8.2) can be replaced by S_N with high probability, one may write

$$
S_N[S_N[S_N[1] + t] + t] \overset{w.h.p.}{=} z_2'.
$$

Here z_2' is the second byte of the faulty stream. Experimental results confirm that at this point on average 240 entries of S are recovered and the rest can be found from the original keystream. Wrong guesses of $S_N[S_N[1]]$ can be eliminated through inconsistency of the collected equations or by comparing the original keystream with the stream generated from the recovered S_N.

We summarize the above strategy of Hoch and Shamir in the form of Algorithm *HS-FaultAttack*.

Input: The non-faulty keystream.
Output: The permutation S_N after the KSA.

1 **for** 2^{16} *iterations* **do**
2   Apply a fault to a single byte of the S_N table;
3   Generate a faulty keystream of length 1024;
4   Identify the fault;
    end
5 For faults affecting $S_N[1]$, form equations of the form (8.3);
6 **repeat**
7   Use the equations to infer the entries of S_N from the known entries;
8   Guess an unknown entry whenever necessary;
    until *all entries of S_N are known* ;
9 Check the recovered state for consistency with the non-faulty keystream;
10 If inconsistent, go to Step 8 to change the guesses and continue;

Algorithm 8.1.1: HS-FaultAttack

The algorithm requires 2^{16} fault injections. Each of the data complexity and the time complexity is of the order of 2^{26}.

8.2 Impossible and Differential Fault Attacks

In [16], two fault attack models were proposed. One of them is called the *impossible fault attack* [16, Section 3] that exploits the Finney cycle. Another one is called the *differential fault attack* [16, Section 4]. We describe both of them in this section.

8.2.1 Impossible Fault Attack

In the impossible fault attack model, the attacker repeatedly injects faults into either register i^G or register j^G, until a Finney state is identified. If at PRGA step r, S_r^G is identified to be in a Finney state, then according to Item

(3) of Theorem 5.1.1, the subsequent 256 bytes produced by decimating the keystream after every 255 bytes reveal the entries of the initial permutation S_r^G shifted cyclically. Since $S_r^G[i_r^G + 1] = 1$, the location of the value 1 and thereby the locations of all other values in S_r^G are automatically determined.

Since the conditions for the Finney state are $j_r^G = i_r^G + 1$ and $S_r^G[j_r^G] = 1$, the probability of falling into a Finney state by a fault is $2^{-8} \times 2^{-8} = -16$, assuming that the faults are independent. In other words, after around 2^{16} fault injections, the internal state is expected to fall into a Finney state.

An obvious problem is how to detect whether a state is a Finney state. It turns out that the number of steps, after which a collision of two bytes of the keystream should occur, gives an indication of whether the state is a Finney state or a normal RC4 state.

There are $N = 256$ possible bytes in the keystream. The probability that no byte is repeated in m successive keystream output bytes of normal RC4 is given by $\frac{N(N-1)\cdots(N-1+m)}{N^m}$ and the probability that there is at least one repetition is $p = 1 - \frac{N(N-1)\cdots(N-1+m)}{N^m}$. Setting $p > \frac{1}{2}$ yields $m \geq 20$. This is nothing but the *birthday paradox* where there are 256 different birthdays and m different people.

For Finney states, the case where the colliding output byte has value 1 can be ignored as it has a very low probability. Assume that the fault is injected in step x. When the output is not 1, a collision of a byte a steps after the fault and another byte b steps after the fault can occur if

1. $S_{x+a}^G[i_{x+a}^G] + S_{x+a}^G[j_{x+a}^G] = S_{x+b}^G[i_{x+b}^G] + S_{x+b}^G[j_{x+b}^G] + 1$ and

2. $i_{x+a}^G \leq S_{x+a}^G[i_{x+a}^G] + S_{x+a}^G[j_{x+a}^G] \leq i_{x+b}^G$.

The probability of Event 1 is $\frac{1}{N}$ and that of Event 2 is $\frac{b-a}{N}$. Thus, the probability that a collision occurs within the first m bytes after the fault injection is approximately given by

$$p' = \sum_{0 \leq a < b < m} \frac{1}{N} \cdot \frac{b-a}{N} \approx \frac{m^3}{3 \cdot 2^{17}}.$$

Conditioning $p' > \frac{1}{2}$ gives $m \geq 58$ and forcing $p' = 1$ yields $m \approx 73$.

Algorithm *DetectFinney* gives a step-by-step description of how to determine whether the state is a Finney state or not.

Since the average number of bytes required to recognize a Finney state is around 2^5 and this cost is incurred for each of the 2^{16} fault injections, the number of keystream bytes required to mount the attack is around 2^{21}. Note that the additional $255 \times 256 \approx 2^{16}$ keystream bytes required to recover the state after detection of the Finney cycle, when added to 2^{21}, do not change the order. The time complexity of the attack is the same as the data complexity.

Input: The keystream after a fault is injected.
Output: Classification of the state (Finney or non-Finney).

1 Process the next keystream byte z_r;
2 **if** *z already appeared at least once (i.e., there has been at least one collision with z_r)* **then**
3   Denote the number of bytes since the last fault by n;
4   **if** $n < 30$ **then**
5 | Return(non-Finney);
     end
6   **if** $30 \leq n < 40$ *and this is the second collision* **then**
7 | Return(non-Finney);
     end
8   **if** $40 \leq n < 50$ *and this is the third collision* **then**
9 | Return(non-Finney);
     end
10   **if** $n > 255$ **then**
11   Vefiry that the 255th preceding byte z_{r-255} is different from the current byte z_r (unless one of z_r and z_{r-255} is 1);
12   Also verify that no other pair (x, z_r) or (z_{r-255}, y) appeared separated by 255 rounds;
13   If such a pair appeared, return(non-Finney);
     end
14   **if** $n > 600$ **then**
15 | Return(Finney);
     end
16   **else**
17 | Go to Step 1;
     end
   end
18 **else**
19 | Go to Step 1;
   end

Algorithm 8.2.1: DetectFinney

8.2.2   Differential Fault Attack

This attack relies on collecting three sets of data with the same unknown key. In each of the following cases, the KSA is performed without any fault.

1. The first set is the first 256 keystream bytes z_r, $r = 1, 2, \ldots, 256$ from normal RC4.

2. For each t in $[0, 255]$, make a fault in $S_N[t]$ and run the PRGA for 30 steps. Thus, we have 256 output streams, each of length 30. Denote these streams by X_t, $t = 0, 1, \ldots, 255$.

3. Run the PRGA for 30 steps. Then for each t in $[0, 255]$, inject a fault in $S_{30}^G[t]$ and starting from that step collect longer keystreams Y_t.

The first non-faulty keystream output byte z_1 depends on the indices $i_1^G (= 1)$, $j_1^G (= S_N[1])$ and $S_N[i_1^G] + S_N[j_1^G]$. Thus, except for $t = i_1^G, j_1^G, S_N[i_1^G] + S_N[j_1^G]$, the first byte of all the other X_t streams are all equal to z_1. Once these three streams are identified, the values of i_1^G, j_1^G and $S_N[i_1^G] + S_N[j_1^G]$ are also known, but which value corresponds to which variable cannot be determined. Since i_1^G is known, we need to differentiate between j_1^G and $S_N[i_1^G] + S_N[j_1^G]$. We continue the same analysis for the following bytes $z_2, z_3, \ldots$.

The technique employed to guess the entries of the permutation is called *cascade guessing*. For example, whenever two consecutive bytes in the stream are the two consecutive j^G values, one may subtract the previous j^G from the current one and derive $S^G[i^G]$. Whenever a contradiction is reached, wrong guesses are eliminated. According to the birthday paradox, a contradiction is expected after an average of less than 20 observed bytes.

If the considered keystream bytes are far from the faults in item 2 above, the identification of the three indices become difficult. One possible solution is to have many sets with faults injected at different times, similar to Y_t in item (3). The complete attack requires injection of 2^{10} faults and inspection of 2^{16} keystream bytes.

8.3   Fault Attack Based on Fork Model

This attack model has been considered in [108, Section 7], where the attacker injects a fault either to j^G or to one of the entries of S^G that are located closely after a fixed index i^G. Repeating this many times, he collects many instances of RC4 that are identical initially but they diverge at a certain point. Let round r be the divergence point. As discussed below, the attacker can iteratively recover all the entries of S_r^G, one by one.

Suppose the attacker wants to recover $S_r^G[u] = v$. He waits until the round r' when the index i^G reaches location u. This happens after at the most $N-1$ steps. With probability at least $(\frac{N-1}{N})^{N-1} \approx \frac{1}{e}$, the index j^G in the intermediate rounds does not visit u, thereby keeping the entry v at u. Given this, due to Corollary 5.2.2, with probability $\frac{2}{N}$ the following holds.

$$
\begin{aligned}
z_{r'} &= r' - S^G_{r'-1}[i^G_{r'}] \\
&= u - S^G_{r'-1}[u] \\
&= u - v.
\end{aligned}
$$

When v does not remain at index u till round r', the probability of which is approximately $1 - e$, the equality $z_{r'} = u - v$ holds due to random association with probability approximately $\frac{1}{N}$. Hence,

$$
\begin{aligned}
P(v = u - z_{r'}) &\approx \frac{1}{e} \cdot \frac{2}{N} + (1 - e) \cdot \frac{1}{N} \\
&= \frac{1}{N}\left(1 + \frac{1}{e}\right).
\end{aligned}
$$

Since the attacker has many instances of RC4 diverging from a single point, he may employ a simple voting mechanism to estimate v as the most frequent value of $u - z_{r'}$. As reported in [108], the complete internal state can be recovered with 80% success probability by injecting around 2^{14} faults.

8.4 Fault Attack with Pseudo-Random Index Stuck

In this section, we describe a technique to recover the RC4 permutation completely when the value of the pseudo-random index j^G is stuck at some value x from a point when $i^G = i_{st}$. This analysis first appeared in [104] and requires $8N$ keystream output bytes suffice to retrieve the permutation in around $O(N)$ time when x and i_{st} are known and in $O(N^3)$ time when they are unknown. Further, if one has a total of 2^{16} keystream bytes after j^G is stuck, then the complete state can be recovered in just 2^8 time complexity, irrespective of whether i^G and j^G are known or unknown.

A fault where the value of the index j^G is stuck at some value x ($0 \le x \le N-1$) during the PRGA, is called a *stuck-at fault*. The corresponding PRGA is termed as StuckPRGA, presented in Algorithm 8.4.1.

Apart from the cryptanalytic significance, studying RC4 with j^G stuck at some value reveals nice combinatorial structures. First, an internal state on any step becomes a restricted permutation over some fixed initial state. Secondly, if one considers consecutive outputs at steps $r, r + 257, r + 514$ and so on, then the resulting keystream sequence

- either consists of the same values (in very few cases), or

- is exactly the subarray (e.g., at indices $y, y + 1, y + 2, \ldots$) of the initial permutation (in most of the cases), or

- is the subarray (e.g., at indices $y, y + 1, y + 3, y + 4, \ldots$) of the initial permutation with a jump in the initial permutation (in very few cases).

Input: Scrambled permutation array $S[0 \ldots N - 1]$, output of KSA.
Output: Pseudo-random keystream bytes z.

Initialization:
$i = j = 0$;

Output Keystream Generation Loop:
$i = i + 1$;
$j = x$;
Swap($S[i]$, $S[j]$);
$t = S[i] + S[j]$;
Output $z = S[t]$;

Algorithm 8.4.1: StuckPRGA

For the remainder of this section we assume, without loss of generality, that

$$S_0^G = \; < a_0, a_1, a_2, \ldots, a_{N-1} > .$$

The subscripts in a_y's follow arithmetic modulo N. For example, a_{-y}, a_{N-y} and in general $a_{\rho N - y}$ for any integer ρ represent the same element. Our goal is to recover $< a_0, a_1, a_2, \ldots, a_{N-1} >$ from the faulty keystream.

8.4.1 Analysis of StuckPRGA

We begin with a few definitions that will be needed in the subsequent analysis.

Definition 8.4.1. (Run) *A run of the RC4 PRGA is defined to be a set of any N consecutive rounds of keystream output byte generation during which the deterministic index i^G takes each value in $\{0, \ldots, N - 1\}$ exactly once.*

Definition 8.4.2. (Successor) *Given a permutation $\mathcal{P}$ of size N, the n-th successor of an element u in $\mathcal{P}$, denoted by $suc^n(u)$, is defined to be the element which appears n locations after u, if we move from left to right in $\mathcal{P}$ in a circular fashion. If $\mathcal{P} = \; < b_0, b_1, b_2, \ldots, b_{N-1} >$, then $suc^n(b_y) = b_{y+n}$.*

Definition 8.4.3. (Rotated Permutation) *Given a permutation $\mathcal{P}$ of size N, the n-rotated permutation, denoted by $rot^n(\mathcal{P})$, is defined to be the permutation obtained by circularly right-shifting $\mathcal{P}$ by n positions. If $\mathcal{P} = \; < b_0, b_1, b_2, \ldots, b_{N-1} >$, then*

$$rot^n(\mathcal{P}) = \; < b_n, b_{n+1}, \ldots, b_{N-1}, b_0, b_1, \ldots, b_{n-2}, b_{n-1} > .$$

Definition 8.4.4. (Candidate Pair) *An ordered pair (u, v) is called a candidate pair, if u appears $N + 1$ rounds after v in the keystream and both u, v come from the same index in the respective permutations.*

Definition 8.4.5. (Conflict Set) *A set $\{(u, v_1), (u, v_2)\}$ of two candidate pairs is called a conflict set, if $v_1 \neq v_2$ and it is not known whether $v_1 = suc^1(u)$ or $v_2 = suc^1(u)$.*

Definition 8.4.6. (Resolved Pair) *A candidate pair (u, v) is called a resolved pair for a permutation $\mathcal{P}$, if it is known that $v = suc^1(u)$.*

Definition 8.4.7. (Resolved Permutation) *A permutation $\mathcal{P}$ of size N is said to be resolved, if for each element u in $\mathcal{P}$, $suc^1(u)$ is known, or in other words, if $N - 1$ many distinct candidate pairs are resolved.*

Since the permutation has N distinct elements, knowledge of $N - 1$ successors for any $N - 1$ elements reveals the successor of the remaining element.

Definition 8.4.8. (Partially Resolved Permutation) *A permutation $\mathcal{P}$ of size N is said to be partially resolved, if for some element u in $\mathcal{P}$, $suc^1(u)$ is not known, or in other words, if less than $N - 1$ many distinct candidate pairs are resolved.*

Definition 8.4.9. *Given two partially resolved permutations $\mathcal{P}_1$ and $\mathcal{P}_2$ of the same N elements, we say $\mathcal{P}_1 > \mathcal{P}_2$ or $\mathcal{P}_1 = \mathcal{P}_2$ or $\mathcal{P}_1 < \mathcal{P}_2$, if the number of resolved pairs for $\mathcal{P}_1$ is more than or equal to or less than the number of resolved pairs for $\mathcal{P}_2$ respectively.*

In the analysis of this section and the next, without loss of generality we assume that j^G is stuck at $x = 0$ from round 1 onwards. Since the index i visits 0 to $N - 1$ cyclically, similar results hold for $x \neq 0$ also, which will be discussed in Section 8.4.3.

Table 8.1 illustrates the structure of the permutation under the above fault model at different rounds of the PRGA.

Proposition 8.4.10. *Suppose the permutation after round ρN of the PRGA, $\rho \geq 0$, is*

$$S_{\rho N}^G = <b_0, b_1, \ldots, b_{N-1}> .$$

Then the permutation after round $\rho N + y$ of the PRGA, $1 \leq y \leq N - 1$, is given by

$$S_{\rho N+y}^G = <b_y, b_0, \ldots, b_{y-1}, b_{y+1}, \ldots, b_{N-1}> .$$

Proof: We prove it by induction on y.
Base Case: When $y = 1$, the deterministic index i^G takes the value 1. So, b_0, b_1 are swapped and

$$S_{\rho N+1}^G = <b_1, b_0, b_2, \ldots, b_{N-1}> .$$

Hence the result holds for $y = 1$.

Round r	i	Bytes of the Permutation S_r^G									Output Index t_r
		0	1	2	3	4	5	...	$N-2$	$N-1$	
0		a_0	a_1	a_2	a_3	a_4	a_5	...	a_{254}	a_{255}	
1	1	a_1	a_0	a_2	a_3	a_4	a_5	...	a_{254}	a_{255}	$a_0 + a_1$
2	2	a_2	a_0	a_1	a_3	a_4	a_5	...	a_{254}	a_{255}	$a_1 + a_2$
3	3	a_3	a_0	a_1	a_2	a_4	a_5	...	a_{254}	a_{255}	$a_2 + a_3$
⋮	⋮	⋮	⋮	⋮	⋮	⋮	⋮	⋮	⋮	⋮	⋮
255	255	a_{255}	a_0	a_1	a_2	a_3	a_4	...	a_{253}	a_{254}	$a_{254} + a_{255}$
256	0	a_{255}	a_0	a_1	a_2	a_3	a_4	...	a_{253}	a_{254}	$2a_{255}$
257	1	a_0	a_{255}	a_1	a_2	a_3	a_4	...	a_{253}	a_{254}	$a_0 + a_{255}$
258	2	a_1	a_{255}	a_0	a_2	a_3	a_4	...	a_{253}	a_{254}	$a_0 + a_1$
259	3	a_2	a_{255}	a_0	a_1	a_3	a_4	...	a_{253}	a_{254}	$a_1 + a_2$
260	4	a_3	a_{255}	a_0	a_1	a_2	a_4	...	a_{253}	a_{254}	$a_2 + a_3$
⋮	⋮	⋮	⋮	⋮	⋮	⋮	⋮	⋮	⋮	⋮	⋮
511	255	a_{254}	a_{255}	a_0	a_1	a_2	a_3	...	a_{252}	a_{253}	$a_{253} + a_{254}$
512	0	a_{254}	a_{255}	a_0	a_1	a_2	a_3	...	a_{252}	a_{253}	$2a_{254}$
513	1	a_{255}	a_{254}	a_0	a_1	a_2	a_3	...	a_{252}	a_{253}	$a_{254} + a_{255}$
514	2	a_0	a_{254}	a_{255}	a_1	a_2	a_3	...	a_{252}	a_{253}	$a_0 + a_{255}$
515	3	a_1	a_{254}	a_{255}	a_0	a_2	a_3	...	a_{252}	a_{253}	$a_0 + a_1$
516	4	a_2	a_{254}	a_{255}	a_0	a_1	a_3	...	a_{252}	a_{253}	$a_1 + a_2$
⋮	⋮	⋮	⋮	⋮	⋮	⋮	⋮	⋮	⋮	⋮	⋮

TABLE 8.1: Evolution of the permutation during StuckPRGA with j^G stuck at 0.

Inductive Case: Suppose for some y, $1 \le y \le N-2$, the result holds, i.e.,

$$S^G_{\rho N+y} = <b_y, b_0, \ldots, b_{y-1}, b_{y+1}, \ldots, b_{N-1}>$$

(*inductive hypothesis*).

Now, in round $\rho N + y + 1$, the index i^G becomes $y+1$ and the other index j^G remains fixed at 0. Thus, the values b_y and b_{y+1} are swapped. Hence,

$$S^G_{\rho N+y+1} = <b_{y+1}, b_0, \ldots, b_y, b_{y+2}, \ldots, b_{N-1}>,$$

i.e., the result also holds for $y+1$.   ∎

Lemma 8.4.11.
(1) *After round $\rho N + y$ of the PRGA, $\rho \ge 0$, $1 \le y \le N-1$, the permutation $S^G_{\rho N+y}$ is given by*

$$< a_{N-\rho+y}, a_{N-\rho}, a_{N-\rho+1}, \ldots, a_{N-\rho+y-1}, a_{N-\rho+y+1}, \ldots, a_{N-\rho-2}, a_{N-\rho-1} >$$

and the permutation $S_{(\rho+1)N}$ after round $\rho N + N$ is the same as the permutation $S_{\rho N + N -1}$ after round $\rho N + N - 1$.
(2) *The index where the keystream output byte is chosen from is given by*

$$t_{\rho N + y} = \begin{cases} a_{N-\rho+y-1} + a_{N-\rho+y} & \text{if } \rho \ge 0, \ 1 \le y \le N-1; \\ 2a_{N-\rho} & \text{if } \rho \ge 1, \ y = 0. \end{cases}$$

Proof: The proof of item (1) will be based on induction on ρ.
Base Case: Take $\rho = 0$. We need to prove that

$$S^G_y = < a_y, a_0, a_1, \ldots, a_{y-1}, a_{y+1}, \ldots, a_{N-2}, a_{N-1} >, \quad 1 \le y \le N-1.$$

This immediately follows from Proposition 8.4.10 above, taking $\rho = 0$.
Inductive Case: Suppose the result holds for some $\rho \ge 0$, i.e., for $1 \le y \le N-1$, the permutation $S^G_{\rho N+y}$ is given by

$$< a_{N-\rho+y}, a_{N-\rho}, a_{N-\rho+1}, \ldots, a_{N-\rho+y-1}, a_{N-\rho+y+1}, \ldots, a_{N-\rho-2}, a_{N-\rho-1} >$$

(*inductive hypothesis*).

Thus, in round $\rho N + N - 1$, we have

$$S^G_{\rho N+N-1} = < a_{N-\rho-1}, a_{N-\rho}, a_{N-\rho+1}, \ldots, a_{N-\rho-2} > .$$

In the next round, i.e. in round $\rho N + N$, the deterministic index i^G becomes 0 which is equal to the value of j^G and hence no swap is involved. Thus,

$$S^G_{(\rho+1)N} = S^G_{\rho N+N-1} = < a_{N-\rho-1}, a_{N-\rho}, a_{N-\rho+1}, \ldots, a_{N-\rho-2} >,$$

which can be rewritten as

$$< b_0, b_1, \ldots, b_{N-1} >, \text{ where } b_y = a_{N-\rho-1+y}, 0 \le y \le N-1.$$

According to Proposition 8.4.10,

$$
\begin{aligned}
S^{G}_{(\rho+1)N+y} &= \; < b_y, b_0, b_1, \ldots, b_{y-1}, b_{y+1}, \ldots, b_{N-2}, b_{N-1} > \\
&= \; < a_{N-(\rho+1)+y}, a_{N-(\rho+1)}, a_{N-(\rho+1)+1}, \ldots, \\
&\qquad a_{N-(\rho+1)+y-1}, a_{N-(\rho+1)+y+1}, \ldots, a_{N-(\rho+1)-1} > .
\end{aligned}
$$

Hence, the result holds for the case $\rho + 1$ also.

Now we prove item (2). In round $\rho N + y$, the value of the deterministic index i^G is $y \; (\mathrm{mod}\, N)$ and that of the index j^G remains fixed at 0. Hence the output is generated from the index

$$
t_{\rho N+y} = S^{G}_{\rho N+y}[y] + S^{G}_{\rho N+y}[0].
$$

Writing the permutation bytes in terms of the a_y's, we get the result. ∎

Theorem 8.4.12. *Consider the two rounds $\rho N + y$ and $(\rho + 1)N + (y + 1)$, $\rho \geq 0, 1 \leq y \leq N - 2$. The two keystream output bytes $z_{\rho N+y}$ and $z_{(\rho+1)N+(y+1)}$ come from the same location $t = a_{N-\rho+y-1} + a_{N-\rho+y}$ in the respective permutations $S^{G}_{\rho N+y}$ and $S^{G}_{(\rho+1)N+(y+1)}$ with the following characteristics.*

1. $t = 0 \iff z_{\rho N+y} = z_{(\rho+1)N+(y+1)} = S^{G}_0[N - \rho + y]$.

2. $t = y + 1 \iff z_{\rho N+y} = suc^2(z_{(\rho+1)N+(y+1)})$ *with respect to S^{G}_0.*

3. $t \in \{0, 1, \ldots, y - 1, y, y + 2, y + 3, \ldots, N - 1\}$
   $\iff z_{\rho N+y} = suc^1(z_{(\rho+1)N+(y+1)})$ *with respect to S^{G}_0.*

Proof: Consider $\rho \geq 0, 1 \leq y \leq N - 2$. From Lemma 8.4.11, we get

$$
\begin{aligned}
t_{\rho N+y} &= t_{(\rho+1)N+(y+1)} \\
&= a_{N-\rho+y-1} + a_{N-\rho+y} \\
&= t \quad (\text{say}).
\end{aligned}
$$

Again from Lemma 8.4.11, we have

$$
\begin{aligned}
S^{G}_{\rho N+y} &= \; < a_{N-\rho+y}, a_{N-\rho}, a_{N-\rho+1}, \ldots, a_{N-\rho+y-1}, a_{N-\rho+y+1}, a_{N-\rho+y+2}, \\
&\qquad \ldots, a_{N-\rho-2}, a_{N-\rho-1} >,
\end{aligned}
$$

and

$$
\begin{aligned}
S^{G}_{(\rho+1)N+(y+1)} &= \; < a_{N-\rho+y}, a_{N-\rho-1}, a_{N-\rho}, \ldots, a_{N-\rho+y-2}, a_{N-\rho+y-1}, \\
&\qquad a_{N-\rho+y+1}, \ldots, a_{N-\rho-3}, a_{N-\rho-2} > .
\end{aligned}
$$

Thus, $t = 0$ if and only if

$$
\begin{aligned}
z_{\rho N+y} &= z_{(\rho+1)N+(y+1)} \\
&= z \quad (\text{say}).
\end{aligned}
$$

And in that case, z reveals $a_{N-\rho+y}$, i.e., the value at index 0 in $S^G_{\rho N+y}$ or equivalently the value at index $N-\rho+y$ in the original permutation S^G_0. This proves item 1.

Of the $N-1$ other possible values of t, if $t = y+1$, then $z_{\rho N+y} = a_{N-\rho+y+1}$ and $z_{(\rho+1)N+(y+1)} = a_{N-\rho+y-1}$; and vice versa. This proves item 2.

If, however, t takes any of the remaining $N-2$ values (other than 0 and $y+1$), then $z_{\rho N+y}$ appears next to $z_{(\rho+1)N+(y+1)}$ in S^G_0; and vice versa. This proves item 3. ∎

Note that the keystream output bytes coming from the same index in two consecutive *runs* are always separated by $N+1$ rounds.

Let us elaborate the pattern considered in the above Theorem. Consider the indices of the keystream output bytes in two consecutive *runs* as follows.

y	run ρ	run $\rho+1$
1	$a_{N-\rho} + a_{N-\rho+1}$	$a_{N-\rho-1} + a_{N-\rho}$
2	$a_{N-\rho+1} + a_{N-\rho+2}$	$a_{N-\rho} + a_{N-\rho+1}$
...	...	...
$N-2$	$a_{N-\rho-3} + a_{N-\rho-2}$	$a_{N-\rho-4} + a_{N-\rho-3}$
$N-1$	$a_{N-\rho-2} + a_{N-\rho-1}$	$a_{N-\rho-3} + a_{N-\rho-2}$
N	$2a_{N-\rho-1}$	$2a_{N-\rho-2}$

Observe that the keystream output selection indices in *run* ρ for $y = 1$ to $N-2$ exactly match with those in *run* $\rho+1$ for $y = 2$ to $N-1$ respectively. Further, as discussed in the proof of Theorem 8.4.12, the permutations in *run* $\rho+1$ for $y = 2$ to $N-1$ are right shifts of the permutations in *run* ρ for $y = 1$ to $N-2$ respectively except at two locations.

8.4.2 State Recovery of RC4 with StuckPRGA

The combinatorial structure identified in Theorem 8.4.12 can be exploited to devise an efficient algorithm *PartResolvePerm* for getting a *partially resolved* permutation from the faulty keystream bytes.

In this algorithm, $Next[u]$ denotes the value that comes immediately after the value u in the permutation S^G_0. If $Next[u] = -1$, it means that the element next to the element u in S^G_0 is not yet determined. The algorithm tallies two consecutive *runs* of the PRGA and fill in the array $Next$ by observing the *candidate pairs*, i.e., the keystream output bytes that come from the same index in the respective permutations. Due to *item* 2 of Theorem 8.4.12, for some u, one may record $suc^2(u)$ as $suc^1(u)$, resulting in some *conflict sets*, i.e., candidate pairs (u, v_1) and (u, v_2) such that $v_1 \neq v_2$. Then it is not known which one of v_1, v_2 is $suc^1(u)$. We keep an array $Conflict$ to contain conflict sets of the form $\{(u, v_1), (u, v_2)\}$. Each entry of $Conflict$ consists of three fields, namely, (i) *value* for storing u, (ii) *first* for storing v_1 and (iii) *second* for storing v_2.

Input: The RN many keystream output bytes from the first
$R(\geq 2)$ *runs* of StuckPRGA.

Output:

1. A partially resolved permutation in the form of an array $Next$.

2. A set of conflict pairs in an array $Conflict$.

```
1  for u = 0 to N − 1 do
2  |  Set Next[u] = −1;
   end
3  NumConflicts = 0;
4  for ρ = 0 to R − 2 do
5  |  for y = 1 to N − 2 do
6  |  |  if z_{ρN+y} = z_{(ρ+1)N+(y+1)} then
7  |  |  |  Set S_0^G[N − ρ + y] = z_{ρN+y};
   |  |  end
8  |  |  else
9  |  |  |  if Next[z_{(ρ+1)N+(y+1)}] = −1 then
10 |  |  |  |  Set Next[z_{(ρ+1)N+(y+1)}] = z_{ρN+y};
   |  |  |  end
11 |  |  |  else if Next[z_{(ρ+1)N+(y+1)}] ≠ z_{ρN+y} then
12 |  |  |  |  Set NumConflicts = NumConflicts + 1;
13 |  |  |  |  Set Conflict[NumConflicts].value = z_{(ρ+1)N+(y+1)};
14 |  |  |  |  Set Conflict[NumConflicts].first = Next[z_{(ρ+1)N+(y+1)}];
15 |  |  |  |  Set Conflict[NumConflicts].second = z_{ρN+y};
   |  |  |  end
   |  |  end
   |  end
   end
```

Algorithm 8.4.2: PartResolvePerm

Remark 8.4.13. *At the point j^G gets stuck, the permutation S^G can be considered to be a random permutation. Thus each byte of the first run, after j^G got stuck, can be assumed to be chosen uniformly at random from $\{0, \ldots, N-1\}$. Since we use sets of two consecutive runs for collecting the* candidate pairs, *the values in each such pair can also be regarded uniformly random for estimating the expected numbers of distinct* candidate pairs *and* conflict sets. *Extensive experimentations also confirm this randomness assumption, which is followed in the technical results in the rest of this section.*

Theorem 8.4.14. *The expected number of unassigned entries in the array* Next *after the execution of the PartResolvePerm algorithm is $N \cdot (\frac{N-1}{N})^{(R-1)(N-2)}$.*

Proof: The *candidate pairs* are of the form

$$(z_{(\rho+1)N+(y+1)}, z_{\rho N+y}), \qquad 0 \le \rho \le R-2, 1 \le y \le N-2.$$

Thus, each distinct value of

$$z_{\rho N+y}, \qquad 0 \le \rho \le R-2, 1 \le y \le N-2,$$

would give rise to a *candidate pair* and hence assign exactly one entry of the array *Next*.

Let $x_u = 1$, if the value u does not occur in any of the $(R-1)(N-2)$ many keystream bytes $z_{\rho N+y}$, $0 \le \rho \le R-2$, $1 \le y \le N-2$; otherwise, let $x_u = 0$, $0 \le u \le N-1$. Hence, the total number of values that did not occur in those keystream bytes is given by $X = \sum_{u=0}^{N-1} x_u$. Assuming that each keystream byte is uniformly randomly distributed in $\{0, \ldots, N-1\}$, we have

$$P(x_u = 1) = \left(\frac{N-1}{N}\right)^{(R-1)(N-2)}.$$

Thus,

$$E(x_u) = \left(\frac{N-1}{N}\right)^{(R-1)(N-2)}$$

and

$$
\begin{aligned}
E(X) &= \sum_{u=0}^{N-1} E(x_u) \\
&= N \cdot \left(\frac{N-1}{N}\right)^{(R-1)(N-2)}.
\end{aligned}
$$

∎

Corollary 8.4.15. *The expected number of distinct* candidate pairs *after the execution of the PartResolvePerm algorithm is $N \cdot \left(1 - (\frac{N-1}{N})^{(R-1)(N-2)}\right)$.*

Theorem 8.4.16. *The expected number of conflict sets after the execution of the PartResolvePerm algorithm is bounded by* $(R-1) \cdot (\frac{N-2}{N})$.

Proof: The *candidate pairs* are of the form $(z_{(\rho+1)N+(y+1)}, z_{\rho N+y})$ and the corresponding output indices are

$$t_{\rho,y} = a_{N-\rho+y-1} + a_{N-\rho+y}, \qquad 0 \le \rho \le R-2, 1 \le y \le N-2.$$

According as item 2 of Theorem 8.4.12, if $t_{\rho,y} = y+1$, then

$$z_{\rho N+y} = suc^2(z_{(\rho+1)N+(y+1)}),$$

but due to Step 10 of the *PartResolvePerm* algorithm, $z_{\rho N+y}$ is wrongly recorded as $suc^1(z_{(\rho+1)N+(y+1)})$. For $0 \le \rho \le R-2$, $1 \le y \le N-2$, let $x_{\rho,y} = 1$ if $t_{\rho,y} = y+1$; otherwise, let $x_{\rho,y} = 0$. Hence, the total number of wrong entries in the array $Next$ after the execution of the *PartResolvePerm* algorithm is given by

$$X = \sum_{\rho=0}^{R-2} \sum_{y=1}^{N-2} x_{\rho,y}.$$

Each wrong entry in $Next$ is a potential contributor to one *conflict set*, i.e., $NumConflicts \le X$. Assuming that each output index is uniformly randomly distributed, we have $P(x_{\rho,y} = 1) = \frac{1}{N}$. Thus, $E(x_y) = \frac{1}{N}$ and

$$
\begin{aligned}
E(X) &= \sum_{\rho=0}^{R-2} \sum_{y=1}^{N-2} E(x_{\rho,y}) \\
&= (R-1) \cdot \left(\frac{N-2}{N} \right).
\end{aligned}
$$

∎

Given a *conflict set* $\{(u, v_1), (u, v_2)\}$, if we are able to locate v_1, v_2 in a *candidate pair*, the conflict is resolved and the exact order of u, v_1, v_2 is known. Using this observation, we can devise an algorithm *ResolveConflicts* which takes as input a partially resolved permutation $\mathcal{P}_1$ over $\mathbb{Z}_N$ and a collection of conflict sets and produces as output another partially resolved permutation $\mathcal{P}_2$ such that $\mathcal{P}_2 \ge \mathcal{P}_1$.

Even after R many *runs* of the PRGA, the permutation may still remain *partially resolved*. Then we may exhaustively fill in the remaining unassigned entries in the array $Next$ to form all possible *resolved permutations*. By running StuckPRGA on each resolved permutation in turn, we can determine its first element and thereby recover the entire permutation.

Lemma 8.4.17. *If the initial permutation S_0^G becomes resolved at any stage of the RC4 PRGA, then S_0^G can be retrieved completely in $O(N)$ average time.*

Proof: Suppose one runs PRGA for M rounds starting with an arbitrary permutation $\mathcal{P}$ over $\mathbb{Z}_N$. Assuming that the keystream output bytes are

Input:

    1. A partially resolved permutation $\mathcal{P}_1$ over $\mathbb{Z}_N$ in the form of an array $Next$.

    2. A set of conflict pairs in an array $Conflict$.

Output: A partially resolved permutation $\mathcal{P}_2 \geq \mathcal{P}_1$ in the form of the array $Next$.

```
1  for u = 1 to NumConflicts do
2      for ρ = 0 to R − 2 do
3          for y = 1 to N − 2 do
4              if Conflict[u].first = z_(ρ+1)N+(y+1)  and
5              Conflict[u].second = z_ρN+y  then
6                  Set Next[Conflict[u].value] = Conflict[u].first;
7                  Set
                   Next[Next[Conflict[u].value]] = Conflict[u].second;
               end
8              if Conflict[u].second = z_(ρ+1)N+(y+1)  and
9              Conflict[u].first = z_ρN+y  then
10                 Set Next[Conflict[u].value] = Conflict[u].second;
11                 Set
                   Next[Next[Conflict[u].value]] = Conflict[u].first;
               end
           end
       end
   end
```

Algorithm 8.4.3: ResolveConflicts

uniformly randomly distributed, the probability that the set of M random keystream bytes obtained by running PRGA on $\mathcal{P}$ would match with the M keystream bytes in hand (obtained by running PRGA on S_0^G) is $\frac{1}{N^M}$. With $N = 256$, a small value of M such as $M = 8$ yields a negligibly small value $\frac{1}{2^{64}}$ (close to 0) of this probability. Thus, running PRGA on $\mathcal{P}$ for only 8 rounds, with almost certainty one would be able to determine if that permutation indeed was the original permutation S_0^G.

Now, suppose $\mathcal{P}$ is a resolved permutation. So for any arbitrary element in $\mathcal{P}$, we know what is its successor. Starting from any element u as the first element, if we write all the elements in sequence, then we get a permutation

$$\mathcal{Q} = <u, suc^1(u), suc^2(u), \ldots, suc^{N-1}(u) >$$

such that

$$\mathcal{P} = rot^n(\mathcal{Q}) \qquad \text{for some } n, \, 0 \le n \le N - 1.$$

We run PRGA starting with the initial permutation once as $\mathcal{Q}$, next as $rot^1(\mathcal{Q})$, next as $rot^2(\mathcal{Q})$, and so on, until the first 8 keystream bytes match with the observed keystream bytes in hand. With at most N such trials, the entire permutation can be constructed. ∎

The above Lemma immediately gives an algorithm *ConstructPerm* to construct S_0^G from a partially resolved permutation.

Input: A partially resolved permutation $\mathcal{P}$ in the form of an array $Next$.

Output: The original permutation S_0^G before the PRGA begins.

1 Set $m =$ the number of unassigned (i.e., -1) entries in the array $Next$;

2 **for** *each possible assignment of those m entries* **do**

3 $\quad$ Get the corresponding *resolved permutation T*;

4 $\quad$ **for** $n = 0$ *to* $N - 1$ **do**

5 $\quad\quad$ Run StuckPRGA starting with $rot^n(T)$ as the initial permutation for 8 rounds and generate the first 8 keystream bytes;

6 $\quad\quad$ **if** *the above 8 keystream bytes match with the first 8 keystream bytes obtained in the actual execution of StuckPRGA* **then**

7 $\quad\quad\quad$ Return $S_0^G = rot^n(T)$;

$\quad\quad$ **end**

$\quad$ **end**

end

Algorithm 8.4.4: ConstructPerm

The procedures *PartResolvePerm*, *ResolveConflicts* and *ConstructPerm* can be combined to devise an efficient algorithm *InvertStuckPRGA* to retrieve the initial permutation S_0^G from the first few *runs* of keystream output bytes generation.

Input: The RN many keystream output bytes from the first
$R(\geq 2)$ *runs* of the PRGA.
Output: The original permutation S_0^G before the PRGA begins.

1 Run *PartResolvePerm* with the given keystream bytes and generate the arrays $Next$ and $Conflict$;
2 Run *ResolveConflicts* on $Next$;
3 Run *ConstructPerm* on updated $Next$;

Algorithm 8.4.5: InvertStuckPRGA

Theorem 8.4.18. *The average case time complexity of the InvertStuckPRGA algorithm is* $O\left(\left(R^2 + \lceil A \rceil !\right)N\right)$, *where* $A = N \cdot \left(\frac{N-1}{N}\right)^{(R-1)(N-2)}$.

Proof: The time complexity of Step 1 in *InvertStuckPRGA* is $O(RN)$, since there are two nested "for" loops in *PartResolvePerm* of $R - 1$ and $N - 2$ many iterations respectively and one execution of the steps inside the "for" loops takes $O(1)$ time.

The time complexity of Step 2 in *InvertStuckPRGA* is $O(R^2 N)$, since from Theorem 8.4.16, the average value of *NumConflicts* is $O(R)$ and resolving each of them in the two nested "for" loops in *ResolveConflicts* takes $O(RN)$ time.

According to Theorem 8.4.14, just before the execution of *ConstructPerm* in Step 3 of *InvertStuckPRGA*, the average value of the number m of unassigned entries in the array $Next$ is given by

$$A = N \cdot \left(\frac{N-1}{N}\right)^{(R-1)(N-2)}.$$

Hence the "for" loop in Step 2 of *ConstructPerm* is iterated $\lceil A \rceil !$ times on the average. Again, from Lemma 8.4.17, the time complexity of each iteration of the "for" loop in Step 2 of *ConstructPerm* is $O(N)$. Hence the overall complexity of Step 3 in *InvertStuckPRGA* is $O(\lceil A \rceil !N)$.

Thus, the time complexity of *InvertStuckPRGA* is $O\left(\left(R^2 + \lceil A \rceil !\right)N\right)$. ∎

Remark 8.4.19. *If* $z_{\rho N + y} = z_{(\rho+1)N + (y+1)}$ *for some* $\rho \geq 0$ *and some* $y \in \{1, \ldots, N - 2\}$, *i.e., if the two values in a candidate pair turn out to be equal, then we would have* $S_0^G[N - \rho + y] = z_{\rho N + y}$, *according to item 1 of Theorem 8.4.12. Once the position of any one entry of a resolved permutation is known, the positions of all other entries are immediately known. If one makes use of this fact in the* PartResolvePerm *algorithm, then rotating* T *in the* ConstructPerm *algorithm is not required at all. One can run* StuckPRGA *for 8 rounds on* T *itself to check whether* $T = S_0^G$ *or not. In that case, the* *average case time complexity of the* InvertStuckPRGA *algorithm would be reduced to* $O\left(R^2 N + \lceil A \rceil!\right)$. *However,* i^G *must be known to have this advantage. Since in general* i^G *would not be known to the attacker (see Section 8.4.3 for details), item 1 of Theorem 8.4.12 is not used in the state recovery strategy.*

The quantity $log_2(\lceil A \rceil!)$ may be interpreted as a measure of uncertainty in resolving the permutation. For $N = 256$, Table 8.2 lists the values of A and $log_2(\lceil A \rceil!)$ for some values of R. Observe that if all the successors are unresolved (e.g. the case $R = 1$), the uncertainty is $log_2(256!) \approx 1683.9961$.

Runs R	Avg. No. of Elements with Unassigned Successors A	Uncertainty in Permutation $log_2(\lceil A \rceil!)$
1	256	1684
2	94.73	491.69
3	35.05	138.09
4	12.97	32.54
5	4.80	6.91
6	1.78	1.00
7	0.66	0.00
8	0.24	0.00

TABLE 8.2: Decrease in uncertainty of resolving the permutation with increasing *runs*

The data indicates that as one considers more number of *runs*, the uncertainty in resolving the permutation is reduced. In fact, the uncertainty starts decreasing when only the first 258 keystream output bytes are available, as z_1 and z_{258} come from the same index $a_0 + a_1$ (see Table 8.1) and form a *candidate pair*. Table 8.2 also shows that theoretically 7 *runs* are enough to shrink the uncertainty in resolving the permutation to zero and recover the permutation in $O(R^2 N)$ time. However, empirical results confirm that 8 *runs* provide a conservative estimate, after which there is no uncertainty in resolving the permutation. Zero uncertainty implies that the permutation obtained after *ResolveConflicts* in Step 2 of *InvertStuckPRGA* algorithm is a *resolved permutation*. In this case, the time complexity of *InvertStuckPRGA* reduces to $8^2 \cdot 256 = 2^{14}$.

The permutation can still be completely recovered, even if fewer numbers of keystream bytes are available. The price to be paid is an increase in time complexity due to exhaustive assignment of successor elements. For example, when $R = 4$, i.e., when we start with the first 1024 keystream output bytes, the average number of permutation elements whose next entries are not determined is $256 \cdot \left(\frac{255}{256}\right)^{3*254} = 12.97$. If we exhaustively fill in these $\lceil 12.97 \rceil = 13$ successor values, we would generate $13! \approx 2^{32}$ possible resolved permutations. The time complexity of complete permutation recovery with 1024 keystream bytes would be around $(4^2 + 2^{32}) \cdot 256 \approx 2^{48}$.

Interestingly, if we go for $R = N$ *runs*, i.e., if we have a total of N^2 ($= 2^{16}$ for $N = 256$) many keystream output bytes, we can construct the permutation in just $O(N)$ time. From Lemma 8.4.11, when $\rho = N$, we have

$$S_{\rho N}^G = <a_0, a_1, a_2, \ldots, a_{N-1}> = S_0^G.$$

In this case, after the first N *runs*, the structure of the permutation, the output indices as well as the keystream output bytes start repeating in the same order. Considering the keystream output bytes coming from a fixed index $a_{y-1} + a_y$, i.e., the values $z_{\rho N + \rho + y}$ for a fixed y, $0 \leq \rho \leq N - 1$, we can readily get a resolved permutation, without requiring to perform exhaustive assignments.

8.4.3   State Recovery when Both Indices Are Unknown

So far we have considered that the value at which j^G is stuck is zero. All the results above can easily be extended when j^G is stuck at any value $x \in \{0, \ldots, N - 1\}$. Suppose r_{st} is the round at which j^G gets stuck at x for the first time and from round r_{st} onwards it remains stuck at x, $r_{st} \geq 1$. The value of the deterministic index i^G at round r_{st} is $i_{st} = r_{st} \bmod N$. Then after $d = (x - i_{st}) \bmod N$ more rounds, i.e., at round $r_{st} + d + 1$, i^G becomes $x + 1$ for the first time after j^G got stuck. We can denote the indices of the permutation and the corresponding values after the end of round $r_{st} + d$ as follows.

Indices	0	1	...	$x-1$	x	$x+1$	...	$N-2$	$N-1$
Values	b_{N-x}	b_{N-x+1}	...	b_{N-1}	b_0	b_1	...	b_{N-2-x}	b_{N-1-x}

Thus, we have the following result.

Proposition 8.4.20. *"The evolution of the permutation from round $r_{st}+d+1$ onwards, given j^G stuck at x from round r_{st}" is analogous to "the x-rotation of the permutation evolving from round 1 onwards, given j^G stuck at 0 from round 1."*

Suppose that the attacker can access the keystream bytes from the point when j^G got stuck. Because of the cyclic pattern mentioned in Proposition 8.4.20, and because of the relative gap of $N + 1$ rounds between the elements of any *candidate pair* (as discussed following Theorem 8.4.12 in Section 8.4.1), the attacker does not need to know r_{st} or i_{st}. If the attacker starts counting the first *run* from the point when he has the first keystream byte at his disposal, he can efficiently recover the permutation at the point when j^G got stuck. In the subsequent analysis, we assume that the attacker does not know

1. the round r_{st} from which j^G got stuck,

2. the value i_{st} of the deterministic index i^G when j^G got stuck and

3. the value x at which j^G remains stuck.

We here sketch the modification corresponding to Step 5 of the algorithm *PartResolvePerm* as well as Step 3 of algorithm *ResolveConflicts*), where y varies from 1 to $N-2$. In Section 8.4.1, j^G was assumed to be stuck at 0 from round $r_{st} = 1$, when i_{st} was also 1. Thus, it was already known in advance that the *candidate pairs* $(z_{(\rho+1)N+(y+1)}, z_{\rho N+y})$, for $y = N-1, N$ (i.e., when the deterministic index i^G takes the values $j^G - 1, j^G$), should be ignored. On the contrary, here both r_{st} and i_{st} are unknown. In order to process the keystream bytes using the algorithms *PartResolvePerm* and *ResolveConflicts*, we need to initialize ρ to 0 at the point j^G gets stuck and from that point onwards the keystream bytes are named as $z_1, z_2, \ldots$ and so on. Given j^G is stuck at an unknown value x, when the deterministic index i^G would take the values $x-1$ and x, the two corresponding *candidate pairs* should be ignored. However, these two cases cannot be identified and eliminated here, as x is not known. Hence, in Step 5 of the algorithm *PartResolvePerm* and Step 3 of algorithm *ResolveConflicts*, we should consider that y varies from 1 to N for each value of $\rho \in \{0, 1, \ldots, R-3\}$ and y varies from 1 to $N-1$ for $\rho = R-2$.

In this modified approach, utilizing all the RN keystream bytes yields a total of $(R-1)N-1$ *candidate pairs*. Following the same line of arguments as in Theorem 8.4.14, we get the expected number of unassigned entries in the array *Next* just before the execution of Step 3 of *InvertStuckPRGA* as

$$A' = N \cdot \left(\frac{N-1}{N}\right)^{(R-1)N-1}$$

which, for large N (such as $N = 256$), is approximately equal to

$$A = N \cdot \left(\frac{N-1}{N}\right)^{(R-1)(N-2)}.$$

Since we get two additional wrong entries in each of the $R-1$ *runs*, the number of *conflict sets* would increase by $2(R-1)$ from the value estimated in Theorem 8.4.16. Thus, the modified upper bound on the expected number of *conflict sets* after the execution of the *PartResolvePerm* algorithm, when the *candidate pairs* are formed by considering all the keystream bytes in each run, is given by $(R-1) \cdot (2 + \frac{N-2}{N})$. Observe that the bound is still $O(R)$ as in Theorem 8.4.16. However, since i_{st} and x are not known, we need to run Step 4 of the *ConstructPerm* Algorithm for each possible value of i_{st} and x in $\{0, \ldots, N-1\}$, until Step 6 of *ConstructPerm* reveals the true initial permutation. Thus, we need at most N^2 executions of Step 4 of *ConstructPerm* for each of the $\lceil A \rceil!$ resolved permutations, where

$$A = N \cdot \left(\frac{N-1}{N}\right)^{(R-1)(N-2)}.$$

Thus, when i_{st} and x are unknown, the average case time complexity of *ConstructPerm* is given by $\lceil A \rceil ! N^3$, and following the same analysis as in Theorem 8.4.18, the average case time complexity of *InvertStuckPRGA* is given by $O\left(R^2 N + \lceil A \rceil ! N^3\right)$. Plugging in $N = 256$ and $R = 8$, we get $A = 0$ and the time complexity becomes

$$8^2 \cdot 256 + 256^3 \approx 2^{25}.$$

8.4.4 Detecting the Stuck-At Fault

The analysis so far assumes that the keystream bytes from the point when j^G got stuck is available to the attacker. In practice, the attacker has access to the keystream output bytes only. By analyzing the keystream, he should be able to determine the interval during which j^G remains stuck at a certain value. After the fault detection, he can run the *InvertStuckPRGA* algorithm on the keystream bytes obtained during that interval.

Theoretically determining the exact point when j^G gets stuck appears to be extremely tedious. However, we describe a simple test which can distinguish between a sequence of normal RC4 keystream bytes when j^G is pseudo-randomly updated and a sequence of RC4 keystream bytes when j^G is stuck. The following theorem shows that the number of *conflict sets* from the *PartResolvePerm* algorithm when j^G is pseudo-randomly updated is much more than that when j^G is stuck at a certain value.

Theorem 8.4.21. *Assume that the index j^G is pseudo-randomly updated in each round of the PRGA. Then the expected number of conflict sets after the execution of the PartResolvePerm algorithm, when the candidate pairs are formed by considering all the keystream bytes in each run, is bounded by* $(R-1) \cdot (N-1) - \frac{N-1}{N}$.

Proof: The *candidate pairs* are of the form $(z_{(\rho+1)N+(y+1)}, z_{\rho N+y})$ and the corresponding output indices are

$$t_{\rho,y} = a_{N-\rho+y-1} + a_{N-\rho+y},$$

where the range of y depends on ρ as follows. The range is $1 \le y \le N$ for $0 \le \rho \le R - 3$ and it is $1 \le y \le N - 1$ for $\rho = R - 2$.

Let $x_{\rho,y} = 1$ if

$$z_{\rho N+y} \ne suc^1(z_{(\rho+1)N+(y+1)});$$

otherwise, let $x_{\rho,y} = 0$. Then the total number of wrong entries in the array *Next* after the execution of the *PartResolvePerm* algorithm is given by

$$X = \sum_{\rho=0}^{R-3} \sum_{y=1}^{N} x_{\rho,y} + \sum_{y=1}^{N-1} x_{R-2,y}.$$

Each wrong entry in $Next$ is a potential contributor to one *conflict set*, i.e.,

$$NumConflicts \leq X.$$

Assuming that the index j^G is pseudo-randomly updated and each output index is uniformly randomly distributed, we have $P(x_{\rho,y} = 1) = \frac{N-1}{N}$. Thus, $E(x_{\rho,y}) = \frac{N-1}{N}$ and

$$
\begin{aligned}
E(X) &= \sum_{\rho=0}^{R-3} \sum_{y=1}^{N} E(x_{\rho,y}) + \sum_{y=1}^{N-1} E(x_{R-2,y}) \\
&= (R-1) \cdot (N-1) - \frac{N-1}{N}.
\end{aligned}
$$

∎

Thus, the attacker can run the *PartResolvePerm* algorithm once on the sequence of available keystream bytes and count the number of *conflict sets*. *Numconflicts* would be $O(R)$ if j^G is stuck during the interval when those keystream bytes were generated; otherwise, *Numconflicts* will be $O(RN)$.

Research Problems

Problem 8.1 Can Algorithm 8.2.1 be improved to detect Finney cycles more efficiently?

Problem 8.2 Can a fault attack be mounted if instead of the whole byte of j, a single of j is stuck?

Problem 8.3 Can a fault attack be mounted if the attacker can choose specific bits of j to inject faults?

Chapter 9

Variants of RC4

We already discussed many cryptanalytic results on RC4 KSA and PRGA in the previous chapters. In order to remove the weaknesses and improve the cipher, many modifications and variants of RC4 have been proposed so far. In this chapter, we outline these works.

9.1 Byte-Oriented Variants

At FSE 2004, two variants of RC4 were proposed. One is VMPC and the other is RC4A. VMPC [192] is a generalization of RC4. The key scheduling of VMPC transforms a secret key K of length l bytes (typically, $16 \leq l \leq 64$) into a permutation S of $\mathbb{Z}_N$ and initializes a variable j. An optional IV of length l' may be used. Another 16-bit variable m is used as a counter.

The name VMPC comes from *Variably Modified Permutation Composition* which is a transformation of permutation P into the permutation Q. A k-level VMPC function is defined as

$$Q[x] = P[P_k[P_{k-1}[\ldots[P_1[P[x]]]\ldots]]], 0 \leq x \leq N - 1,$$

where $P_i[x] = (P[x] + i) \bmod N$. VMPC cipher uses this transformation in its PRGA to output the keystream bytes and update S through shuffle-exchanges.

RC4A [139] tries to increase the security of RC4 by using two S-boxes S_1 and S_2. The key scheduling of RC4A is the same as the RC4 KSA, except that it uses two different secret keys two construct the two S-boxes. Hence we do not give the RC4A KSA separately. The keystream generation part is in the same line of RC4 PRGA. Two pseudo-random indices j_1 and j_2 are used corresponding to S_1 and S_2 respectively to update them through shuffle-exchanges. The only modification is that the index $S_1[i] + S_1[j]$ evaluated on S_1 produces output from S_2 and vice-versa. The RC4A PRGA is given in Algorithm 9.1.3

RC4A uses fewer CPU cycles per keystream output byte than RC4. To produce two successive output bytes, the index i is incremented once in case of RC4A, but twice in RC4.

The first distinguishing attacks on VMPC and RC4A appeared in [113]. According to [113], the distinguisher for VMPC requires around 2^{54} keystream

Input:

    1. Secret key array $\kappa[0 \ldots (l-1)]$.

    2. (Optional) Initialization Vector $IV[0 \ldots (l-1)]$.

Output: Scrambled permutation array $S[0 \ldots N-1]$.

Initialization:
for $i = 0, \ldots, N-1$ **do**
    | $S[i] = i;$
    | $j = 0;$
end

Scrambling:
for $m = 0, \ldots, 3N-1$ **do**
    | $i = m \bmod N;$
    | $j = S[j + S[i] + \kappa[m \bmod l]];$
    | Swap$(S[i], S[j]);$
end

Scrambling with IV (if IV is used):
for $m = 0, \ldots, 3N-1$ **do**
    | $i = m \bmod N;$
    | $j = S[j + S[i] + IV[m \bmod l']];$
    | Swap$(S[i], S[j]);$
end

Algorithm 9.1.1: VMPC KSA

Input:

    1. A key-dependent scrambled permutation $S[0 \ldots N-1]$.

    2. A key-dependent index j.

Output: Pseudo-random keystream bytes z.

Initialization:
$i = 0;$

Output Keystream Generation Loop:
$j = S[j + S[i]];$
$t = S[i] + S[j];$
Output $z = S[S[S[j]] + 1];$
Swap$(S[i], S[j]);$
$i = i + 1;$

Algorithm 9.1.2: VMPC PRGA

Input: Two key-dependent scrambled permutations $S_1[0 \ldots N-1]$ and
$\quad\quad$ $S_2[0 \ldots N-1]$.
Output: Pseudo-random keystream bytes.

Initialization:
$i = j_1 = j_2 = 0;$

Output Keystream Generation Loop:
$i = i + 1;$
$j_1 = j_1 + S_1[i];$
$\mathrm{Swap}(S_1[i], S_1[j_1]);$
$t_1 = S_1[i] + S_1[j_1];$
$Output = S_2[t_1];$
$j_2 = j_2 + S_2[i];$
$\mathrm{Swap}(S_2[i], S_2[j_2]);$
$t_2 = S_2[i] + S_2[j_2];$
$Output = S_1[t_2];$

Algorithm 9.1.3: RC4A PRGA

---

bytes and that for RC4A requires 2^{58} keystream bytes. Shortly, better attacks based on the non-uniformity of the first few keystream bytes were reported in [182]. These improved distinguishers for VMPC and RC4A require 2^{38} and 2^{23} keystream bytes respectively. Subsequently, a corrected version [114] of the paper [113] appeared and reported a data complexity of 2^{40} keystream bytes for distinguishing VMPC from random stream. It is not surprising that the *ABTAB* distinguisher [109] for RC4 described in Section 6.3.1 also works for RC4A.

Several variants of RC4 and RC4-like ciphers have been described in [118], such as Chameleon and RC4B. These are less popular in terms of practical use and we omit their descriptions here.

9.2 Word-Oriented Variants

In [118, Chapter 6], two new ciphers, named Sheet Bend and Bowline, were developed by expanding RC4 to 32 bits. The work [126] proposed a generalization of RC4 with an aim to expand RC4 to 32/64 bits with a state size much smaller than 2^{32} or 2^{64}. The new algorithm is called RC4(n, m), where $N = 2^n$ is the size of the state array in words, m is the word size in bits, $n \le m$. Later, the name NGG was adopted for this cipher after the initials of its designers. NGG KSA and PRGA update the indices i, j in the same way as do the RC4 KSA and PRGA respectively. In the NGG KSA, the array S is initialized to a precomputed random array a. Then, $S[i]$ and $S[j]$

are swapped and the sum of these two elements (modulo $M = 2^m$) is assigned
to $S[i]$.

Input:

    1. Secret key array $K[0 \ldots N - 1]$.

    2. Precomputed random array $a[0 \ldots N - 1]$.

Output: Scrambled array $S[0 \ldots N - 1]$.

Initialization:
for $i = 0, \ldots, N - 1$ **do**
    $S[i] = a_i$;
    $j = 0$;
end

Scrambling:
for $i = 0, \ldots, N - 1$ **do**
    $j = (j + S[i] + K[i]) \bmod N$;
    Swap($S[i], S[j]$);
    $S[i] = (S[i] + S[j]) \bmod M$;
end

Algorithm 9.2.1: NGG KSA

In the PRGA phase, a pseudo-random element is sent to output and immediately after that it is changed by an addition (modulo M) of two other elements from the array S.

Input: Key-dependent scrambled array $S[0 \ldots N - 1]$.
Output: Pseudo-random keystream bytes z.

Initialization:
$i = j = 0$;

Output Keystream Generation Loop:
$i = (i + 1) \bmod N$;
$j = (j + S[i]) \bmod N$;
Swap($S[i], S[j]$);
Output $z = S[((S[i] + S[j]) \bmod M) \bmod N]$;
$S[((S[i] + S[j]) \bmod M) \bmod N] = (S[i] + S[j]) \bmod M$;

Algorithm 9.2.2: NGG PRGA

Another version of NGG was introduced in [63] and came to be known as the GGHN cipher. In GGHN, in addition to i, j, a third variable k is used to increase the security of the cipher. k is initialized in the KSA and is key dependent. Number of KSA loop repetitions r depends on the parameters n, m. For example, for $n = 8$ and $m = 32$, r is taken to be 20.

Input:

    1. Secret key array $K[0 \ldots N-1]$.

    2. Precomputed random array $a[0 \ldots N-1]$.

Output:

    1. Scrambled array $S[0 \ldots N-1]$.

    2. Key-dependent secret variable k.

Initialization:
for $i = 0, \ldots, N-1$ **do**
    $S[i] = a_i;$
    $j = k = 0;$
end

Scrambling:
repeat
    for $i = 0, \ldots, N-1$ **do**
        $j = (j + S[i] + \kappa[i \bmod l]) \bmod N;$
        Swap$(S[i], S[j]);$
        $S[i] = (S[i] + S[j]) \bmod M;$
        $k = (k + S[i]) \bmod M;$
    end
until r *iterations* ;

Algorithm 9.2.3: GGHN KSA

In the PRGA, k is used for updating S as well as for masking the output.

Input:

    1. Key-dependent scrambled array $S[0 \ldots N-1]$.

    2. Key-dependent secret variable k.

Output: Pseudo-random keystream bytes z.

Initialization:
$i = j = 0;$

Output Keystream Generation Loop:
$i = (i+1) \bmod N;$
$j = (j + S[i]) \bmod N;$
$k = (k + S[j]) \bmod M;$
Output $z = S\left[(S[i] + S[j]) \bmod N + k)\right] \bmod M;$
$S\left[(S[i] + S[j]) \bmod N\right] = (k + S[i]) \bmod M;$

Algorithm 9.2.4: GGHN PRGA

A distinguished attack on NGG using only 100 keystream words appeared in [186]. Later, GGHN was attacked in [183] that built a distinguisher based on a bias in the first two words of the keystream, associated with approximately 2^{30} secret keys.

In 2005, an RC4-like stream cipher Py (pronounced Roo) was submitted [17] to the eSTREAM project. The cipher produces two 32-bit words as output at each step. A building block of the cipher is *rolling arrays*. The designers of Py define the rolling array to be a vector whose entries are cyclically rotated by one place at every step. Py uses one word variable s and two rolling arrays P and Y, where P is a permutation over $\mathbb{Z}_{256}$, indexed by $0, \ldots, 255$, and Y is a 260-word array of 32-bit words, indexed by $-3, \ldots, 256$. The key schedule of Py initializes the variable s and the array Y from the secret key. We omit the key scheduling here and discuss only the keystream generation. Interested readers may look into [17] for further details.

In [17], a variant of Py, called Py6, was also presented, where the permutation P is reduced to be over $\mathbb{Z}_{64}$ (i.e., each entry is of 6 bits) and the size of Y is reduced to 68 entries, each being a 32-bit word. The speed of Py6 is essentially the same as Py. The smaller size of the internal state allows only faster key and IV setup.

In 2006, a distinguishing attack on Py was reported [140]. It requires $2^{84.7}$ randomly chosen key/IVs and the first 24 output bytes for each key. To resist this attack, the designers of Py came up with a variant, called Pypy [18]. However, a series of papers [77, 189–191] revealed several weaknesses of the Py family of ciphers and the cipher could not reach the final round of eSTREAM. Again the designers of Py came up with some variants [19], tweaking the IV setup, but these variants were also attacked [151–154].

Input:

1. Key-dependent variable s.

2. Key-dependent scrambled array $Y[-3\ldots256]$.

Output: Pseudo-random keystream words z.

Swap and Rotate P:
Swap($P[0]$,$P[Y[185]$ & $0xFF]$);
Rotate(P);

Update s:
$s+=Y[P[72]]-Y[P[239]]$;
$s=ROTL32\,(s,((P[116]+18)\ \&\ 31))$;

Output 8 bytes (from LSB to MSB order):
Output $(ROTL32(s,25)\oplus Y[256])+Y[P[26]]$;
Output $(s\oplus Y[-1])+Y[P[208]]$;

Update and Rotate Y:
$Y[-3]=(ROTL32(s,14)\oplus Y[-3])+Y[P[153]]$;
Rotate(Y);

Algorithm 9.2.5: Py PRGA (Single step)

9.3 RC4-Based Hash

In [28], a hash-function based on RC4 was proposed. This hash function, named RC4-Hash, produces a variable length output from 16 bytes to 64 bytes.

RC4-Hash is parameterized by l and works as a function

$$RCH_l : \{0,1\}^{<2^{64}} \to \{0,1\}^{8l}.$$

The message M (with length at most $2^{64}-1$) is padded as follows:

$$pad(M) = bin_8(l)\|M\|1\|0^k\|bin_{64}(|M|),$$

where $k \geq 0$ is the least integer such that

$$8+|M|+1+k+64 = 0 \bmod 512.$$

One may write

$$pad(M) = M_1\|\cdots\|M_t,$$

where each M_i is of size 512 bits. S^{IV} is a permutation of $\mathbb{Z}_{256}$ and for $i \in [0, 255]$, $r(i)$ gives a number in $[0, 63]$. S^{IV} and $r(i)$'s are specified in [28,

Appendix]. Let $(S_0, j_0) = (S^{IV}, 0)$ be the initial state. A compression function C defined in Algorithm 9.3.1 is invoked iteratively as follows:

$$(S_0, j_0) \overset{M_1}{\to} (S_1, j_1) \overset{M_2}{\to} \cdots (S_{t-1}, j_{t-1}) \overset{M_t}{\to} (S_t, j_t),$$

where $(S, j) \overset{m}{\to} (S', j')$ means

$$C((S, j), m) = (S', j').$$

Input:

    1. $S[0 \ldots 255]$: a permutation over $\mathbb{Z}_{256}$.

    2. j: an integer in $\mathbb{Z}_{256}$.

    3. m: a 512-bit binary string.

Output:

    1. The scrambled permutation $S[0 \ldots 255]$.

    2. Updated j.

for $i = 0$ *to* 255 **do**
    | $j = (j + S[i] + m[r(i)]) \bmod 256$;
    | Swap($S[i]$,$S[j]$);
end
Return(S, j);

Algorithm 9.3.1: RC4-Hash: The compression function $C((S, j), m)$

After t iterations, the following two post-processing operations are performed in sequence.

1. Compute $S_{t+1} = S_0 \circ S_t$ and $j_{t+1} = j_t$.

2. The final hash value is given by

$$RCH_l(M) = HBG_l(OWT(S_{t+1}, j_{t+1})),$$

where OWT and HBG_l are defined in Algorithms 9.3.2 and 9.3.3 respectively.

Later, collisions for RC4-Hash were reported in [76]. This work reveals that finding a set of $2m$ colliding messages has an expected cost of $2^7 + m \cdot 2^8$ many compression function evaluations.

Input:

1. $S[0\ldots255]$: a permutation over $\mathbb{Z}_{256}$.

2. j: an integer in $\mathbb{Z}_{256}$.

Output:

1. The scrambled permutation $S[0\ldots255]$.

2. Updated j.

for $i = 0$ *to 511* do
   $j = (j + S[i \bmod 256]) \bmod 256$;
   Swap($S[i],S[j]$);
end
$Temp2 = S$;
$S = Temp1 \circ Temp2 \circ Temp1$;
Return(S,j);

Algorithm 9.3.2: RC4-Hash: The function $OWT((S,j))$

Input:

1. $S[0\ldots255]$: a permutation over $\mathbb{Z}_{256}$.

2. j: an integer in $\mathbb{Z}_{256}$.

Output: An l-byte hash value.

for $i = 1$ *to l* do
   $j = (j + S[i]) \bmod 256$;
   Swap($S[i],S[j]$);
   $Output = S[(S[i] + S[j]) \bmod 256]$;
end

Algorithm 9.3.3: RC4-Hash: The function $HBG_l((S,j))$

9.4 RC4$^+$

In all the variants we discussed so far, the design is modified to a great extent relative to RC4. In [101], a new variant called RC4$^+$ is presented that keeps the RC4 structure as it is and adds a few more operations to strengthen the cipher. This design attempts to exploit the good points of RC4 and then provide some additional features for a better security margin. The existing literature on RC4 reveals that in spite of having a very simple description, the cipher possesses nice combinatorial structures in the shuffle-exchange paradigm. The design of [101] retains this elegant property of RC4 and at the same time removes the existing weaknesses of RC4.

Section 9.4.1 focuses on the design of the new key scheduling. After a summary of the existing weaknesses of the RC4 KSA, the description of the modified KSA is presented. It is also argued that the new algorithm KSA$^+$ circumvents the weaknesses of the standard RC4 KSA. Section 9.4.2 explains the modification of the RC4 PRGA, called PRGA$^+$. Logical arguments as well as empirical evidence are provided to support the security conjectures of the new design. Finally, Section 9.4.3 presents a study on the software performance of both the KSA$^+$ and PRGA$^+$.

9.4.1 KSA$^+$: Modifications to RC4 KSA

First, let us review the important weaknesses of RC4 KSA, that the new design attempts to get rid of.

1. Roos' biases, i.e., biases of $S_N[y]$ toward f_y for initial values of y are described in detail in Section 3.1.1. In summary, the probability $P(S_N[y] = f_y)$ decreases from 0.37 for $y = 0$ to 0.006 for $y = 48$ and then settles down to 0.0039 ($\approx \frac{1}{256}$).

2. In addition to $S_N[y]$, the nested entries $S_N[S_N[y]]$, $S_N[S_N[S_N[y]]]$, and so on, are also biased to f_y. In particular, they proved that $P(S_N[S_N[y]] = f_y)$ decreases from 0.137 for $y = 0$ to 0.018 for $y = 31$ and then slowly settles down to 0.0039 (beyond $y = 48$). We discuss these results in Section 3.1.3.

3. Recall from Section 4.5.1 that the the inverse permutation entries $S_N^{-1}[y]$, $S_N^{-1}[S_N^{-1}[y]]$, and so on are biased to j_{y+1}, and in turn, to f_y.

4. The above biases and related results can be used to recover the secret key from the final permutation S_N after the KSA, as discussed in Chapter 4 in detail.

5. In RC4 KSA, the update rule is given by $j = (j + S[i] + K[i])$. In Section 3.1.2, it has been demonstrated that for a certain class of update

functions which update j as a function of "the permutation S and j in the previous round" and "the secret key K," one can always construct explicit functions of the key bytes which the permutation at every stage of the KSA would be biased to.

6. Each permutation byte after the KSA is significantly biased (either positively or negatively) toward many values in the range $0, \ldots, N-1$, independent of the secret key. For each y, $0 \le y \le N-2$, $P(S_N[y] = v)$ is maximum at $v = y+1$ and this maximum probability ranges approximately between $\frac{1}{N}(1 + \frac{1}{3})$ and $\frac{1}{N}(1 + \frac{1}{5})$ for different values of y, with $N = 256$. We describe these absolute value biases in Section 3.2.

7. As shown in Section 3.3, the expectation E_v of the number of times each value v in the permutation is visited by the indices i, j is not uniform. E_v decreases from 3.0 to 1.37, as v increases from 0 to 255.

8. Known IV is prepended with the secret key when RC4 is used in WEP and WPA protocols. This mode of use is vulnerable to practical attacks which we cover in Chapter 7.

Let the permutation after the KSA$^+$ be denoted by S_{N+}. The motivations behind the new design according to [101] are as follows.

1. Removal of the existing weaknesses of the KSA.

2. Generation of a random-looking S_{N+} after the key scheduling so that identification of any non-uniformity in it becomes difficult.

3. Recovering the secret key from S_{N+} should be of the same order as exhaustive search.

4. It should be hard to get two secret keys k, k' and two initialization vectors iv, iv' (with at least one of the events $k \ne k'$ and $iv \ne iv'$ holding) that can result in the same S_{N+} after KSA$^+$.

The KSA$^+$ consists of three-layers of key scheduling. The initialization and the basic scrambling in the first layer are the same as those of the original RC4 KSA.

Initialization
For $i = 0, \ldots, N-1$
$\quad S[i] = i;$
$j = 0;$

Layer 1: Basic Scrambling
For $i = 0, \ldots, N-1$
$\quad j = (j + S[i] + K[i]);$
$\quad \mathrm{Swap}(S[i], S[j]);$

In the second layer, the permutation is scrambled further using IVs. According to [73], stream ciphers using IVs shorter than the key may be vulnerable against the Time Memory Trade-Off attack [68]. Keeping this in mind, the IV size in KSA$^+$ is kept the same as the secret key length. The deterministic index i moves first from the middle down to the left end and then from the middle up to the right end. Two copies of an l-byte IV, denoted by an array $iv[0, \ldots, l-1]$, are used as follows. The IV bytes are stored from index $\frac{N}{2} - 1$ down to $\frac{N}{2} - l$ in order from the least significant byte to the most significant byte. The same IV bytes are repeated from index $\frac{N}{2}$ up to $\frac{N}{2} + l - 1$. The former copy is used in the leftward movement of the index i and the latter during the rightward movement. N is assumed to be even, which is the case in standard RC4. For simplicity, an array IV of length N can be declared with $IV[y] = 0$ for those indices which do not contain any IV byte.

$$
IV[y] = \begin{cases} iv[\frac{N}{2} - 1 - y] & \text{for } \frac{N}{2} - l \le y \le \frac{N}{2} - 1; \\ iv[y - \frac{N}{2}] & \text{for } \frac{N}{2} \le y \le \frac{N}{2} + l - 1; \\ 0 & \text{otherwise.} \end{cases}
$$

For $N = 256$ and $l = 16$, this scheme places $16 \times 2 = 32$ many bytes in the middle of the IV array spanning from index 112 to 143. Use of IVs in the above manner has two implications.

1. Repeating the IV bytes creates a dependency so that one cannot choose all the 32 bytes freely to find some weakness in the system, since one byte at the left corresponds to one byte at the right, when viewed symmetrically from the middle of an N-byte array.

2. Moreover, in two different directions, the key bytes are added with the IV bytes in an opposite order. Apart from the $2l$ many operations involving the IV, the rest of $N - 2l$ many operations are without the involvement of IV in Layer 2. This helps in covering the IV values, and chosen IV kind of attacks will be hard to mount.

Layer 2: Scrambling with IV	Layer 3: Zigzag Scrambling
For $i = \frac{N}{2} - 1$ down to 0 $\quad j = (j + S[i]) \oplus (K[i] + IV[i]);$ $\quad \text{Swap}(S[i], S[j]);$ For $i = \frac{N}{2}, \ldots, N-1$ $\quad j = (j + S[i]) \oplus (K[i] + IV[i]);$ $\quad \text{Swap}(S[i], S[j]);$	For $y = 0, \ldots, N-1$ $\quad$ If $y \equiv 0 \bmod 2$ then $\qquad i = \frac{y}{2};$ $\quad$ Else $\qquad i = N - \frac{y+1}{2};$ $\quad j = (j + S[i] + K[i]);$ $\quad \text{Swap}(S[i], S[j]);$

In the third and final layer, more scrambling is performed in a zigzag fashion, where the deterministic index i takes values in the following order: 0, 255, 1, 254, 2, 253, $\ldots$, 125, 130, 126, 129, 127, 128. In general, for y varying from 0 to $N - 1$ in steps of 1, we can write

$$
i = \begin{cases} \frac{y}{2} & \text{if } y \equiv 0 \bmod 2; \\ N - \frac{y+1}{2} & \text{if } y \equiv 1 \bmod 2. \end{cases}
$$

We write the steps of KSA$^+$ in Algorithm 9.4.1.

More scrambling steps definitely increases the cost of the cipher. The running time of the KSA$^+$ is around three times that of RC4 KSA, since there

---

Input: Secret key array $K[0 \ldots N-1]$.
Output: Scrambled permutation array $S[0 \ldots N-1]$.

Initialization:
for $i = 0, \ldots, N-1$ **do**
   | $S[i] = i$;
   | $j = 0$;
end

1   Perform scrambling layer 1: basic scrambling as in RC4 KSA;
2   Perform scrambling layer 2: scrambling with IV;
3   Perform scrambling layer 3: zigzag scrambling;

Algorithm 9.4.1: KSA$^+$ of RC4$^+$

are three similar scrambling layers instead of one, each with N iterations. As the key scheduling is run only once, this does not affect the performance of the cipher much.

Next we discuss how the new KSA$^+$ avoids many weaknesses of the original KSA.

Removal of secret key correlation with the permutation bytes

After Layer 1, the values in the first quarter of the permutation are biased to linear combinations of the secret key bytes. In Layer 1, the deterministic index i is moved from the middle to the left end so that the above biases fade away. This takes care of Item (1) in the list of the weaknesses of the RC4 KSA. A similar operation is performed in the second half of the permutation to get rid of the biases of the inverse permutation as described in Item (4). Next, the XOR operation helps further to wipe out these biases. The biases involving the nested indexing mentioned in Item (3) and Item (4) arise due to the biases of direct indexing. So, the removal of the biases at the direct indices of S_N and S_N^{-1} automatically eliminates those at the nested indices also.

The bias of Item (2) are generalization of the biases of Item (1). They originate from the incremental update of j which helps to form a recursive equation involving the key bytes. In the new design, the bit-by-bit XOR operation as well as the zigzag scrambling in Layer 3 prevents one from forming such recursive equations connecting the key bytes and the permutation bytes.

Extensive experimentation has not revealed any correlation between the permutation entries $S_{N+}[y]$ with f_y. Also, the nested entries $S_{N+}[S_{N+}[y]]$, $S_{N+}[S_{N+}[S_{N+}[y]]]$ etc. are also found to be unbiased.

Item (7) is about the movement frequency of the permutation entries. The following experimental results show that such weaknesses of RC4 KSA are absent in the new design. The data is generated by averaging over 100 million runs of KSA$^+$ with 16 bytes key in each run. We find that as v increases from

0 to 255, E_v decreases from 4.99 to 3.31 after the end of Layer 2 and from 6.99 to 5.31 after the end of Layer 3. Table 9.1 demonstrates the individual as well as the incremental effect of each of Layer 2 and Layer 3, when they act upon the identity permutation S_0 and the permutation S_N obtained after Layer 1. In the table, Lr means Layer r, $r = 1, 2, 3$. The first row labeled "Theory" corresponds to the values obtained using Theorem 3.3.2. The data illustrate that the effect of Layer 2 or Layer 3 on the identity permutation S_0 is similar to that of Layer 1. However, after Layer 1 is over (when we have somewhat random permutation S_N, the output of RC4 KSA), each of Layer 2 and Layer 3 individually enforces each entry of the permutation to be visited uniformly (approximately twice on average). Thus, each layer incrementally moves the graph of E_v versus v in the positive Y-direction approximately by an amount of 2, as is illustrated in Figure 9.1.

		avg	sd	max	min
KSA$^+$ L1	Theory	2.0025	0.4664	3.0000	1.3700
(RC4 KSA)	Experiment	2.0000	0.4655	2.9959	1.3686
KSA$^+$ L2	L2 on S_0	2.0000	0.4658	2.9965	1.3683
(Experiment)	L2 on S_N	2.0000	0.0231	2.0401	1.9418
	L1 + L2	4.0000	0.4716	4.9962	3.3103
KSA$^+$ L3)	L3 on S_0	2.0000	0.4660	3.0000	1.3676
(Experiment)	L3 on S_N	2.0000	0.0006	2.0016	1.9988
	L1 + L2 + L3	6.0000	0.4715	6.9962	5.3116

TABLE 9.1: Average, standard deviation, maximum and minimum of the expectations E_v over all v between 0 and 255.

Uniform values of the expectations can be achieved easily with normal RC4, by keeping a count of how many times each element is touched and performing additional swaps involving the elements that have been touched fewer number of times. However, this will require additional space and time. In normal RC4, many permutation elements are touched only once (especially those toward the right end of the permutation), leaking information on j in the inverse permutation. The target, here, is to prevent this by increasing the number of times each element is touched, without keeping any additional space such as a counter. Experimental data in Table 9.1 as well as Figure 9.1 show that this purpose is served using the new strategy.

How random is S_{N+}?

Now we present experimental evidence of how the biases of Item (5) in RC4 KSA are removed. We compare the probabilities $P(S[u] = v)$ for $0 \leq u, v \leq 255$ from standard KSA and the KSA$^+$. All the experiments are performed with 100 million runs with randomly chosen secret keys of length 16 bytes and null IV.

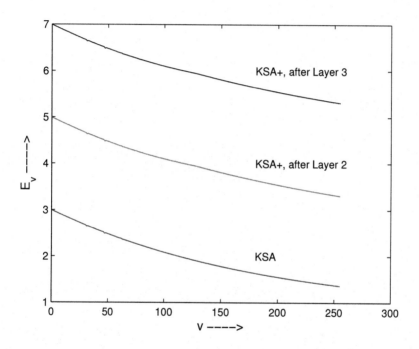

FIGURE 9.1: KSA$^+$: E_v versus v, $0 \leq v \leq 255$.

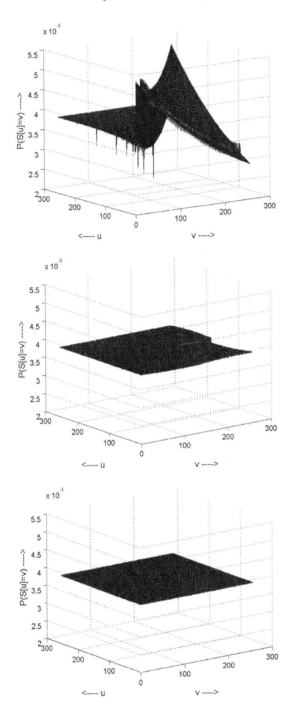

FIGURE 9.2: $P(S[u] = v)$ versus $0 \leq u, v \leq 255$, after each of the three layers of KSA$^+$.

One may see from the graphs in Figure 9.2 that from the end of Layer 1 (top) to the end of Layer 2 (middle), $P(S[u] = v)$ is flattened to a large extent. However, some non-uniformities still remain. After Layer 3 (bottom), the graph becomes completely flat, indicating that there is no bias in the probabilities $P(S[u] = v)$. The maximum and the minimum values of the probabilities as well as the standard deviations in Table 9.2 elaborate this fact further. Recall that $\frac{1}{N} = 0.003906$ for $N = 256$.

		avg	*sd*	*max*	*min*
RC4 KSA	Theory	0.003901	0.000445	0.005325	0.002878
	Experiment	0.003906	0.000448	0.005347	0.002444
KSA+ (Experiment)	After Layer 2	0.003906	0.000023	0.003983	0.003803
	After Layer 3	0.003906	0.000006	0.003934	0.003879

TABLE 9.2: Average, standard deviation, maximum and minimum of the probabilities $P(S[u] = v)$ over all u and v between 0 and 255.

In [110, Page 67], it was mentioned that the RC4 KSA needs to be executed approximately six times in order to get rid of these biases. Whereas, in the new design, the key scheduling time is of the order of three times that of RC4 KSA.

On introducing the IVs

The IV-mode attacks, mentioned in Item (6) in the list of the weaknesses of the RC4 KSA, work because in the original RC4, IVs are either prepended or appended with the secret key. However, in Layer 2 of the KSA+, the IVs are used in the middle and also the corresponding key bytes are added in the update of j. In this layer, $2l$ many operations involve the IV, but $N - 2l$ many operations do not. In addition to such use of IVs, a third layer of zigzag scrambling is performed where no use of IV is made. This almost eliminates the possibility of WEP-like chosen IV attack once the key scheduling is complete.

SSL protocol generates the encryption keys of RC4 by hashing the secret key and the IV together, using both MD5 and SHA-1 hashes. Hence, different sessions have unrelated keys [147], and the WEP attack [52, 175] can be bypassed. Since the KSA+ is believed to be free from the IV-weaknesses, it can be used without any hashing. Thus, the cost of hashing can be engaged in the extra operations in Layer 2 and Layer 3. This conforms to the design motivation of keeping the basic structure of RC4 KSA and still avoids the weaknesses.

As claimed in [101], extensive experimentation did not reveal any weakness with null IVs as well as with randomly chosen IVs.

On retaining the standard KSA in Layer 1

One may argue that Layer 1 is unnecessary and Layers 2, 3 would have taken care of all the existing weaknesses of RC4. While this might be true, Layers 2 and 3, when operated on identity permutation, might introduce some new weaknesses not yet known. It is a fact that RC4 KSA has some weaknesses, but it also reduces the correlation of the secret key with the permutation entries and other biases, at least to some extent compared to the beginning of the KSA. In the process, the permutation is randomized to a certain degree. The structure of RC4 KSA is simple and elegant and easy to analyze. First, this KSA is allowed to run over the identity permutation, so that the exact biases that are to be removed in the subsequent layers are identified. In summary, the KSA$^+$ keeps the good features of RC4 KSA, and removes only the bad ones.

9.4.2   PRGA$^+$: Modifications to RC4 PRGA

The main cryptanalytic results about PRGA that are taken into consideration in designing PRGA$^+$ are listed below.

1. Correlations between the keystream output bytes and the secret key, discussed in Sections 5.5 and 5.6.

2. Distinguishing attacks described in Chapter 6.

3. State recovery from the keystream addressed in Section 5.8.

4. Key recovery in the WEP mode (these exploit the weaknesses of both the KSA and the PRGA) presented in Chapter 7.

KSA$^+$ is designed in such a manner that one cannot get secret key correlations from the permutation bytes. This guarantees that the keystream output bytes, which are some combination of the permutation entries, cannot have any correlation with the secret key. As argued in Section 9.4.1, IVs in KSA^+ cannot be easily exploited to mount an attack. Thus, only two weaknesses, enlisted in Item (2) and (3) above, need to be targeted in the design of PRGA$^+$.

Recall that for any byte b, b_R^n (respectively b_L^n) denotes the byte after right (respectively left) shifting b by n bits. For $r \geq 1$, the permutation, the indices i, j and the keystream output byte after round r of PRGA$^+$ are denoted by S_r^G, i_r^G, j_r^G and z_r respectively, in the same way as in RC4 PRGA.

The main idea behind the design of PRGA$^+$ is masking the output byte such that it does not directly come out from any permutation entry.

- Two entries from the permutation are added modulo 256 (a nonlinear operation) and then the outcome is XOR-ed with a third entry (for masking non-uniformity).

- Incorporating the bytes $S[t'], S[t'']$, in addition to the existing $S[t]$ in RC4, makes the running time of PRGA$^+$ little more than that of RC4 PRGA (see Section 9.4.3 for details). The evolution of the permutation S in PRGA$^+$ stays exactly the same as in RC4 PRGA.

- A constant value $0xAA$ (equivalent to 10101010 in binary) is used in t'. Without this, if j^G becomes 0 in rounds 256, 512, ... (i.e., when $i^G = 0$), then t and t' in such a round become equal with probability 1, giving an internal bias.

Input: Key-dependent scrambled permutation array $S[0 \ldots N-1]$.
Output: Pseudo-random keystream bytes z.

Initialization:
$i = j = 0$;

Output Keystream Generation Loop:
$i = i + 1$;
$j = j + S[i]$;
Swap($S[i], S[j]$);
$t = S[i] + S[j]$;
$t' = (S[i_R^3 \oplus j_L^5] + S[i_L^5 \oplus j_R^3]) \oplus 0xAA$;
$t'' = j + S[j]$;
Output $z = (S[t] + S[t']) \oplus S[t'']$;

Algorithm 9.4.2: PRGA$^+$ of RC4$^+$

Resisting state recovery attacks

We already explained in Chapter 5.8 the basic idea of cryptanalysis in [116], the best known state recovery attack on RC4. Corresponding to a window of $w + 1$ keystream output bytes, one may assume that all the j^G's are known, i.e., $j_r^G, j_{r+1}^G, \ldots, j_{r+w}^G$ are known. Thus w many permutation entries $S_r^G[i_r^G]$ would be available from $j_{r+1}^G - j_r^G$. Then w many equations of the form $S_r^{G^{-1}}[z_r] = S_r^G[i_r^G] + S_r^G[j_r^G]$ can be formed where each equation contains only two unknowns.

PRGA$^+$ does not allow the strategy of [116] to work, since $S^G[S^G[i^G] + S^G[j^G]]$ is not exposed directly; rather it is masked by several other quantities. To form the equations as given in [116], one first needs to guess $S^G[t], S^G[t'], S^G[t'']$. However, looking at the value of z, there is no other option than to go for all the possible choices. The same permutation structure of S in RC4$^+$ could be similarly exploited to get the good patterns [116, Section 3], but introducing the additional output indices t', t'', the non-detectability of such a pattern in the keystream is ensured and thus the idea of [116, Section 4] would not work.

Information on permutation entries is also leaked in the keystream via the

Jenkins' Correlation or Glimpse Theorem [78,108] discussed in Section 5.2. It states that during any PRGA round, $P(S^G[j^G] = i^G - z) = P(S^G[i^G] = j^G - z) \approx \frac{2}{N}$. As we saw in the proof of Theorem 5.2.1, given $i^G = S[i^G] + S[j^G]$, the event $S^G[j^G] = i^G - z$ holds with probability 1. To obtain Glimpse-like such biases in PRGA^+, one needs to have more assumptions of the above form. Thus, Glimpse-like biases of PRGA^+, if they at all exist, would be much weaker.

Resisting distinguishing attacks

In [107], it was proved that $P(z_2 = 0) = \frac{2}{N}$ instead of the uniformly random case of $\frac{1}{N}$. We discuss this distinguishing attack in detail in Section 6.2.2. As we saw in the proof of Theorem 6.2.3, the bias originates from the fact that when $S_N[2] = 0$ and $S_N[1] \neq 2$ after the KSA, the second keystream output byte z_2 takes the value 0 with probability 1. Based on this, a ciphertext-only attack can be mounted in broadcast mode. This kind of situation does not arise in the new design. When 100 million secret keys of length 16 bytes are generated and 1024 rounds of PRGA^+ are executed for each such key, the empirical evidence indicates that $P(z_r = v) = \frac{1}{N}$, $1 \leq r \leq 1024, 0 \leq v \leq N - 1$.

In the work [139], it was reported that $P(z_1 = z_2) = \frac{1}{N} - \frac{1}{N^2}$, which leads to a distinguishing attack. Even after extensive experimentation, such bias in the keystream output bytes of PRGA^+ is not observed. The same experiment described above also supports that $P(z_r = z_{r+1})$ is uniformly distributed for $1 \leq r \leq 1023$.

In [109], it has been shown that getting strings of pattern $ABTAB$ (A, B are bytes and T is a string of bytes of small length Δ, say $\Delta \leq 16$) are more frequent in RC4 keystream than in random stream. In a uniformly random keystream, the probability of getting such pattern irrespective of the length of T is $\frac{1}{N^2}$, whereas for RC4, the probability of such an event is $\frac{1}{N^2}(1 + \frac{e^{\frac{-4-8\Delta}{N}}}{N}) > \frac{1}{N^2}$. We have already presented this distinguishing attack in Section 6.3.1. This result is based on the fact that the permutation entries in locations that affect the swaps and the selection of keystream output bytes in both pairs of rounds that are Δ-round apart, remain unchanged during the intermediate rounds with high probability. The permutation in PRGA^+ evolves in the same way as RC4 PRGA, but the keystream output generation in PRGA^+ is different. This does not allow the pattern AB to propagate down the keystream with significant probability even for smaller interval lengths (Δ). Thus, the source of RC4's ABTAB bias is present in the permutation of RC4^+ also, but it is not revealed in the keystream. The simulation on PRGA^+ also confirms that the keystream is free from such biases.

9.4.3 Performance Evaluation

The performance of the new design has been tested using the *eSTREAM testing framework* [26]. The C-implementation of the testing framework was installed in a machine with Intel(R) Pentium(R) 4 CPU, 2.8 GHz Processor Clock, 512 MB DDR RAM on Ubuntu 7.10 (Linux 2.6.22-15-generic) OS. A benchmark implementation of RC4 is available within the test suite. RC4$^+$, that incorporates both KSA$^+$ and PRGA$^+$, has been implemented maintaining the API compliance of the suite. Test vectors were generated in the NESSIE [127] format.

The results presented here correspond to tests with a 16-byte secret key and null IV using the *gcc_default_O3-ual-ofp* compiler. As per the test, RC4 KSA took 16944.70 cycles/setup, whereas the KSA$^+$ of RC4$^+$ took 49823.69 cycles/setup. The stream encryption speed for RC4 and RC4$^+$ turned out to be 14.39 cycles/byte and 24.51 cycles/byte respectively. In the testing framework, the cycles are measured using the RDTSC (Read Time Stamp Counter) instruction of X86 assembly language. This instruction returns a 64-bit value in registers EDX:EAX that represents the count of ticks from processor reset.

Thus, the running time of our KSA$^+$ is approximately $\frac{49823.69}{16944.70} = 2.94$ times than that of RC4 KSA and the running time of one round of our PRGA$^+$ is approximately $\frac{24.51}{14.39} = 1.70$ times than that of RC4 PRGA.

Research Problems

Problem 9.1 Investigate whether the technique of finding colliding key pairs described in Section 3.4 (Chapter 3) is extendible to RC4-Hash.

Problem 9.2 Perform a detailed study to check if there a distinguisher for RC4$^+$?

Problem 9.3 Give an optimized software implementation of RC4$^+$ to achieve good speed.

Problem 9.4 Perform a comparative study of different RC4-like stream ciphers in terms of the security as well as the speed.

Chapter 10

Stream Cipher HC-128

HC-128 [187] can be considered as the next level of evolution in the design of RC4-like software stream ciphers. While the design philosophy of RC4 is simplicity, lots of complicated operations and functions are used in HC-128 to increase the security. After several rounds of evaluations by eminent cryptologists throughout the world for several years, HC-128 made its way into the final eSTREAM [42] Portfolio (revision 1 in September 2008) in the software category. Some cryptanalytic results have been discovered very recently, but they do not pose any immediate threat to the use of the cipher. In this chapter, we give a description of HC-128 along with exposition of the known results.

10.1 Description of HC-128

Apart from standard notations, HC-128 uses the symbol $\boxminus$ to denote subtraction modulo 512. Each word (or array element) is of 32 bits. Two tables P and Q containing 512 words are used as internal states of HC-128. A 128-bit secret key $K[0, \ldots, 3]$ and a 128-bit initialization vector $IV[0, \ldots, 3]$ are used for key scheduling. The key and IV are expanded into an array W of 1280 words. Let the keystream word generated at the t-th step be denoted by s_t, $t = 0, 1, 2, \ldots$.

HC-128 uses the following six functions.

$$
\begin{aligned}
f_1(x) &= (x \ggg 7) \oplus (x \ggg 18) \oplus (x \gg 3), \\
f_2(x) &= (x \ggg 17) \oplus (x \ggg 19) \oplus (x \gg 10), \\
g_1(x, y, z) &= \big((x \ggg 10) \oplus (z \ggg 23)\big) + (y \ggg 8), \\
g_2(x, y, z) &= \big((x \lll 10) \oplus (z \lll 23)\big) + (y \lll 8), \\
h_1(x) &= Q[x^{(0)}] + Q[256 + x^{(2)}], \\
h_2(x) &= P[x^{(0)}] + P[256 + x^{(2)}],
\end{aligned}
$$

where $x^{(0)}, x^{(1)}, x^{(2)}$ and $x^{(3)}$ are the four bytes of a 32-bit word $x = x^{(3)} \| x^{(2)} \| x^{(1)} \| x^{(0)}$, the order from the least to the most significant byte order being from right to left.

233

The key and IV setups of HC-128 are performed in the key scheduling part and given in Algorithm 10.1.1.

Input:

  1. Secret key array $K[0 \ldots 3]$.

  2. Initialization Vector $IV[0 \ldots 3]$.

Output: The internal state arrays $P[0 \ldots 511]$ and $Q[0 \ldots 511]$.

for $i = 0, \ldots, 3$ do
  | $K[i + 4] = K[i]$;
  | $IV[i + 4] = IV[i]$;
end
for $i = 8, \ldots, 15$ do
  | $W[i] = K[i]$;
end
for $i = 0, \ldots, 7$ do
  | $W[i] = IV[i - 8]$;
end
for $i = 16, \ldots, 1279$ do
  | $W[i] = f_2(W[i - 2]) + W[i - 7] + f_1(W[i - 15]) + W[i - 16] + i$;
end
for $i = 0, \ldots, 511$ do
  | $P[i] = W[i + 256]$;
end
for $i = 0, \ldots, 511$ do
  | $Q[i] = W[i + 768]$;
end
for $i = 0, \ldots, 511$ do
  | $P[i] = \big(P[i] + g_1(P[i \boxminus 3], P[i \boxminus 10], P[i \boxminus 511])\big) \oplus h_1(P[i \boxminus 12])$;
end
for $i = 0, \ldots, 511$ do
  | $Q[i] = \big(Q[i] + g_2(Q[i \boxminus 3], Q[i \boxminus 10], Q[i \boxminus 511])\big) \oplus h_2(Q[i \boxminus 12])$;
end

Algorithm 10.1.1: HC-128 KSA

The keystream generation algorithm is presented in Algorithm 10.1.2.

It is not hard to see that the key and IV setup algorithm can be reversed, if both the P, Q arrays are available after the key scheduling. This has also been pointed out in [97].

Since P is used to denote an array in HC-128, we use *Prob* to denote probability in this chapter.

Input: Two key-dependent internal state arrays:

  1. $P[0 \ldots 511]$;

  2. $Q[0 \ldots 511]$;

Output: Pseudo-random keystream words s_t's.

$i = 0$;
repeat
  | $j = i \bmod 512$;
  | **if** $(i \bmod 1024) < 512$ **then**
  | | $P[j] = P[j] + g_1(P[j \boxminus 3], P[j \boxminus 10], P[j \boxminus 511])$;
  | | $s_i = h_1(P[j \boxminus 12]) \oplus P[j]$;
  | **end**
  | **else**
  | | $Q[j] = Q[j] + g_2(Q[j \boxminus 3], Q[j \boxminus 10], Q[j \boxminus 511])$;
  | | $s_i = h_2(Q[j \boxminus 12]) \oplus Q[j]$;
  | **end**
  | $i = i + 1$;
until *enough keystream bits are generated* ;

Algorithm 10.1.2: HC-128 PRGA

10.2   Linear Approximation of Feedback Functions

The analysis in this section is based on [105]. Two functions g_1, g_2 used in HC-128 are of similar kind. The two addition ("+") operations in g_1 or g_2 are believed to be a source of high nonlinearity. However, good linear approximation can be found by using the result of linear approximation of the addition of three integers.

Linear approximations of modulo-2^n addition of k many n-bit integers have been studied in [172]. For $k = 2$, the probability of the equality of XOR and modulo-2^n sum in the b-th least significant bit tends to $\frac{1}{2}$ as b increases. Below, the case for $k = 3$ is briefly discussed, an instance of which would be used in approximating g_1, g_2.

Recall that the b-th least significant bit of an n-bit word w is given by $[w]^b$, $0 \le b \le n - 1$, i.e.,

$$w = ([w]^{n-1}, [w]^{n-2}, \ldots, [w]^1, [w]^0).$$

This notation is also extended to $[w]^b$, where $b > n - 1$. In that case, $[w]^b$ means $[w]^{b \bmod n}$.

Let X, Y, Z be three n-bit integers;

$$S = (X + Y + Z) \bmod 2^n$$

is the sum (truncated to n bits),

$$T = X \oplus Y \oplus Z$$

is their bitwise XOR. Let C_b denote the carry bit generated in the b-th step of the addition of X, Y and Z. C_b can take the values $0, 1$ and 2, since three bits are involved in the addition. For the LSB addition, C_{-1} is assumed to be 0. Let

$$\rho_{b,v} = Prob(C_b = v), \qquad b \geq -1, v \in \{0,1,2\}.$$

We can write

$$
\begin{aligned}
Prob(S_b = T_b) &=& Prob(C_{b-1} = 0 \text{ or } 2) \\
&=& \rho_{b-1,0} + \rho_{b-1,2} \\
&=& 1 - \rho_{b-1,1}.
\end{aligned}
$$

One can easily show the following recurrences.

1. $\rho_{b+1,0} = \frac{1}{2}\rho_{b,0} + \frac{1}{8}\rho_{b,1}$.

2. $\rho_{b+1,1} = \frac{1}{2}\rho_{b,0} + \frac{3}{4}\rho_{b,1} + \frac{1}{2}\rho_{b,2}$.

3. $\rho_{b+1,2} = \frac{1}{8}\rho_{b,1} + \frac{1}{2}\rho_{b,2}$.

The solution gives $\rho_{b,1} = \frac{2}{3}(1 - \frac{1}{4^{b+1}})$ and so we can state the following result.

Proposition 10.2.1. *For $0 \leq b \leq n-1$, $Prob([S]^b = [T]^b) = \frac{1}{3}(1 + \frac{1}{2^{2b-1}})$.*

An immediate approximation to Proposition 10.2.1 is given in the following corollary.

Corollary 10.2.2. *For $0 \leq b \leq n-1$, $Prob([S]^b = [T]^b) = p_b$, where*

$$
p_b = \begin{cases}
1 & \text{if } b = 0; \\
\frac{1}{2} & \text{if } b = 1; \\
\frac{1}{3} \text{ (approximately)} & \text{if } 2 \leq b \leq n-1.
\end{cases}
$$

During the keystream generation part of HC-128, the array P is updated as

$$P[i \bmod 512] = P[i \bmod 512] + g_1(P[i \boxminus 3], P[i \boxminus 10], P[i \boxminus 511]),$$

where

$$g_1(x, y, z) = ((x \ggg 10) \oplus (z \ggg 23)) + (y \ggg 8).$$

Thus, we can restate the update rule as

$$
\begin{aligned}
P_{up}[i \bmod 512] &=& P[i \bmod 512] + ((P[i \boxminus 3] \ggg 10) \oplus (P[i \boxminus 511] \ggg 23)) \\
&& + (P[i \boxminus 10] \ggg 8). \quad (10.1)
\end{aligned}
$$

In consistence with the notations in [187, Section 4], one may write the keystream generation step as follows. For $0 \leq i \bmod 1024 < 512$,

$$
s_i = \begin{cases} h_1(P[i \boxminus 12]) \oplus P_{up}[i \bmod 512], & \text{for } 0 \leq i \bmod 512 \leq 11; \\ h_1(P_{up}[i \boxminus 12]) \oplus P_{up}[i \bmod 512], & \text{for } 12 \leq i \bmod 512 \leq 511. \end{cases}
$$
(10.2)

Let P'_{up} denote the updated array P, when the two "+" operators are replaced by "$\oplus$" in the right-hand side of Equation (10.1). Then for $0 \leq b \leq n-1$, the b-th bit of the updated words of the array P is given by

$$
\left[P'_{up}[i \bmod 512] \right]^b = \left[P[i \bmod 512] \right]^b \oplus \left[P[i \boxminus 3] \right]^{10+b}
$$
$$
\oplus \left[P[i \boxminus 511] \right]^{23+b} \oplus \left[P[i \boxminus 10] \right]^{8+b}.
$$

Define

$$
s'_i = \begin{cases} h_1(P[i \boxminus 12]) \oplus P'_{up}[i \bmod 512], & \text{for } 0 \leq i \bmod 512 \leq 11; \\ h_1(P_{up}[i \boxminus 12]) \oplus P'_{up}[i \bmod 512], & \text{for } 12 \leq i \bmod 512 \leq 511. \end{cases}
$$

Each s'_i can be treated as the "distorted" value of s_i due the XOR-approximation of the sum of three integers in Equation (10.1). From Corollary 10.2.2 of Proposition 10.2.1 and the definition of s''_i's, one can immediately note the following result.

Lemma 10.2.3. *For* $0 \leq i \bmod 1024 < 512$ *and* $0 \leq b \leq n-1$,
$$
Prob\left(\left[s'_i\right]^b = \left[s_i\right]^b \right) = Prob\left(\left[P'_{up}[i \bmod 512]\right]^b = \left[P_{up}[i \bmod 512]\right]^b \right) = p_b.
$$

Now, 32-bit integers ζ_i's can be constructed as estimates of the keystream words s_i's as follows.

$$
[\zeta_i]^b = \begin{cases} \left[s'_i\right]^b & \text{if } b = 0, 1; \\ 1 \oplus \left[s'_i\right]^b & \text{if } 2 \leq b < 32. \end{cases}
$$

Thus, one has the following result.

Theorem 10.2.4. *The expected number of bits where the two 32-bit integers* s_i *and* ζ_i *match is approximately 21.5.*

Proof: Let $m_b = 1$, if $[s_i]^b = [\zeta_i]^b$; otherwise, let $m_b = 0$, $0 \leq b \leq 31$. Hence, the total number of matches is given by

$$
M = \sum_{b=0}^{31} m_b.
$$

By linearity of expectation,

$$
E(M) = \sum_{b=0}^{31} E(m_b)
$$
$$
= \sum_{b=0}^{31} Prob(m_b = 1).
$$

From Lemma 10.2.3 and the construction of ζ_i, we have

$$Prob(m_b = 1) = \begin{cases} p_b & \text{if } b = 0, 1; \\ 1 - p_b & \text{if } 2 \leq b < 32. \end{cases}$$

This gives

$$\begin{aligned} E(M) &\approx 1 + \frac{1}{2} + 30 \cdot (1 - \frac{1}{3}) \\ &= 21.5. \end{aligned}$$

∎

Theorem 10.2.4 shows the correlation of the HC-128 keystream word s_i with its linear approximation ζ_i.

10.3   Distinguishing Attacks on HC-128

Linear approximation of the feedback functions g_1, g_2 can be used to mount distinguishing attacks on HC-128. Primarily, there are three distinguishers for HC-128. Since addition and XOR-operation exactly match for the least significant bit (LSB), no approximation is needed to devise a distinguisher based on the LSB's of the keystream words. Such a distinguisher was proposed by Wu, the designer himself, in [187]. We discuss this in Section 10.3.1. Subsequently, in [105], extension of this distinguisher to 30 other bits were presented. We discuss this work in Section 10.3.2. Both the above distinguishers work across two blocks of HC-128 keystream, where each block consists of 512 consecutive keystream words. In Section 10.3.3, we present the distinguisher of [106] that is spread over three blocks of the keystream.

10.3.1   Wu's LSB-Based Distinguisher

This result appeared in the original design proposal [187]. Though the keystream words of HC-128 are generated using both the arrays P and Q, the updates of P and Q arrays are independent. For 512 many iterations, the array P is updated with the older values from P itself. For the next 512 many iterations, the array Q is updated with the older values of Q. In this way, alternating updates of P and Q continue. Table 10.1 illustrates how the keystream words s_i's are related to the array elements $P[i]$'s and $Q[i]$'s.

In general, for $0 \leq (i \bmod 1024) < 512$, the keystream output word of HC-128 is produced as

$$s_i = h_1(P[i \boxminus 12]) \oplus P[i \bmod 512],$$

following an update of $P[i \bmod 512]$ via addition of $g_1(P[i\boxminus 3], P[i\boxminus 10], P[i\boxminus$

Old P array:	$P[0]$	$P[1]$	...	$P[511]$
Keystream:	s_0	s_1	...	s_{511}
Intermediate Q array:	$Q[0]$	$Q[1]$	...	$Q[511]$
Keystream:	s_{512}	s_{513}	...	s_{1023}
New P array:	$P[0]$	$P[1]$	...	$P[511]$
Keystream:	s_{1024}	s_{1025}	...	s_{1535}
New Q array:	$Q[0]$	$Q[1]$	...	$Q[511]$
Keystream:	s_{1536}	s_{1537}	...	s_{2047}

TABLE 10.1: Evolution of the arrays P and Q and the corresponding keystream words.

511]) to itself, as formulated in Equations (10.1) and (10.2) of Section 10.2. Let z_i denote $P[i \boxminus 12]$ at the i-th step. From Equation (10.2), it is easy to see that any occurrence of $P[i \bmod 512]$ can be replaced by $s_i \oplus h_1(z_i)$. Performing this replacement in Equation (10.1), we get, for $10 \le i \bmod 512 < 511$,

$$s_i \oplus h_1(z_i) = \left(s_{i-1024} \oplus h_1'(z_{i-1024})\right) + g_1\left(s_{i-3} \oplus h_1(z_{i-3}), s_{i-10}\right.$$
$$\left. \oplus h_1(z_{i-10}), s_{i-1023} \oplus h_1'(z_{i-1023})\right). \tag{10.3}$$

Since $h_1(.)$ and $h_1'(.)$ are related to two P arrays at two different 1024 size blocks that act as two different S-boxes, $h_1(.)$ and $h_1'(.)$ are treated as two different functions.

Inside the function g_1, there are three rotations, one XOR and one addition and outside g_1 there is one more addition. Since the LSB of the XOR of any number of words equals the LSB of the sum of those words, one can write Equation (10.3) as

$$[s_i]^0 \oplus [s_{i-1024}]^0 \oplus [s_{i-3}]^{10} \oplus [s_{i-10}]^8 \oplus [s_{i-1023}]^{23}$$
$$= [h_1(z_i)]^0 \oplus [h_1'(z_{i-1024})]^0 \oplus [h_1(z_{i-3})]^{10} \oplus [h_1(z_{i-10})]^8 \oplus [h_1'(z_{i-1023})]^{23}.$$

Thus, for $1024\tau + 10 \le j < i < 1024\tau + 511$,
$$[s_i]^0 \oplus [s_{i-1024}]^0 \oplus [s_{i-3}]^{10} \oplus [s_{i-10}]^8 \oplus [s_{i-1023}]^{23}$$
$$= [s_j]^0 \oplus [s_{j-1024}]^0 \oplus [s_{j-3}]^{10} \oplus [s_{j-10}]^8 \oplus [s_{j-1023}]^{23}, \text{ if and only if } H(\xi_i) = H(\xi_j), \text{ where}$$

$$H(\xi_i) = [h_1(z_i)]^0 \oplus [h_1'(z_{i-1024})]^0 \oplus [h_1(z_{i-3})]^{10} \oplus [h_1(z_{i-10})]^8 \oplus [h_1'(z_{i-1023})]^{23}.$$

Here $\xi_i = (z_i, z_{i-1024}, z_{i-3}, z_{i-10}, z_{i-1023})$ is an 80-bit input and $H(.)$ can be considered as a random 80-bit-to-1-bit S-box.

The following result gives the collision probability for a general random m-bit-to-n-bit S-box.

Theorem 10.3.1. *Let H be an m-bit-to-n-bit S-box and all those n-bit elements are randomly generated, where $m \ge n$. Let x_1 and x_2 be two m-bit random inputs to H. Then $H(x_1) = H(x_2)$ with probability $2^{-m} + 2^{-n} - 2^{-m-n}$.*

Proof: If $x_1 = x_2$ (this happens with probability 2^{-m}), then $H(x_1) = H(x_2)$ happens with probability 1. If $x_1 \neq x_2$ (this happens with probability $1 - 2^{-m}$), then $H(x_1) = H(x_2)$ happens with probability 2^{-n}. Thus, $Prob\,(H(x_1) = H(x_2)) = 2^{-m} \cdot 1 + (1 - 2^{-m}) \cdot 2^{-n}$. ∎

For HC-128 S-box, whose outputs are $H(\xi_i)$ and $H(\xi_j)$, we have $m = 80$ and $n = 1$. According to Theorem 10.3.1, $Prob\,(H(\xi_i) = H(\xi_j)) = \frac{1}{2} + 2^{-81}$. Hence, for $1024\tau + 10 \leq j < i < 1024\tau + 511$,

$$Prob\Big([s_i]^0 \oplus [s_{i-1024}]^0 \oplus [s_{i-3}]^{10} \oplus [s_{i-10}]^8 \oplus [s_{i-1023}]^{23}$$
$$= [s_j]^0 \oplus [s_{j-1024}]^0 \oplus [s_{j-3}]^{10} \oplus [s_{j-10}]^8 \oplus [s_{j-1023}]^{23}\Big) = \tfrac{1}{2} + 2^{-81}.$$

Thus, one can mount a distinguisher based on the equality of the least significant bits of the keystream word combinations $s_i \oplus s_{i-1024} \oplus (s_{i-3} \ggg 10) \oplus (s_{i-10} \ggg 8) \oplus (s_{i-1023} \ggg 23)$ and $s_j \oplus s_{j-1024} \oplus (s_{j-3} \ggg 10) \oplus (s_{j-10} \ggg 8) \oplus (s_{j-1023} \ggg 23)$. According to Section 6.1, this distinguisher requires approximately 2^{164} many equations of the above form or a total of 2^{156} many keystream words for a success probability 0.9772.

10.3.2 Extension of Wu's Distinguisher to Other Bits

It has been commented in [187] that the distinguishing attack exploiting the other 31 bits would not be effective due to the use of modulo addition. In contrary to the belief of the designer of HC-128, a recent work [105] has shown that the distinguisher works for all the bits (except one) in the keystream words. We present this extension in this Section.

There exist biases in the equality of 31 out of the 32 bits (except the second least significant bit) of the word combinations

$$s_i \oplus s_{i-1024} \oplus (s_{i-3} \ggg 10) \oplus (s_{i-10} \ggg 8) \oplus (s_{i-1023} \ggg 23)$$

and $s_j \oplus s_{j-1024} \oplus (s_{j-3} \ggg 10) \oplus (s_{j-10} \ggg 8) \oplus (s_{j-1023} \ggg 23)$.

Thus, we have a distinguisher for each of those 31 bits separately.

The analysis generalizes the idea of Section 10.3.1 by applying Corollary 10.2.2. Refer to the visualization of the array P as explained in Table 10.1 of Section 10.3.1 and focus on Equation (10.3):

$$s_i \oplus h_1(z_i) = \big(s_{i-1024} \oplus h_1'(z_{i-1024})\big) + g_1\big(s_{i-3} \oplus h_1(z_{i-3}), s_{i-10}$$
$$\oplus h_1(z_{i-10}), s_{i-1023} \oplus h_1'(z_{i-1023})\big).$$

Again let us replace the two "+" operations (one inside and one outside) g_1 by "$\oplus$." Then as per the discussion following Corollary 10.2.2, one can write, for $10 \leq i \bmod 1024 < 511$,

$$[s_i]^b \oplus [s_{i-1024}]^b \oplus [s_{i-3}]^{10+b} \oplus [s_{i-10}]^{8+b} \oplus [s_{i-1023}]^{23+b}$$
$$= [h_1(z_i)]^b \oplus [h_1'(z_{i-1024})]^b \oplus [h_1(z_{i-3})^{10+b}]$$
$$\oplus [h_1(z_{i-10})]^{8+b} \oplus [h_1'(z_{i-1023})]^{23+b}$$

holds with probability $p_0 = 1$ for $b = 0$, with probability $p_1 = \frac{1}{2}$ for $b = 1$ and with probability $p_b \approx \frac{1}{3}$ for $2 \le b \le 31$. In short, for $0 \le b \le 31$,

$$Prob\left([\psi_i]^b = H_b(\xi_i)\right) = p_b,$$

where the 32-bit integer ψ_i is constructed as:

$$[\psi_i]^b = [s_i]^b \oplus [s_{i-1024}]^b \oplus [s_{i-3}]^{10+b} \oplus [s_{i-10}]^{8+b} \oplus [s_{i-1023}]^{23+b}$$

and

$$
\begin{aligned}
H_b(\xi_i) &= [h_1(z_i)]^b \oplus [h_1'(z_{i-1024})]^b \oplus [h_1(z_{i-3})]^{10+b} \\
&\oplus [h_1(z_{i-10})]^{8+b} \oplus [h_1'(z_{i-1023})]^{23+b}.
\end{aligned}
$$

Note that $\xi_i = (z_i, z_{i-1024}, z_{i-3}, z_{i-10}, z_{i-1023})$ is an 80-bit input and each $H_b(.)$, for $0 \le b \le 31$, is a random 80-bit-to-1-bit S-box. Clearly, for $0 \le b \le 31$, $Prob\left([\psi_i]^b = H_b(\xi_i) \oplus 1\right) = 1 - p_b$.

Thus, one can state the following technical result.

Lemma 10.3.2. *For $1024\tau + 10 \le j < i < 1024\tau + 511$ and $0 \le b \le 31$,*

$$Prob\left([\psi_i]^b \oplus [\psi_j]^b = H_b(\xi_i) \oplus H_b(\xi_j)\right) = q_b$$

where

$$
q_b = \begin{cases}
1 & \text{if } b = 0; \\
\frac{1}{2} & \text{if } b = 1; \\
\frac{5}{9} \text{ (approximately)} & \text{if } 2 \le b \le 31.
\end{cases}
$$

Proof:

$$
\begin{aligned}
& Prob\left([\psi_i]^b \oplus [\psi_j]^b = H_b(\xi_i) \oplus H_b(\xi_j)\right) \\
={} & Prob\left([\psi_i]^b = H_b(\xi_i)\right) \cdot Prob\left([\psi_j]^b = H_b(\xi_j)\right) \\
& + Prob\left([\psi_i]^b = H_b(\xi_i) \oplus 1\right) \cdot Prob\left([\psi_j]^b = H_b(\xi_j) \oplus 1\right) \\
={} & p_b \cdot p_b + (1 - p_b) \cdot (1 - p_b).
\end{aligned}
$$

Substituting the values of p_b from Corollary 10.2.2, we get the result. ∎

For $0 \le b \le 31$, one can write

$$Prob\left([\psi_i]^b \oplus [\psi_j]^b = H_b(\xi_i) \oplus H_b(\xi_j) \oplus 1\right) = 1 - q_b.$$

For a given b, all the $H_b(\xi_i)$'s are the outputs of the same random secret 80-bit-to-1-bit S-box $H_b(.)$. Substituting $m = 80$ and $n = 1$ in Theorem 10.3.1 gives the following corollary.

Corollary 10.3.3. *For $1024\tau + 10 \le j < i < 1024\tau + 511$ and $0 \le b \le 31$,*

$$Prob\left(H_b(\xi_i) = H_b(\xi_j)\right) = \frac{1}{2} + 2^{-81}.$$

Obviously, $Prob\left(H_b(\xi_i) = H_b(\xi_j) \oplus 1\right) = \frac{1}{2} - 2^{-81}$.

The above results can be combined to formulate the following theorem.

Theorem 10.3.4. *For* $1024\tau + 10 \le j < i < 1024\tau + 511$,

$$Prob\left([\psi_i]^b = [\psi_j]^b\right) = r_b,$$

where

$$r_b = \begin{cases} \frac{1}{2} + 2^{-81} & \text{if } b = 0; \\ \frac{1}{2} & \text{if } b = 1; \\ \frac{1}{2} + \frac{2^{-81}}{9} \text{ (approximately)} & \text{if } 2 \le b \le 31. \end{cases}$$

Proof:

$$
\begin{aligned}
& Prob\left([\psi_i]^b = [\psi_j]^b\right) \\
= \quad & Prob\left([\psi_i]^b \oplus [\psi_j]^b = H_b(\xi_i) \oplus H_b(\xi_j)\right) \cdot Prob\left(H_b(\xi_i) = H_b(\xi_j)\right) + \\
& Prob\left([\psi_i]^b \oplus [\psi_j]^b = H_b(\xi_i) \oplus H_b(\xi_j) \oplus 1\right) \cdot Prob\left(H_b(\xi_i) = H_b(\xi_j) \oplus 1\right).
\end{aligned}
$$

Substitute values from Lemma 10.3.2 and Corollary 10.3.3 to get the final expression. ∎

For the special case of $b = 0$, there exists a distinguisher based on the bias $\frac{1}{2} + 2^{-81}$ in the equality of the LSB's of ψ_i and ψ_j. This is exactly the same as Wu's distinguisher described in the previous section. The above results show that we can also mount a distinguisher of around the same order for each of the 30 bits corresponding to $b = 2, 3, \ldots, 31$ based on the bias $\frac{1}{2} + \frac{2^{-81}}{9}$.

Two random 32-bit integers are expected to match in 16 bit positions. It is easy to show that if one performs a bitwise comparison of the 32-bit elements

$$
\begin{aligned}
\psi_i \quad &= \quad ([\psi_i]^{31}, [\psi_i]^{30}, \ldots, [\psi_i]^0) \\
\text{and } \psi_j \quad &= \quad ([\psi_j]^{31}, [\psi_j]^{30}, \ldots, [\psi_j]^0)
\end{aligned}
$$

in HC-128, where $1024\tau + 10 \le j < i < 1024\tau + 511$, then the expected number of matches between the corresponding bits is more than 16, and to be specific, is approximately $16 + \frac{13}{12} \cdot 2^{-79}$.

Theorem 10.3.5. *For* $1024\tau + 10 \le j < i < 1024\tau + 511$, *the expected number of bits where the two 32-bit integers* ψ_i *and* ψ_j *match is approximately* $16 + \frac{13}{12} \cdot 2^{-79}$.

Proof: Let $m_b = 1$, if $[\psi_i]^b = [\psi_j]^b$; otherwise, let $m_b = 0$, $0 \le b \le 31$. Hence, the total number of matches is given by

$$M = \sum_{b=0}^{31} m_b.$$

From Theorem 10.3.4, we have $Prob(m_b = 1) = r_b$. Hence, $E(m_b) = r_b$ and by linearity of expectation,

$$E(M) = \sum_{b=0}^{31} E(m_b)$$

$$= \sum_{b=0}^{31} r_b.$$

Substituting the values of r_b's from Theorem 10.3.4, we get

$$E(M) \approx 16 + \frac{13}{3} \cdot 2^{-81}.$$

∎

Along the same line of the above idea, a new word-based distinguisher counting the number of zeros in created sets of (32) bits has recently been reported in [173]. This work claims to require around $2^{152.537}$ such sets to mount the distinguisher.

10.3.3   A Distinguisher Spread over Three 512-Word Blocks

In this section, we are going to describe the distinguisher recently reported in [106].

Once again, refer to the visualization of the array P as explained in Table 10.1. The keystream generation occurs in blocks of 512 words. If $B_1, B_2, B_3, B_4, \ldots$ denote successive blocks, the array P is updated in blocks $B_1, B_3, B_5, \ldots$ and the array Q is updated in blocks $B_2, B_4, B_6, \ldots$, and so on. Without loss of generality, for some fixed t, consider a block B_{2t+1} of 512 keystream word generation in which the array P is updated. The symbol P without any subscript or superscript denotes the *updated* array in the current block B_{2t+1}. Let P_{-1} and P_{-2} denote the *updated* arrays in blocks B_{2t-1} and B_{2t-3} respectively.

From Equation (10.1) (see Section 10.2), using the equality of XOR and sum for the least significant bit, one can write, for $10 \le i \bmod 512 < 511$,

$$[P[i]]^0 = [P_{-1}[i]]^0 \oplus [P[i \boxminus 3]]^{10} \oplus [P_{-1}[i \boxminus 511]]^{23} \oplus [P[i \boxminus 10]]^8. \quad (10.4)$$

Similarly, we can write

$$[P_{-1}[i]]^0 = [P_{-2}[i]]^0 \oplus [P_{-1}[i \boxminus 3]]^{10} \oplus [P_{-2}[i \boxminus 511]]^{23} \oplus [P_{-1}[i \boxminus 10]]^8. \quad (10.5)$$

XOR-ing both sides of Equation (10.4) and Equation (10.5) and rearranging terms, we have, for $10 \le i \bmod 512 < 511$,

$$\begin{aligned}
[P[i]]^0 \oplus [P_{-2}[i]]^0 = & \left([P[i \boxminus 3]]^{10} \oplus [P_{-1}[i \boxminus 3]]^{10} \right) \\
& \oplus \left([P[i \boxminus 10]]^8 \oplus [P_{-1}[i \boxminus 10]]^8 \right) \\
& \oplus \left([P_{-1}[i \boxminus 511]]^{23} \oplus [P_{-2}[i \boxminus 511]]^{23} \right).
\end{aligned} \quad (10.6)$$

As in Section 10.2, the primed array name denotes the updated array, when the two "+" operators are replaced by "$\oplus$" in the update rule (Equation (10.1)). Hence,

$$\left[P'[i \boxminus 3]\right]^{10} = \left[P_{-1}[i \boxminus 3]\right]^{10} \oplus \left[P[i \boxminus 6]\right]^{20} \oplus \left[P_{-1}[i \boxminus 514]\right]^{33} \oplus \left[P[i \boxminus 13]\right]^{18},$$
$$(10.7)$$

$$\left[P'[i \boxminus 10]\right]^{8} = \left[P_{-1}[i \boxminus 10]\right]^{8} \oplus \left[P[i \boxminus 13]\right]^{18} \oplus \left[P_{-1}[i \boxminus 521]\right]^{31} \oplus \left[P[i \boxminus 20]\right]^{16},$$
$$(10.8)$$

and $\left[P'_{-1}[i \boxminus 511]\right]^{23}$

$$= \left[P_{-2}[i \boxminus 511]\right]^{23} \oplus \left[P_{-1}[i \boxminus 514]\right]^{33} \oplus \left[P_{-2}[i \boxminus 1022]\right]^{46} \oplus \left[P_{-1}[i \boxminus 521]\right]^{31}.$$
$$(10.9)$$

Equations (10.7), (10.8) and (10.9) hold in ranges $13 \le i \bmod 512 < 514$, $20 \le i \bmod 512 < 521$ and $9 \le i \bmod 512 < 510$ respectively.

XOR-ing both sides of Equations (10.6), (10.7), (10.8) and (10.9), and rearranging terms, one gets, for $20 \le i \bmod 512 < 510$,

$$\left[P[i]\right]^{0} \oplus \left[P_{-2}[i]\right]^{0} \oplus \left[P[i \boxminus 6]\right]^{20} \oplus \left[P[i \boxminus 20]\right]^{16} \oplus \left[P_{-2}[i \boxminus 510]\right]^{14}$$
$$= \left(\left[P[i \boxminus 3]\right]^{10} \oplus \left[P'[i \boxminus 3]\right]^{10}\right)$$
$$\oplus \left(\left[P[i \boxminus 10]\right]^{8} \oplus \left[P'[i \boxminus 10]\right]^{8}\right)$$
$$\oplus \left(\left[P_{-1}[i \boxminus 511]\right]^{23} \oplus \left[P'_{-1}[i \boxminus 511]\right]^{23}\right).$$
$$(10.10)$$

Lemma 10.3.6. *For* $1024\tau + 20 \le i < 1024\tau + 510$,
$$Prob\left(\left[P[i]\right]^{0} \oplus \left[P_{-2}[i]\right]^{0} \oplus \left[P[i \boxminus 6]\right]^{20} \oplus \left[P[i \boxminus 20]\right]^{16} \oplus \left[P_{-2}[i \boxminus 510]\right]^{14} = 0\right)$$
$$\approx \frac{13}{27}.$$

Proof: From Equation (10.10), we can write the above probability as

$$Prob(\lambda_1 \oplus \lambda_2 \oplus \lambda_3 = 0)$$

where

$$\lambda_1 = \left[P[i \boxminus 3]\right]^{10} \oplus \left[P'[i \boxminus 3]\right]^{10},$$
$$\lambda_2 = \left[P[i \boxminus 10]\right]^{8} \oplus \left[P'[i \boxminus 10]\right]^{8},$$
$$\lambda_3 = \left[P_{-1}[i \boxminus 511]\right]^{23} \oplus \left[P'_{-1}[i \boxminus 511]\right]^{23}.$$

Now, from Lemma 10.2.3 (see Section 10.2), we get

$$Prob(\lambda_1 = 0) = Prob(\lambda_2 = 0) = Prob(\lambda_3 = 0) \approx \frac{1}{3}.$$

Now, XOR of three terms yield 0, if either all three terms are 0 or exactly one is 0 and the other two are 1's. Hence,

$$Prob(\lambda_1 \oplus \lambda_2 \oplus \lambda_3 = 0) \approx \left(\frac{1}{3}\right)^3 + \binom{3}{1}\left(\frac{1}{3}\right)\left(\frac{2}{3}\right)^2$$

$$= \frac{13}{27}.$$

∎

As in Sections 10.3.1 and 10.3.2, let $P[i \boxminus 12]$ at the i-th step be denoted by z_i and replace $P[i \bmod 512]$ by $s_i \oplus h_1(z_i)$. Thus, for $1024\tau + 20 \le i < 1024\tau + 510$, the event

$$\left([P[i]]^0 \oplus [P_{-2}[i]]^0 \oplus [P[i \boxminus 6]]^{20} \oplus [P[i \boxminus 20]]^{16} \oplus [P_{-2}[i \boxminus 510]]^{14} = 0\right)$$

can be alternatively written as $(\chi_i = \mathcal{H}(\mathcal{Z}_i))$, where

$$\chi_i = s_i^0 \oplus s_{i-2048}^0 \oplus s_{i-6}^{20} \oplus s_{i-20}^{16} \oplus s_{i-2046}^{14} \text{ and}$$

$$\mathcal{H}(\mathcal{Z}_i) = h_1(z_i)^0 \oplus h_1''(z_{i-2048})^0 \oplus h_1(z_{i-6})^{20} \oplus h_1(z_{i-20})^{16} \oplus h_1''(z_{i-2046})^{14}.$$

Here $\mathcal{Z}_i = (z_i, z_{i-2048}, z_{i-6}, z_{i-20}, z_{i-2046})$ is an 80-bit input and $\mathcal{H}(.)$ can be considered as a random 80-bit-to-1-bit S-box. Note that $h_1(.)$ and $h_1''(.)$ indicate two different functions, since they are related to two P arrays at two different blocks and thus act as two different S-boxes. With this formulation, Lemma 10.3.6 can be restated as

$$Prob((\chi_i = \mathcal{H}(\mathcal{Z}_i)) \approx \frac{13}{27}. \tag{10.11}$$

Further, from Theorem 10.3.1, for $1024\tau + 20 \le j < i < 1024\tau + 510$,

$$Prob\left(\mathcal{H}(\mathcal{Z}_i) = \mathcal{H}(\mathcal{Z}_j)\right) = \frac{1}{2} + 2^{-81}. \tag{10.12}$$

Theorem 10.3.7. *For $1024\tau + 20 \le j < i < 1024\tau + 510$,*

$$Prob\left(\chi_i = \chi_j\right) \approx \frac{1}{2} + \frac{1}{36^2 8^1},$$

where $\chi_u = s_u^0 \oplus s_{u-2048}^0 \oplus s_{u-6}^{20} \oplus s_{u-20}^{16} \oplus s_{u-2046}^{14}$.

Proof: We need to combine the probability expressions of Equations (10.11) and (10.12). For $1024\tau + 20 \le j < i < 1024\tau + 510$,

$$Prob\left(\chi_i \oplus \chi_j = \mathcal{H}(\mathcal{Z}_i) \oplus \mathcal{H}(\mathcal{Z}_j)\right)$$
$$= Prob\left(\chi_i = \mathcal{H}(\mathcal{Z}_i)\right) \cdot Prob\left(\chi_j = \mathcal{H}(\mathcal{Z}_j)\right) +$$
$$Prob\left(\chi_i = \mathcal{H}(\mathcal{Z}_i) \oplus 1\right) \cdot Prob\left(\chi_j = \mathcal{H}(\mathcal{Z}_j) \oplus 1\right)$$
$$\approx \left(\frac{13}{27}\right)^2 + \left(1 - \frac{13}{27}\right)^2 = \frac{365}{36}.$$

$$Prob\,(\chi_i = \chi_j)$$
$$= \;\; Prob\,(\chi_i \oplus \chi_j = \mathcal{H}(\mathcal{Z}_i) \oplus \mathcal{H}(\mathcal{Z}_j)) \cdot Prob\,(\mathcal{H}(\mathcal{Z}_i) = \mathcal{H}(\mathcal{Z}_j)) \,+$$
$$\quad Prob\,(\chi_i \oplus \chi_j = \mathcal{H}(\mathcal{Z}_i) \oplus \mathcal{H}(\mathcal{Z}_j) \oplus 1) \cdot Prob\,(\mathcal{H}(\mathcal{Z}_i) = \mathcal{H}(\mathcal{Z}_j) \oplus 1)$$
$$\approx \;\; \frac{365}{3^6} \cdot \left(\frac{1}{2} + 2^{-81}\right) + \left(1 - \frac{365}{3^6}\right) \cdot \left(\frac{1}{2} - 2^{-81}\right) = \frac{1}{2} + \frac{1}{3^6 2^{81}}.$$

∎

The above analysis shows that there exist biases in the event $(\chi_i = \chi_j)$, which can be used to mount a distinguishing attack. The probability of the above event can be written as $p_\chi(1 + q_\chi)$, where $p_\chi = \frac{1}{2}$ and $q_\chi = \frac{1}{3^6 2^{80}}$. According to Section 6.1, one would require $\frac{4^2}{p_\chi q_\chi^2} = 3^{12} 2^{165}$ pairs of keystream word combinations of the form (χ_i, χ_j) for a success probability 0.9772. Since each block of $512 = 2^9$ keystream words provides approximately $\binom{512}{2} \approx 2^{17}$ many pairs, the required number of keystream words is approximately $3^{12} 2^{156} \approx 2^{19} 2^{156} = 2^{175}$.

The distinguisher presented in Theorem 10.3.7 may be extended to other bits in a similar line as in Section 10.3.2. However, the bias would be weaker than that presented in Theorem 10.3.7 and so the extended distinguisher would require more keystream words.

10.4   Collisions in h_1, h_2 and State Leakage in Keystream

The previous sections concentrated on the functions g_1, g_2. Here, in a different direction, we study the other two functions h_1, h_2. Without loss of generality, let us focus on the keystream block corresponding to the P array, i.e., the block of 512 rounds where P is updated in each round and Q remains constant. As j runs from 0 to 511, let the corresponding output $h_1(P[j \boxminus 12]) \oplus P[j]$ be denoted by s_j. Recall that $h_1(x) = Q[x^{(0)}] + Q[256 + x^{(2)}]$. The results presented this section are in terms of the function h_1. The same analysis holds for the function h_2 in the other keystream block.

In [40], it has been observed that

$$Prob(s_j \oplus s_{j+1} = P[j] \oplus P[j+1]) \approx 2^{-16}, \tag{10.13}$$

where s_j, s_{j+1} are two consecutive keystream output words. A more detailed study of this observation is studied [105] and a sharper association is found, which gives twice the above probability. In this section, we essentially present this and related results of [105].

The following technical result shows that XOR of two words of P is leaked in the keystream words, if the corresponding values of $h_1(.)$ collide.

Lemma 10.4.1. *For $0 \le u \neq v \le 511$, $s_u \oplus s_v = P[u] \oplus P[v]$ if and only if $h_1(P[u \boxminus 12]) = h_1(P[v \boxminus 12])$.*

Proof: We have $s_u = h_1(P[u \boxminus 12]) \oplus P[u]$ and $s_v = h_1(P[v \boxminus 12]) \oplus P[v]$. Thus,

$$s_u \oplus s_v = \big(h_1(P[u \boxminus 12]) \oplus h_1(P[v \boxminus 12])\big) \oplus (P[u] \oplus P[v]).$$

The term $\big(h_1(P[u \boxminus 12]) \oplus h_1(P[v \boxminus 12])\big)$ vanishes if and only if $s_u \oplus s_v = P[u] \oplus P[v]$.   ∎

Assume that the array P from which the input to the function h_1 is selected and the array Q from which the output of h_1 is chosen contain uniformly distributed 32-bit elements. The notations h and U are used in Lemma 10.4.2, which is in a more general setting than just HC-128; these may be considered to model h_1 and Q respectively for HC-128.

Lemma 10.4.2. *Let $h(x) = U[x^{(0)}] + U[x^{(2)} + 2^m]$ be an n-bit to n-bit mapping, where each entry of the array U is an n-bit number, U contains 2^{m+1} many elements and $x^{(0)}$ and $x^{(2)}$ are two disjoint m-bit segments from the n-bit input x. Suppose x and x' are two n-bit random inputs to h. Assume that the entries of U are distributed uniformly at random. Then for any collection of t distinct bit positions $\{b_1, b_2, \ldots, b_t\} \subseteq \{0, 1, \ldots, n-1\}$, $0 \leq t \leq n$,*

$$Prob\left(\bigwedge_{l=0}^{t} \left([h(x)]^{b_l} = [h(x')]^{b_l}\right)\right) = \gamma_{m,t}$$

where $\gamma_{m,t} = 2^{-2m} + (1 - 2^{-2m})2^{-t}$.

Proof: Each value $v \in [0, 2^n - 1]$ can be expressed as the modulo 2^n sum of two integers $a, b \in [0, 2^n - 1]$ in exactly 2^n ways, since for each choice of $a \in [0, 2^n - 1]$, the choice of b is fixed as $b = (v - a) \bmod 2^n$. It is given that each $U[\alpha]$ is uniformly distributed over $[0, 2^n - 1]$. Thus, for uniformly random $\alpha, \beta \in [0, 2^{m+1} - 1]$, the sum $(U[\alpha] + U[\beta]) \bmod 2^n$ is also uniformly distributed and so is any collection of t bits of the sum.

Now, $h(x) = U[x^{(0)}] + U[x^{(2)} + 2^m]$ and $h(x') = U[x'^{(0)}] + U[x'^{(2)} + 2^m]$ can be equal in bit positions $b_1, b_2, \ldots, b_t$ in the following two ways.

Case I: $x^{(0)} = x'^{(0)}$ and $x^{(2)} = x'^{(2)}$. This happens with probability $2^{-m} \cdot 2^{-m}$.

Case II: The event "$x^{(0)} = x'^{(0)}$ and $x^{(2)} = x'^{(2)}$" does not happen and still $\big(U[x^{(0)}] + U[x^{(2)} + 2^m]\big)$ and $\big(U[x'^{(0)}] + U[x'^{(2)} + 2^m]\big)$ match in the t bits due to random association. This happens with probability $(1 - 2^{-2m}) \cdot 2^{-t}$.

Adding the two components, we get the result.   ∎

$\gamma_{m,t}$ depends only on m, t and not on n, as is expected from the particular form of $h_1(.)$ that uses only $2m$ bits from its n-bit random input. For HC-128, we have $n = 32$ and $m = 8$. For the equality of the whole output word of $h_1(.)$, one needs to set $t = n = 32$. Thus, $Prob(h_1(x) = h_1(x')) = \gamma_{8,32} =$

$2^{-16} + 2^{-32} - 2^{-48} \approx 0.0000152590$, which is slightly greater than 2^{-16}. Note that if one checks the equality of two n-bit random integers, the probability of that event turns out to be 2^{-n} only, which is as low as 2^{-32} for $n = 32$.

Now the result given in [40] can be formalized as

Theorem 10.4.3. *In HC-128, consider a block of 512 many keystream words corresponding to the array P. For $0 \le u \ne v \le 511$,*

$$Prob\left((s_u \oplus s_v) = (P[u] \oplus P[v])\right) = \gamma_{8,32} > 2^{-16}.$$

Proof: The result follows from Lemma 10.4.1 and Lemma 10.4.2.   ∎

A sharper result which gives twice the above probability is as follows.

Theorem 10.4.4. *In HC-128, consider a block of 512 many keystream words corresponding to the array P. For any u, v, $0 \le u \ne v < 500$, if*

$$\left((s_u^{(0)} = s_v^{(0)}) \ \& \ (s_u^{(2)} = s_v^{(2)})\right),$$

then

$$Prob\left((s_{u+12} \oplus s_{v+12}) = (P[u+12] \oplus P[v+12])\right) \approx \frac{1}{2^{15}}.$$

Proof: From Lemma 10.4.1, $s_u^{(b)} \oplus s_v^{(b)} = P[u]^{(b)} \oplus P[v]^{(b)}$ if and only if

$$h_1(P[u \boxminus 12])^{(b)} = h_1(P[v \boxminus 12])^{(b)}, \quad \text{for } b = 0, 1, 2, 3.$$

Given that $s_u^{(0)} = s_v^{(0)}$ and $s_u^{(2)} = s_v^{(2)}$, we have $P[u]^{(0)} = P[v]^{(0)}$ and $P[u]^{(2)} = P[v]^{(2)}$ if and only if $h_1(P[u \boxminus 12])^{(0)} = h_1(P[v \boxminus 12])^{(0)}$ and $h_1(P[u \boxminus 12])^{(2)} = h_1(P[v \boxminus 12])^{(2)}$. Thus,

$$Prob\left(P[u]^{(0)} = P[v]^{(0)} \ \& \ P[u]^{(2)} = P[v]^{(2)} \mid s_u^{(0)} = s_v^{(0)} \ \& \ s_u^{(2)} = s_v^{(2)}\right)$$
$$= Prob\left(h_1(P[u \boxminus 12])^{(0)} = h_1(P[v \boxminus 12])^{(0)} \ \& \right.$$
$$\left. h_1(P[u \boxminus 12])^{(2)} = h_1(P[v \boxminus 12])^{(2)}\right)$$
$$= \gamma_{8,16} \quad \text{(from Lemma 10.4.2)}$$
$$\approx \frac{1}{2^{15}}.$$

By definition, $h_1(x) = Q[x^{(0)}] + Q[256 + x^{(2)}]$. So the equalities $P[u]^{(0)} = P[v]^{(0)}$ and $P[u]^{(2)} = P[v]^{(2)}$ give $h_1(P[u]) = h_1(P[v])$ and this in turn gives $s_{u+12} \oplus s_{v+12} = P[u+12] \oplus P[v+12]$ by Lemma 10.4.1.   ∎

The event $(s_u^{(0)} = s_v^{(0)} \ \& \ s_u^{(2)} = s_v^{(2)})$ occurs with probability 2^{-16} for a random stream. If the probability of this event in HC-128 is away from 2^{-16}, we would immediately have a distinguisher. The experimental data does not reveal any observable bias for this event. However, it would be interesting to investigate if there exists a bias of very small order for this or a similar event in HC-128.

The Jenkins' Correlation or Glimpse Theorem (see Section 5.2) is an important result on the weakness of RC4 stream cipher. This result quantifies the leakage of state information into the keystream of RC4. The leakage probability is twice the random association 2^{-8}. Theorem 10.4.4 is a Glimpse-like theorem on HC-128 that shows the leakage of state information into the keystream with a probability $\approx 2^{-15}$, that is much more than 2^{-31} (two times the random association 2^{-32}). This probability is in fact two times the square-root of the random association. However, in RC4, the Jenkins' Correlation turns the key-state correlations into key-keystream correlations that lead to practical attacks. On the other hand, in case of HC-128, no key-state correlations have been found so far. So the state leakage in HC-128 keystream, as of now, does not lead to any real attack on HC-128.

10.5 Constructing Full-State Given Only Half-State Information

Recently, it has been shown in [137] that the knowledge of any one of the two internal state arrays of HC-128 along with the knowledge of 2048 keystream words is sufficient to construct the other state array completely in 2^{42} time complexity. Though this analysis does not lead to any attack on HC-128, it reveals a nice combinatorial property of HC-128 keystream generation algorithm. Moreover, the secret key being derivable from any state [97], only half of the state is now sufficient to determine the key uniquely. This analysis can also serve as a general model to study stream ciphers that have a similar dual state internal structure. We present this analysis in this section.

There are two internal state arrays of HC-128, P and Q, each containing 512 many 32-bit words. The keystream is generated in blocks of 512 words. Within a block, one of these arrays gets updated and the keystream word is produced by XOR-ing the updated entry with the sum of two words from the other array. The role of the two arrays is reversed after every block of 512 keystream words generation.

10.5.1 Formulation of Half-State Exposure Analysis

Without loss of generality, let us consider four consecutive blocks B_1, B_2, B_3 and B_4 of keystream generation such that Q is updated in blocks B_1 and B_3 and P is updated in blocks B_2 and B_4. Suppose the keystream words corresponding to all of these four blocks are known. Henceforth, by the symbols P and Q, we will denote the arrays after the completion of block B_1 and before the start of block B_2. After the completion of block B_2, Q remains unchanged and P is updated to, say, P_N. After the completion of block B_3, Q would again be updated to, say, Q_N.

Block B_1:	Block B_2:	Block B_3:
P unchanged,	P updated to P_N,	P_N unchanged,
Q updated.	Q unchanged.	Q updated to Q_N.
(Q denotes the updated array)		

Block B_4, that is not shown in the diagram, would only be used for verifying whether the reconstruction is correct or not. Algorithm 10.5.1 (called *Reconst>ructState*) takes as inputs the 512 words of the array P and (assuming that the 2048 keystream words corresponding to the four blocks B_1, B_2, B_3 and B_4 are known) produces as output 512 words of the array Q_N. Since the update of an array depends only on itself, it turns out that from block B_3 onwards the complete state becomes known.

In B_1 and B_3, Q is updated. Let Q denote the updated array after the completion of block B_1 and let Q_N be the new array after Q is updated in block B_3. In B_1, P remains unchanged and in B_2, it is updated to P_N. Let $s_{b,i}$ denote the i-th keystream word produced in block B_b, $1 \le b \le 4$, $0 \le i \le 511$.

The update of P (or Q) depends only on itself, i.e.,

$$P_N[i] = \begin{cases} P[i] + g_1(P[509+i], P[502+i], P[i+1]), & \text{for } 0 \le i \le 2; \\ P[i] + g_1(P_N[i-3], P[502+i], P[i+1]), & \text{for } 3 \le i \le 9; \\ P[i] + g_1(P_N[i-3], P_N[i-10], P[i+1]), & \text{for } 10 \le i \le 510; \\ P[i] + g_1(P_N[i-3], P_N[i-10], P_N[i-511]), & \text{for } i = 511. \end{cases}$$
$$(10.14)$$

Thus, if one knows the 512 words of P (or Q) corresponding to any one block, then one can easily derive the complete P (or Q) array corresponding to any subsequent block.

Consider that the keystream words $s_{b,i}$, $1 \le b \le 4$, $0 \le i \le 511$, are observable. A state reconstruction problem can be formulated as follows.

> Given the partial state information $P[0 \dots 511]$, reconstruct the complete state ($P_N[0 \dots 511], Q_N[0 \dots 511]$).

Since the update of each of P and Q depends only on P and Q respectively, once we determine P_N and Q_N, we essentially recover the complete state information for all subsequent steps.

10.5.2 State Reconstruction Strategy

The state reconstruction proceeds in five phases. The **First Phase** would be to determine P_N from P using (10.14).

The keystream generation of block B_2 follows the equation

$$s_{2,i} = \begin{cases} h_1(P[500+i]) \oplus P_N[i], & \text{for } 0 \le i \le 11; \\ h_1(P_N[i-12]) \oplus P_N[i], & \text{for } 12 \le i \le 511. \end{cases} \qquad (10.15)$$

Since $h_1(x) = Q[x^{(0)}] + Q[256 + x^{(2)}]$, we can rewrite (10.15) as

$$Q[l_i] + Q[u_i] = s_{2,i} \oplus P_N[i], \qquad (10.16)$$

where

$$\begin{array}{ll} \text{for } 0 \le i \le 11, & l_i = (P[500 + i])^{(0)}, u_i = 256 + (P[500 + i])^{(2)} \\ \text{\& for } 12 \le i \le 511, & l_i = (P_N[i - 12])^{(0)}, u_i = 256 + (P_N[i - 12])^{(2)}. \end{array} \Bigg\}$$

$$(10.17)$$

Here l_i, u_i and the right-hand side $s_{2,i} \oplus P_N[i]$ of system (10.16) of equations are known for all $i = 0, 1, \ldots, 511$. Thus, there are 512 equations in ≤ 512 unknowns. Simply applying Gauss elimination would require a complexity of $512^3 = 2^{27}$. However, according to Lemma 10.5.1, a unique solution does not exist for any such system and hence a different approach is required to solve the system.

Lemma 10.5.1. *Suppose $r + s$ linear equations are formed using variables from the set $\{x_1, x_2, \ldots, x_r, y_1, y_2, \ldots, y_s\}$. If each equation is of the form $x_i + y_j = b_{ij}$ for some i in $[1, r]$ and some j in $[1, s]$, where b_{ij}'s are all known, then such a system does not have a unique solution.*

Proof: Consider the $(r+s) \times (r+s)$ coefficient matrix A of the system with the columns denoted by $C_1, \ldots, C_{r+s}$, such that the first r columns $C_1, \ldots, C_r$ correspond to the variables $x_1, \ldots, x_r$ and the last s columns $C_{r+1}, \ldots, C_{r+s}$ correspond to the variables $y_1, \ldots, y_s$. Every row of A has the entry 1 in exactly two places and the entry 0 elsewhere. The first 1 in each row appears in one of the columns $C_1, \ldots, C_r$ and the second 1 in one of the columns $C_{r+1}, \ldots, C_{r+s}$. After the elementary column transformations $C_1 \leftarrow C_1 + \ldots + C_r$ and $C_{r+1} \leftarrow C_{r+1} + \ldots + C_{r+s}$, the two columns C_1 and C_{r+1} has 1's in all the rows and hence become identical. This implies that the matrix is not of full rank and hence a unique solution does not exist for the system. ∎

The left-hand side of every equation in system (10.16) is of the form $Q[l_i] + Q[u_i]$, where $0 \le l_i \le 255$ and $256 \le u_i \le 511$. Taking $r = s = 256$, $x_i = Q[i - 1]$, $1 \le i \le 256$ and $y_j = Q[255 + j]$, $1 \le j \le 256$, we see that Lemma 10.5.1 directly applies to this system, establishing the non-existence of a unique solution. At this stage, one could remove the redundant rows to find a linear space which contains the solution. However, it is not clear how many variables need to be guessed to arrive at the final solution. A graph theoretic approach has been used in [137] to derive the entries of the array Q efficiently, by guessing the value of only a single variable.

Definition 10.5.2. *System (10.16) of 512 equations can be represented in the form of a bipartite graph $G = (V_1, V_2, E)$, where $V_1 = \{0, \ldots, 255\}$, $V_2 = \{256, \ldots, 511\}$ and for each term $Q[l_i] + Q[u_i]$ of (10.16), there is an edge $\{l_i, u_i\} \in E$, $l_i \in V_1$ and $u_i \in V_2$. Thus, $|E| = 512$ (counting repeated edges, if any). We call such a graph G, with the vertices as the indices of one internal array of HC-128, as the index graph of the state of HC-128.*

Lemma 10.5.3. *Let M be the size of the largest connected component of the index graph G corresponding to block B_2. Then M out of 512 words of the array Q can be derived in 2^{32} search complexity.*

Proof: Consider any one of the 512 equations of System (10.16). Since the sum $Q[l_i] + Q[u_i]$ is known, knowledge of one of $Q[l_i]$, $Q[u_i]$ reveals the other. Thus, if we know one word of Q at any index of a connected component, we can immediately derive the words of Q at all the indices of the same component. Since this holds for each connected component, we can guess any one 32-bit word in the largest connected component correctly in 2^{32} attempts and thereby the result follows. ∎

Since the arrays P, Q and the keystream of HC-128 are assumed to be random, the *index graph* G can be considered to be a random bipartite graph. Theoretical analysis of the size distribution of the connected components of random finite graphs is a vast area of research in applied probability and there have been several works [29, 65, 85, 125, 176] in this direction under different graph models. In [176], the model considered is a bipartite graph $G(n_1, n_2, T)$ with n_1 vertices in the first part, n_2 vertices in the second one and the graph is constructed by T independent trials. Each of them consists of drawing an edge which joins two vertices chosen independently of each other from distinct parts. This coincides with the index graph model of Definition 10.5.2 with $n_1 = |V_1|$, $n_2 = |V_2|$ and $T = |E|$.

In general, let $n_1 \geq n_2$, $\alpha = \frac{n_2}{n_1}$, $\beta = (1-\alpha)\ln n_1$, $n = n_1 + n_2$. Let $\xi_{n_1,n_2,T}$ and $\chi_{n_1,n_2,T}$ respectively denote the number of isolated vertices and the number of connected components in $G(n_1, n_2, T)$. We have the following result from [176].

Proposition 10.5.4. *If $n \to \infty$ and $(1+\alpha)T = n\ln n + Xn + o(n)$, where X is a fixed number, then $Prob\left(\chi_{n_1,n_2,T} = \xi_{n_1,n_2,T} + 1\right) \to 1$ and for any $k = 0, 1, 2, \ldots$, $Prob\left(\xi_{n_1,n_2,T} = k\right) - \frac{\lambda^k e^{-\lambda}}{k!} \to 0$, where $\lambda = \frac{e^{-X}(1+e^{-\beta})}{1+\alpha}$.*

In other words, if n is sufficiently large and n_1, n_2, T are related by $(1+\alpha)T = n\ln n + Xn + o(n)$, then the graph contains one giant connected component and isolated vertices whose number follows a Poisson distribution with parameter λ given above.

Corollary 10.5.5. *If M is the size of the largest component of the index graph G, then the mean and standard deviation of M is respectively given by $E(M) \approx 442.59$ and $sd(M) \approx 8.33$.*

Proof: For the index graph of HC-128, $n_1 = n_2 = 256$, $n = n_1 + n_2 = 512$, $T = 512$, $\alpha = \frac{n_2}{n_1} = 1$, $\beta = (1-\alpha)\ln n_1 = 0$. The relation $(1+\alpha)T = n\ln n + Xn + o(n)$ is equivalent to $\frac{(1+\alpha)}{n}T = \ln n + X + \frac{o(n)}{n}$. As $n \to \infty$, the ratio $\frac{o(n)}{n} \to 0$ and hence $X \to \frac{(1+\alpha)}{n}T - \ln n$. Substituting $\alpha = 1$, $T = 512$ and $n = 512$, we get $X \approx -4.24$. By Proposition 10.5.4, the limiting distribution of the random variable $\xi_{n_1,n_2,T}$ is Poisson with mean (as well as variance) $\lambda = \frac{e^{-X}(1+e^{-\beta})}{1+\alpha} \approx e^{4.24} \approx 69.41$. Moreover, in the limit, $\chi_{n_1,n_2,T} = \xi_{n_1,n_2,T} + 1$ and this implies that all the vertices except the isolated ones would be in a single giant component. So $M = n - \xi_{n_1,n_2,T}$ and the expectation $E(M) = n - E(\xi_{n_1,n_2,T}) = n - \lambda \approx 512 - 69.41 = 442.59$.

Again, the variance $Var(M) = Var(n - \xi_{n_1,n_2,T}) = Var(\xi_{n_1,n_2,T}) = \lambda$, giving $sd(M) = sd(\xi_{n_1,n_2,T}) = \sqrt{\lambda} \approx 8.33$. ∎

Simulation with 10 million trials, each time with 1024 consecutive words of keystream generation for the complete arrays P and Q, gives the average of the number $\xi_{n_1,n_2,T}$ of isolated vertices of the index graph of the state of HC-128 as 69.02 with a standard deviation of 6.41. These values closely match with the theoretical estimates of the mean $\lambda \approx 69.41$ and standard deviation $\sqrt{\lambda} \approx 8.33$ of $\xi_{n_1,n_2,T}$ derived in Corollary 10.5.5.

From Corollary 10.5.5, theoretical estimates of the mean and standard deviation of the size M of the largest component is 442.59 and 8.33 respectively. From the same simulation described above, the empirical average and standard deviation of M are found to be $407.91 \approx 408$ and 9.17 respectively.

The small gap between the theoretical estimate and the empirical result arises from the fact that the theory takes n as infinity, but its value in the context of HC-128 is 512. In the limit when $n \to \infty$, each vertex is either an isolated one or part of the single giant component. In practice, on the other hand, except for the isolated vertices (≈ 69 in number) and the vertices of the giant component (≈ 408 in number), the remaining few ($\approx 512 - 69 - 408 = 35$ in number) vertices form some small components. However, the low (9.17) empirical standard deviation of M implies that the empirical estimate 408 of $E(M)$ is robust. We will see later that as a consequence of Theorem 10.5.7, any $M > 200$ is sufficient for our purpose.

If $C = \{y_1, y_2, \ldots, y_M\}$ be the largest component of G, then we can guess the word corresponding to any fixed index, say y_1. As explained in the proof of Lemma 10.5.3, each guess of $Q[y_1]$ uniquely determines the values of $Q[y_2], \ldots, Q[y_M]$. According to Corollary 10.5.5 and the discussion following it, we can guess around 408 words of Q in this method. This is the **Second Phase** of the recovery algorithm.

The following result, known as the *Propagation Theorem*, can be used to determine the remaining unknown words.

Theorem 10.5.6 (Propagation Theorem). *If $Q[y]$ is known for some y in $[0, 499]$, then $m = \lfloor \frac{511-y}{12} \rfloor$ more words of Q, namely, $Q[y + 12], Q[y + 24], \ldots, Q[y + 12m]$, can all be determined from $Q[y]$ in a time complexity that is linear in the size of Q.*

Proof: Consider block B_1. Following the notation in Section 10.5.1, the equation for keystream generation is
$$s_{1,i} = h_2(Q[i - 12]) \oplus Q[i], \text{ for } 12 \leq i \leq 511.$$
Written in another way, it becomes
$$Q[i] = s_{1,i} \oplus \left(P\left[(Q[i - 12])^{(0)}\right] + P\left[256 + (Q[i - 12])^{(2)}\right] \right).$$
Setting $y = i - 12$, we have, for $0 \leq y \leq 499$,

$$Q[y + 12] = s_{1,y+12} \oplus \left(P\left([Q[y]]^{(0)}\right) + P\left[256 + (Q[y])^{(2)}\right] \right) \qquad (10.18)$$

This is a recursive equation, in which all s_1 values and the array P are completely known. Clearly, if we know one $Q[y]$, we know all subsequent $Q[y + 12k]$, for $k = 1, 2, \ldots$, as long as $y + 12k \leq 511$. This means $k \leq \frac{511-y}{12}$. The number m of words of Q that can be determined is then the maximum allowable value of k, i.e., $m = \lfloor \frac{511-y}{12} \rfloor$. ∎

By recursively applying (10.18) to the words of the Q array determined from the maximum size connected component of the index graph, we derive many of around $104(= 512 - 408)$ unknown words in the array. This is the **Third Phase** of the recovery algorithm. If we imagine the words initially labeled as "known" or "unknown," then this step can be visualized as propagation of the "known" labels in the forward direction. Even after this step, some words remain unknown. However, as Theorem 10.5.7 implies, through this propagation, all the words $Q[500], Q[501], \ldots, Q[511]$ can be "known" with probability almost 1.

Theorem 10.5.7. *After the Third Phase, the expected number of unknown words among $Q[500]$, $Q[501]$, $\ldots$, $Q[511]$ is approximately $8 \cdot (1 - \frac{43}{512})^M + 4 \cdot (1 - \frac{42}{512})^M$, where M is the size of the largest component of the index graph G.*

Proof: After the Second Phase, exactly M words $Q[y_1], Q[y_2], \ldots, Q[y_M]$ are known corresponding to the distinct indices $y_1, y_2, \ldots, y_M$ in the largest component C of size M in G. Since G is a random bipartite graph, each of indices $y_1, y_2, \ldots y_M$ can be considered to be drawn from the set $\{0, 1, \ldots, 511\}$ uniformly at random (without replacement). We partition this sample space into 12 disjoint residue classes modulo 12, denoted by, $[0], [1], \ldots, [11]$. Then, each of the indices $y_1, y_2, \ldots, y_M$ can be considered to be drawn from the set $\{[0], [1], \ldots, [11]\}$ (this time with replacement; this is a reasonable approximation because $M \gg 12$) with probabilities proportional to the sizes of the residue classes. Thus, for $1 \leq j \leq M$, $Prob(y_j \in [r]) = \frac{43}{512}$ if $0 \leq r \leq 7$ and $\frac{42}{512}$ if $8 \leq r \leq 11$.

Let $m_r = 1$, if none of $y_1, y_2, \ldots, y_M$ are from $[r]$; otherwise, let $m_r = 0$. Hence, the total number of residue classes from which no index is selected is $Y = \sum_{r=0}^{11} m_r$. Now, in the Third Phase, we propagate the known labels in the forward direction using (10.18) (see Theorem 10.5.6, the Propagation Theorem). The indices $\{500, 501, \ldots, 511\}$ are to the extreme right end of the array Q and hence they also form the set of "last" indices where the propagation eventually stops. Further, each index in the set $\{500, 501, \ldots, 511\}$ belongs to exactly one of the sets $[r]$. Hence, the number of unknown words among $Q[500], Q[501], \ldots, Q[511]$ is also given by Y.

We have,

$$E(m_r) = Prob(m_r = 1) = \begin{cases} (1 - \frac{43}{512})^M & \text{for } 0 \leq r \leq 7; \\ (1 - \frac{42}{512})^M & \text{for } 8 \leq r \leq 11. \end{cases}$$

Thus, $E(Y) = \sum_{r=0}^{11} E(m_r) = 8 \cdot (1 - \frac{43}{512})^M + 4 \cdot (1 - \frac{42}{512})^M$.   ∎

Substituting M by its theoretical mean estimate 443 as well as by its empirical mean estimate 408 yields $E(Y) \approx 0$. In fact, for any $M > 200$, the expression $(1 - \frac{43}{512})^M + 4 \cdot (1 - \frac{42}{512})^M$ for $E(Y)$ becomes vanishingly small. The experimental data also supports that in every instance, none of the words $Q[500], Q[501], \ldots, Q[511]$ remains unknown.

Next, the following result can be used to determine the entire Q_N array.

Theorem 10.5.8. *Suppose the complete array P_N and the 12 words $Q[500]$, $Q[501]$, ..., $Q[511]$ from the array Q are known. Then the entire Q_N array can be reconstructed in a time complexity linear in the size of Q.*

Proof: Following the notation in Section 10.5.1, the equation for the keystream generation of the first 12 steps of block B_3 is $s_{3,i} = h_2(Q[500 + i]) \oplus Q_N[i]$, $0 \le i \le 11$. Expanding $h_2(.)$, we get, for $0 \le i \le 11$,

$$Q_N[i] = s_{3,i} \oplus \left(P_N \left[(Q[500 + i])^{(0)} \right] + P_N \left[256 + (Q[500 + i])^{(2)} \right] \right).$$

Thus, we can determine

$$Q_N[0], Q_N[1], \ldots Q_N[11]$$

from

$$Q[500], Q[501], \ldots Q[511].$$

Now, applying Theorem 10.5.6 on these first 12 words of Q_N, we can determine all the words of Q_N in linear time (in size of Q).   ∎

Applying Theorem 10.5.8 constitute the **Fourth Phase** of the algorithm.

After Q_N is derived, its correctness needs to be verified. For this, one can update P_N as in block B_4 and generate 512 keystream words (with this P_N and the derived Q_N). If the generated keystream words entirely match with the observed keystream words of block B_4, then the guess can be assumed to be correct. This verification is the **Fifth** (and final) **Phase** of the algorithm. If a mismatch is found, the procedure needs to be repeated with the next guess, i.e., with another possible value in $[0, 2^{32} - 1]$ of the word $Q[y_1]$.

After Q_N is correctly determined, the words of the Q array for all the succeeding blocks can be deterministically computed from the update rule for Q.

The above discussion is formalized in Algorithm 10.5.1, called *Reconstruct-State*.

Theorem 10.5.9. *The data complexity of Algorithm 10.5.1 is 2^{16} and its time complexity is 2^{42}.*

Proof: For the First Phase, we do not need any keystream word. For each of the Second, Third, Fourth and Fifth Phases, we need a separate block

Input: $P[0\ldots511]$.
Output: $P_N[0\ldots511], Q_N[0\ldots511]$.

First Phase:

1   **for** $i \leftarrow 0$ **to** 511 **do**
2 | Determine $P_N[i]$ using (10.14);
    end

Second Phase:

3   Form a bipartite graph $G = (V_1, V_2, E)$ as follows;
4   $V_1 \leftarrow \{0, \ldots, 255\};\ V_2 \leftarrow \{256, \ldots, 511\};\ E \leftarrow \emptyset$;
5   **for** $i \leftarrow 0$ **to** 511 **do**
6 | Determine l_i and u_i using (10.17);
7 | $E \leftarrow E \cup \{l_i, u_i\}$;
    end
8   Find all connected components of G;
9   Let $C = \{y_1, y_2, \ldots, y_M\}$ be the largest component with size M;
10   Guess $Q[y_1]$ and thereby determine $Q[y_2], \ldots, Q[y_M]$ from (10.16);
     and for each such guess of $Q[y_1]$, repeat the three Phases below;

Third Phase:

11   **for** $j \leftarrow 1$ **to** M **do**
12 | $y \leftarrow y_j$;
13 | **while** $y \leq 499$ **do**
14 | **if** $Q[y+12]$ *is still unknown* **then**
15 | $Q[y+12] \leftarrow s_{1,y+12} \oplus \left(P\left[(Q[y])^{(0)}\right] + P\left[256 + (Q[y])^{(2)}\right] \right)$;
       end
16 | $y \leftarrow y + 12$;
     end
    end

Fourth Phase:

17   **for** $i \leftarrow 0$ **to** 11 **do**
18 | $Q_N[i] \leftarrow s_{3,i} \oplus \left(P_N\left[(Q[500+i])^{(0)}\right] + P_N\left[256 + (Q[500+i])^{(2)}\right] \right)$;
20 | $y \leftarrow i$;
21 | **while** $y \leq 499$ **do**
22 | $Q_N[y+12] \leftarrow s_{3,y+12} \oplus \left(P_N\left[(Q_N[y])^{(0)}\right] + P_N\left[256 + (Q_N[y])^{(2)}\right] \right)$;
23 | $y \leftarrow y + 12$;
     end
    end

Fifth Phase:

24   With the new Q_N, generate 512 keystream words by updating P_N;
25   Verify correctness of the guess in Step 10 by matching these keystream
     words with the observed keystream words of block B_4;

Algorithm 10.5.1: ReconstructState

of 512 keystream words. Thus, the required amount of data is $4 \cdot 512 = 2^{11}$ no. of 32 $(= 2^5)$-bit keystream words.

From Step 1 in the First Phase up to Step 7 of the Second Phase, the total time required is linear in the size of P (or Q), i.e., of complexity 2^9. Step 8 in the Second Phase of Algorithm 10.5.1 can be performed through depth-first search which requires $O(|V_1| + |V_2| + |E|)$ time complexity. For $|V_1| = 256$, $|V_2| = 256$ and $|E| = 512$, the value turns out to be 2^{10}. After this, the guess in Step 10 of Algorithm 10.5.1 consumes 2^{32} time and for each such guess, the complete Phases 3, 4 and 5 together take time that is linear in the size of the array Q, i.e., of complexity 2^9. Thus, the total time required is $2^9 + 2^{10} + 2^{32} \cdot 2^9 < 2^{42}$. ■

Note that for system (10.16) of equations, one must verify the solution by first generating some keystream words and then matching them with the observed keystream, as is done in the Fifth Phase of Algorithm 10.5.1. During Step 10 in the Second Phase, one may exploit the cycles of the largest component to verify correctness of the guess. If the guessed value of a variable in a cycle does not match with the value of the variable derived when the cycle is closed, we can discard that guess. However, in the worst case, all the 2^{32} guesses have to be tried and if there is no conflict in a cycle, the guess has to be verified by keystream matching. Thus, it is not clear if there is any significant advantage by detecting and exploiting the cycles.

10.6 Design Modification with Respect to Known Observations

This section is also based on [137] which advocates two design goals.

- to guard against the available analysis in literature and

- not to sacrifice the speed in the process.

Thus, an attempt is made to keep the same structure as the original HC-128 with minimal changes.

$h_1(.)$ as well as $h_2(.)$ makes use of only 16 bits from the 32-bit input and this fact leads to all the vulnerabilities discussed so far in this Chapter. Thus, the form of $h_1(.), h_2(.)$ must be modified so as to incorporate all the 32 bits of their inputs. In the new versions of these functions (equation (10.19)), XOR-ing the entire input with the existing output (sum of two array entries) is suggested. However, certain precautions may need to be taken so that other security threats do not come into play.

h_1 and h_2 are modified as follows.

$$\left. \begin{aligned} h_{N1}(x) &= (Q[x^{(0)}] + Q[256 + x^{(2)}]) \oplus x. \\ h_{N2}(x) &= (P[x^{(0)}] + P[256 + x^{(2)}]) \oplus x. \end{aligned} \right\} \tag{10.19}$$

The update functions g_1 and g_2 also need to be modified with the twin motivation of preserving the internal state as well as making sure that the randomness of the keystream is ensured. The following modifications are proposed in [137].

$$\left. \begin{aligned} g_{N1}(x,y,z) &= \left((x \ggg 10) \oplus (z \ggg 23)\right) + Q[(y \gg 7) \wedge 1FF]. \\ g_{N2}(x,y,z) &= \left((x \lll 10) \oplus (z \lll 23)\right) + P[(y \gg 7) \wedge 1FF]. \end{aligned} \right\} \quad (10.20)$$

f_1 and f_2 are kept the same as in original HC-128.

A randomly chosen word from the Q array is included in the update of the P array elements and a randomly chosen word from the P array while updating the Q array elements. This would ensure that each new block of the P (or Q) array is dependent on the previous block of Q(or P) array. Thus, the analysis of Section 10.5 would not apply and the internal state would be preserved, even if half the internal state elements are known.

Likewise, in the equation of the distinguisher proposed by the designer [187, Section 4], the term $P[i \boxminus 10]$ would get replaced by some random term of Q array. With this replacement, it is not obvious how a similar distinguishing attack can be mounted. The similar situation will happen for the distinguishers proposed in [105].

Now consider the fault attack of [84]. It assumes that if a fault occurs at $Q[f]$ in the block in which P is updated, then $Q[f]$ is not referenced until step $f - 1$ of the next block (in which Q would be updated). This assumption does not hold for the new design due to the nesting use of P and Q in the updates of one another (equation (10.20)). Thus, on the modified design, the fault position recovery algorithm given in [84, Section 4.2] would not work immediately. In particular, Lemma 1 and Lemma 2 of [84] would not hold on the modified cipher. However, the security claim of the modified HC-128, as in any stream cipher, is always a conjecture.

10.6.1 Performance Evaluation

The performance of the new design has been evaluated using the eSTREAM testing framework [26]. The C-implementation of the testing framework was installed in a machine with Intel(R) Pentium(R) D CPU, 2.8 GHz Processor Clock, 2048 KB Cache Size, 1 GB DDR RAM on Ubuntu 7.04 (Linux 2.6.20-17-generic) OS. A benchmark implementation of HC-128 and HC-256 [188] is available within the test suite. The modified version of HC-128 was implemented, maintaining the API compliance of the suite. Test vectors were generated in the NESSIE [127] format. The results presented below correspond to tests with null IV using the gcc-3.4_prescott_O3-ofp compiler.

	HC-128	New Proposal	HC-256
Stream Encryption (cycles/byte)	4.13	4.29	4.88

The encryption speed of the proposed design is of the same order as that of the original HC-128. Also observe that the extra array element access in the new update rules (Equation (10.20)) as compared to the original update rules does not affect the performance much. HC-128 was designed as a lightweight version of HC-256. The idea of cross-referencing each other in the update rules of P and Q has also been used in the design of HC-256 and that is why the half state exposure does not reveal the full state in case of HC-256. However, the modification to HC-128 removes the known weaknesses of HC-128 but keeps the speed much better than HC-256, with only a little reduction in speed compared to HC-128.

Research Problems

Problem 10.1 Is there a distinguishing attack on HC-128, that requires less than 2^{128} keystream words?

Problem 10.2 Check the feasibility of a state recovery attack on HC-128.

Problem 10.3 Do a parallelized implementation of HC-128 and estimate the improvement in speed.

Chapter 11

Conclusion

In the preceding chapters we have presented a detailed study of the RC4 KSA and PRGA and their variants. In this short final chapter, we summarize the current status in RC4 research.

- The design is nice and simple.

- It invites a lot of cryptanalytic results.

- The cipher is well studied.

- The cipher requires discarding some amount of initial keystream bytes.

- To date, RC4 is quite safe as 128-bit stream cipher.

- Different approaches to incorporate IV and MAC in RC4 may be possible, but one needs to be careful regarding the security issues.

- Hardware at the speed of one byte per clock is available [112, 155].

$RC4^+$ is a modified version of RC4 for a better security margin. Even with the technical arguments and empirical evidences, the security claim of $RC4^+$ is a conjecture, as is the case with many of the existing stream ciphers. No immediate weakness of the new design could be observed, and the cipher is subject to further analysis.

11.1 Safe Use of RC4

With respect to all the analysis so far, we can enlist the following precautions that need to be considered before using RC4.

1. Always throw few initial keystream output bytes. Typically, throwing 1024 bytes removes most of the weaknesses.

2. Avoid broadcasting under different keys, if the initial output bytes are not thrown.

3. Do not append or prepend IV. Mix it in a non-obvious complex manner.

4. Probably, one should mask the output as in RC4$^+$ to avoid state recovery attacks.

Appendix A

A Sample C Implementation of RC4

```c
#include <stdio.h>
#include <stdlib.h>
#include <time.h>

unsigned char key[256], K[256], S[256];
unsigned int key_len = 16, i, j;        // 128 bit key

void rc4_keygen() {
        for(i = 0; i < key_len; i++) {
                key[i] = (unsigned char) (256.0*rand()/(RAND_MAX+1.0));
        }
}

void rc4_expandkey() {
        for(i = 0; i < 256; i++) {
                K[i] = key[i % key_len];
        }
}

void swap(unsigned char *s, unsigned int i, unsigned int j) {
        unsigned char temp = s[i];
        s[i] = s[j];
        s[j] = temp;
}

void rc4_ksa() {
        for (i = 0; i < 256; i++) {
                S[i] = i;
        }
        for (i = j = 0; i < 256; i++) {
                j = (j + K[i] + S[i]) & 0xFF; // Fast modulo 256 operation
                swap(S, i, j);
        }
        i = j = 0;
}

unsigned char rc4_prga() {
        i = (i + 1) & 0xFF;
        j = (j + S[i]) & 0xFF;
        swap(S, i, j);
        return(S[(S[i] + S[j]) & 0xFF]);
}

int main() {
        int round, keystream_len = 1024;
        srand(time(NULL));
        rc4_keygen();
        rc4_expandkey();
        rc4_ksa();
        for(round = 1; round <= keystream_len; round ++) {
                printf("%x", rc4_prga());
        }
}
```

263

Appendix B

Verified Test Vectors of RC4

Secret Key (128 bit, in Hex)	Keystream (First 16 bytes, in Hex)
00000000000000000000000000000000	de188941a3375d3a8a061e67576e926d
ffffffffffffffffffffffffffffffff	6d252f2470531bb0394b93b4c46fdd9c
01010101010101010101010101010101	06080e0e18202929393349576676783
d764c8cce93255c4478d7aa05d83f3ea	c86e2d580e675554423642f33a6468e9
2a83e82681a22df7a04329387f7f2cd5	960c0b913786a18411b0e9e4a7499bbc
e38248bb72dd1350b2994e61ad0f9509	6cca71a62bb276d9d9ebc853970fe9fe
b540a80fbb1c787c244e0e189ee9deff	50a75801240ece2030f43a90e55319d6
993a6e57d103ef8affc253c841689fd2	45b1ddd8eb16da566c435742f3370d43

Bibliography

[1] C. Adams and S. Lloyd. *Understanding PKI: Concepts, Standards, and Deployment Considerations*. Addison-Wesley Professional. Second edition, November 16, 2002.

[2] http://csrc.nist.gov/CryptoToolkit/aes/rijndael [last accessed on April 30, 2011].

[3] http://www.iaik.tugraz.at/research/krypto/AES [last accessed on April 30, 2011].

[4] http://www2.mat.dtu.dk/people/Lars.R.Knudsen/aes.html [last accessed on April 30, 2011].

[5] M. Akgün, P. Kavak and H. Demirci. New Results on the Key Scheduling Algorithm of RC4. Preprint, received on May 19, 2008 by email. INDOCRYPT 2008, pages 40–52, vol. 5365, Lecture Notes in Computer Science, Springer.

[6] Anonymous (email id: nobody@jpunix.com). Thank you Bob Anderson. Cypherpunks mailing list, September 9, 1994. Available at http://cypherpunks.venona.com/date/1994/09/msg00304.html [last accessed on January 1, 2009].

[7] S. Babbage. A Space/Time Tradeoff in Exhaustive Search Attacks on Stream Ciphers. European Convention on Security and Detection, IEE Conference Publication No. 408, May 1995.

[8] R. Basu, S. Ganguly, S. Maitra and G. Paul. RC4 Keystream Always Leaks Information about the Hidden Index j. Proceedings of the State of the Art of Stream Ciphers (SASC), special Workshop hosted by ECRYPT, the European Network of Excellence in Cryptography, February 13–14, 2008, Lausanne, Switzerland, pages 233–247.

[9] R. Basu, S. Ganguly, S. Maitra and G. Paul. A Complete Characterization of the Evolution of RC4 Pseudo Random Generation Algorithm. *Journal of Mathematical Cryptology*, pages 257–289, vol. 2, no. 3, October, 2008. This is an extended version of the paper [8].

[10] R. Basu, S. Maitra, G. Paul and T. Talukdar. On Some Sequences of the Secret Pseudo-Random Index j in RC4 Key Scheduling. Applied Algebra,

Algebraic Algorithms, and Error Correcting Codes (AAECC) 2009, pages 137–148, vol. 5527, Lecture Notes in Computer Science, Springer.

[11] C. H. Bennett and G. Brassard. Quantum Cryptography: Public key distribution and coin tossing. Proceedings of the IEEE International Conference on Computers, Systems, and Signal Processing, Bangalore, India, 1984, page 175–179.

[12] D. J. Bernstein, J. Buchmann and E. Dahmen (Eds.). *Post-Quantum Cryptography*. Springer, 2009.

[13] E. Biham and O. Dunkelman. Crypanalysis of the A5/1 GSM stream cipher. INDOCRYPT 2000, pages 43–51, vol. 1977, Lecture Notes in Computer Science, Springer.

[14] E. Biham and O. Dunkelman. Differential Cryptanalysis in Stream Ciphers. IACR Eprint Server, eprint.iacr.org, number 2007/218, June 6, 2007. [last accessed on April 30, 2011].

[15] E. Biham and Y. Carmeli. Efficient Reconstruction of RC4 Keys from Internal States. FSE 2008, pages 270–288, vol. 5086, Lecture Notes in Computer Science, Springer.

[16] E. Biham, L. Granboulan and P. Q. Nguyen. Impossible Fault Analysis of RC4 and Differential Fault Analysis of RC4. FSE 2005, pages 359–367, vol. 3557, Lecture Notes in Computer Science, Springer.

[17] E. Biham and J. Seberry. Py: A Fast and Secure Stream Cipher using Rolling Arrays. April, 2005. Available at `http://www.ecrypt.eu.org/stream/pyp2.html` [last accessed on April 30, 2011].

[18] E. Biham and J. Seberry. Pypy: Another version of Py. June, 2006. Available at `http://www.ecrypt.eu.org/stream/pyp2.html` [last accessed on April 30, 2011].

[19] E. Biham and J. Seberry. Tweaking the IV Setup of the Py Family of Stream Ciphers - The ciphers TPy, TPypy, and TPy6. January, 2007. Available at `http://www.ecrypt.eu.org/stream/pyp2.html` [last accessed on April 30, 2011].

[20] A. Biryukov, A. Shamir and D. Wagner. Real Time Cryptanalysis of A5/1 on a PC. FSE 2000, pages 1–18, vol. 1978, Lecture Notes in Computer Science, Springer.

[21] R. E. Blahut. *Principles and Practice of Information Theory*. Addisson-Wesley, 1983.

[22] SIG Bluetooth. Bluetooth specification. Available at http://www.bluetooth.com [last accessed on April 30, 2011].

[23] C. Blundo and P. D'Arco. The Key Establishment Problem. Foundations of Security Analysis and Design II 2004, page 44–90, vol. 2946, Lecture Notes in Computer Science, Springer.

[24] D. Bouwmeester, A. Ekert and A. Zeilinger. *The Physics of Quantum Information: Quantum Cryptography, Quantum Teleportation, Quantum Computation.* Springer. First edition, 2000.

[25] C. Boyd and A. Mathuria. *Protocols for Authentication and Key Establishment.* Springer, 2003.

[26] C. D. Cannire. eSTREAM testing framework. Available at `http://www.ecrypt.eu.org/stream/perf` [last accessed on April 30, 2011].

[27] G. J. Chaitin. On the Length of Programs for Computing Finite Binary Sequences. *Journal of the ACM*, pages 547–570, vol. 13, 1966.

[28] D. Chang, K. C. Gupta, M. Nandi. RC4-Hash: A New Hash Function Based on RC4. INDOCRYPT 2006, pages 80–94, vol. 4329, Lecture Notes in Computer Science, Springer.

[29] C. Cooper and A. Frieze. The Size of the Largest Strongly Connected Component of a Random Digraph with a Given Degree Sequence. *Combinatorics, Probability and Computing*, vol. 13, no. 3, 2004, pages 319–337.

[30] T. H. Cormen, C. E. Leiserson, R. L. Rivest and C. Stein. *Introduction to Algorithms.* The MIT Press. Second Edition, September 1, 2001.

[31] T. Cover and J. A. Thomas. *Elements of Information Theory.* Wiley series in Telecommunication. Wiley, 1991.

[32] T. W. Cusick and P. Stanica. Cryptographic Boolean Functions and Applications. Academic Press, March 26, 2009.

[33] J. Daemen and P. Kitsos. The self-synchronizing stream cipher MOUSTIQUE. Available at `http://www.ecrypt.eu.org/stream/p3ciphers/mosquito/mosquito_p3.pdf` [last accessed on April 30, 2011].

[34] J. Daemen and V. Rijmen. The Block Cipher Rijndael. Smart Card Research and Applications 2000, pages 288–296, vol. 1820, Lecture Notes in Computer Science, Springer.

[35] B. Debraize, I. M. Corbella. Fault Analysis of the Stream Cipher SNOW 3G. FDTC 2009.

[36] D. Deutsch. Quantum Theory, the Church-Turing Principle and the Universal Quantum Computer. Proceedings of the Royal Society of London, pages 97–117, vol. 400, 1985.

[37] W. Diffie and M. E. Hellman. Multiuser Cryptographic Techniques. Federal Information Processing Standard Conference Proceedings, pages 109–112, vol. 45, 1976.

[38] W. Diffie and M. E. Hellman. New Directions in Cryptography. *IEEE Transactions on Information Theory*, pages 644–654, vol. 22, 1976.

[39] W. Du and M. J. Atallah. Secure multi-party computation problems and their applications: A review and open problems. Proceedings of the New Security Paradigms Workshop, pages 11–20, Coudcroft, New Mexico, USA, September 11–13, 2001.

[40] O. Dunkelman. A small observation on HC-128. Available at `http://www.ecrypt.eu.org/stream/phorum/read.php?1,1143` Date: November 14, 2007. [last accessed on April 30, 2011].

[41] P. Ekdahl and T. Johansson. Another attack on A5/1. *IEEE Transcation on Information Theory*, pages 284–289, vol. 49, no. 1, 2003.

[42] eSTREAM, the ECRYPT Stream Cipher Project. `http://www.ecrypt.eu.org/stream` [last accessed on April 30, 2011].

[43] ETSI/SAGE Specification Version 1.1. Specification of the 3GPP Confidentiality and Integrity Algorithms UEA2 & UIA2, Document 2: SNOW 3G Specification. September 6, 2006.

[44] ETSI/SAGE Specification Version 1.4. Specification of the 3GPP Confidentiality and Integrity Algorithms 128-EEA3 & 128-EIA3, Document 2: ZUC Specification. July 30, 2010.

[45] Federal Information Processing Standards (FIPS). Publication 46, January 15, 1977.

[46] Federal Information Processing Standards (FIPS). Publication 197, 2001.

[47] J. Feghhi and P. Williams. *Digital Certificates: Applied Internet Security.* Addison-Wesley Professional. October 9, 1998.

[48] N. Ferguson. Michael: an improved MIC for 802.11 WEP. IEEE doc. 802.11-02/020r0, Januaru 2002.

[49] R. P. Feynman. Simulating Physics with Computers. *International Journal of Theoretical Physics*, pages 467–488, vol. 21, no. 6/7, 1982.

[50] H. Finney. An RC4 cycle that can't happen. Post in sci.crypt, September 1994.

[51] S. R. Fluhrer and D. A. McGrew. Statistical Analysis of the Alleged RC4 Keystream Generator. FSE 2000, pages 19–30, vol. 1978, Lecture Notes in Computer Science, Springer.

[52] S. R. Fluhrer, I. Mantin and A. Shamir. Weaknesses in the Key Scheduling Algorithm of RC4. SAC 2001, pages 1–24, vol. 2259, Lecture Notes in Computer Science, Springer.

[53] S. R. Fluhrer and S. Lucks. Analysis of E_0 encryption system. SAC 2001, pages 38–48, vol. 2259, Lecture Notes in Computer Science, Springer.

[54] M. Gangné. Identity-based Encryption: a Survey. *CryptoBytes*, pages 10–19, vol. 6, no. 1, 2003.

[55] O. Goldreich. *Foundations of Cryptography: Volume 1, Basic Tools*, Cambridge University Press. First edition. January 18, 2007.

[56] S. Goldwasser, S. Micali and C. Rackoff. The Knowledge Complexity of Interactive Proof Systems. Proceedings of the 17th Annual ACM Symposium on Theory of Computing (STOC), pages 291–304, May 6–8, 1985, Providence, Rhode Island, USA.

[57] S. Goldwasser, S. Micali and C. Rackoff. The Knowledge Complexity of Interactive Proof Systems. *SIAM Journal on Computing*, pages 186–208, vol. 18, no. 1, 1989. This is an extended version of [56].

[58] J. Golic. Linear statistical weakness of alleged RC4 keystream generator. EUROCRYPT 1997, pages 226–238, vol. 1233, Lecture Notes in Computer Science, Springer.

[59] J. Golic. Cryptanalysis of alleged A5 stream cipher. EUROCRYPT 1997, pages 239–255, vol. 1233, Lecture Notes in Computer Science, Springer, 1997.

[60] J. D. Golic, V. Bagini and G. Morgari. Linear Cryptanalysis of Bluetooth stream cipher. EUROCRYPT 2002, pages 238–255, vol. 2332, Lecture Notes in Computer Science, Springer, 2002.

[61] J. Golic and G. Morgari. Iterative Probabilistic Reconstruction of RC4 Internal States. IACR Eprint Server, eprint.iacr.org, number 2008/348, August 8, 2008 [last accessed on April 30, 2011].

[62] S. W. Golomb. *Shift Register Sequences*. Aegean Park Press. Second edition, 1982.

[63] G. Gong, K. C. Gupta, M. Hell and Y. Nawaz. Towards a General RC4-Like Keystream Generator. CISC 2005, pages 162–174, vol. 3822, Lecture Notes in Computer Science, Springer.

[64] D. Hankerson, A. Menezes and S. Vanstone. *Guide to Elliptic Curve Cryptography*. Springer, 2004.

[65] J. Hansen and J. Jaworski. Large components of bipartite random mappings. *Random Structures & Algorithms*, vol. 17, no. 3–4, October 2000, pages 317–342.

[66] M. Hell, T. Johansson and W. Meier. Grain – A Stream Cipher for Constrained Environments. *International Journal of Wireless and Mobile Computing*, Special Issue on Security of Computer Network and Mobile Systems, 2006.

[67] M. Hell, T. Johansson, A. Maximov and W. Meier. A Stream Cipher Proposal: Grain-128. ISIT 2006, Seattle, USA.

[68] M. E. Hellman. A Cryptanalytic Time-Memory Trade-Off. *IEEE Transactions on Information Theory*, page 401–406, vol. 26, issue 4, July 1980.

[69] M. Hermelin and K. Nyberg. Correlation properties of Bluetooth combiner. ICISC 1999, pages 17–29, vol. 1787, Lecture Notes in Computer Science, Springer, 2000.

[70] I. N. Herstein. *Abstract Algebra*. Wiley. Third Edition, January 1, 1996.

[71] J. J. Hoch and A. Shamir. Fault Analysis of Stream Ciphers. CHES 2004, pages 240–253, vol. 3156, Lecture Notes in Computer Science, Springer.

[72] R. V. Hogg, A. Craig and J. W. McKean. *Introduction to Mathematical Statistics*. Prentice Hall. Sixth Edition, June 27, 2004.

[73] J. Hong and P. Sarkar. New Applications of Time Memory Data Trade-offs. ASIACRYPT 2005, pages 353–372, vol. 3788, Lecture Notes in Computer Science, Springer.

[74] E. Horowitz and S. Sahni. *Fundamentals of Data Structures*. W.H. Freeman & Company, 1982.

[75] R. Housley, D. Whiting and N. Ferguson. Alternate temporal key hash. IEEE doc. 802.11-02/282r2, April 2002.

[76] S. Indesteege and B. Preneel. Collisions for RC4-Hash. ISC 2008, pages 355–366, vol. 5222, Lecture Notes in Computer Science, Springer.

[77] T. Isobe, T. Ohigashi, H. Kuwakado and M. Morii. How to Break Py and Pypy by a Chosen-IV Attack. 2006. Available at http://www.ecrypt.eu.org/stream/pyp2.html [last accessed on April 30, 2011].

[78] R. J. Jenkins. ISAAC and RC4. 1996. Available at http://burtleburtle.net/bob/rand/isaac.html [last accessed on April 30, 2011].

[79] P. Judge and M. Ammar. Security Issues and Solutions in Multicast Content Distribution: A Survey. *IEEE Network*, January/February 2003.

[80] D. Kahn. *The Codebreakers: The Comprehensive History of Secret Communication from Ancient Times to the Internet*. Scribner, 1996.

[81] J. Katz and Y. Lindell. *Introduction to Modern Cryptography*. CRC Press, 2007.

[82] A. Kerckhoffs. La cryptographie militaire. *Journal des sciences militaires*, vol. IX, pages 5–83, January 1883 and pages 161–191, February 1883.

[83] S. Khazaei and W. Meier. On Reconstruction of RC4 Keys from Internal States. Accepted in Mathematical Methods in Computer Science (MMICS), December 17–19, 2008, Karlsruhe, Germany.

[84] A. Kircanski and A. M. Youssef. Differential Fault Analysis of HC-128. Africacrypt 2010, pages 360–377, vol. 6055, Lecture Notes in Computer Science, Springer.

[85] I. B. Kalugin. The number of components in a random bipartite graph. *Diskretnaya Matematika*, vol. 1, no. 3, 1989, pages 62–70.

[86] A. Klein. Attacks on the RC4 stream cipher. *Designs, Codes and Cryptography*, pages 269–286, vol. 48, no. 3, September, 2008.

[87] L. R. Knudsen, W. Meier, B. Preneel, V. Rijmen and S. Verdoolaege. Analysis Methods for (Alleged) RCA. ASIACRYPT 1998, pages 327–341, vol. 1514, Lecture Notes in Computer Science, Springer.

[88] D. E. Knuth. *The Art of Computer Programming*, vol. 2 (Seminumerical Algorithms), Addison-Wesley Publishing Company. Third edition, 1997.

[89] D. E. Knuth. *The Art of Computer Programming*, vol. 3 (Sorting and Searching), Addison-Wesley Publishing Company. Second edition, 1998.

[90] N. Koblitz. Elliptic Curve Cryptosystems. *Mathematics of Computation*, pages 203–209, vol. 48, 1987.

[91] A. N. Kolmogorov. Three Approaches to the Quantitative Definition of Information. *Problems of Information Transmission*, pages 1–7, vol. 1, no. 1, 1965.

[92] LAN/MAN Standard Committee. ANSI/IEEE standard 802.11b: Wireless LAN Medium Access Control (MAC) and Physical Layer (phy) Specifications. 1999.

[93] LAN/MAN Standard Committee. ANSI/IEEE standard 802.11i: Amendment 6: Wireless LAN Medium Access Control (MAC) and Physical Layer (phy) Specifications. Draft 3, 2003.

[94] LAN/MAN Standard Committee. ANSI/IEEE standard 802.11i: Amendment 6: Wireless LAN Medium Access Control (MAC) and Physical Layer (phy) Specifications, 2004.

[95] LAN/MAN Standard Committee. ANSI/IEEE standard 802.1X: Port Based Network Access Control, 2001.

[96] R. Lidl and H. Niederreiter. *Finite Fields*. Cambridge University Press. Second edition, 1997.

[97] Y. Liu and T. Qin. The key and IV setup of the stream ciphers HC-256 and HC-128. International Conference on Networks Security, Wireless Communications and Trusted Computing, pages 430–433, 2009.

[98] Y. Lu, W. Meier and S. Vaudenay. The Conditional Correlation Attack: A Practical Attack on Bluetooth Encryption. CRYPTO 2005, pages 97–117, vol. 3621, Lecture Notes in Computer Science, Springer, 2001.

[99] M. Luby and C. Rackoff. How to Construct Pseudorandom Permutations and Pseudorandom Functions. *SIAM Journal on Computing*, pages 373–386, vol. 17, 1988.

[100] S. Maitra, G. Paul and S. Sen Gupta. Attack on Broadcast RC4 Revisited. FSE 2011, pages 199–217, vol. 6733, Lecture Notes in Computer Science, Springer.

[101] S. Maitra and G. Paul. Analysis of RC4 and Proposal of Additional Layers for Better Security Margin. INDOCRYPT 2008, pages 27–39, vol. 5365, Lecture Notes in Computer Science, Springer.

[102] S. Maitra and G. Paul. Analysis of RC4 and Proposal of Additional Layers for Better Security Margin (Full Version). IACR Eprint Server, eprint.iacr.org, number 2008/396, September 19, 2008. This is an extended version of [101].

[103] S. Maitra and G. Paul. New Form of Permutation Bias and Secret Key Leakage in Keystream Bytes of RC4. FSE 2008, pages 253–269, vol. 5086, Lecture Notes in Computer Science, Springer.

[104] S. Maitra and G. Paul. Recovering RC4 Permutation from 2048 Keystream Bytes if j is Stuck. ACISP 2008, pages 306–320, vol. 5107, Lecture Notes in Computer Science, Springer.

[105] S. Maitra, G. Paul and S. Raizada. Some Observations on HC-128. Proceedings of the International Workshop on Coding and Cryptography (WCC), May 10–15, 2009, Ullensvang, Norway, pages 527–539.

[106] S. Maitra, G. Paul, S. Raizada, S. Sen and R. Sengupta. Some Observations on HC-128. *Designs, Codes and Cryptography*, pages 231–245, vol. 59, no. 1–3, April 2011. This is an extended version of [105].

[107] I. Mantin and A. Shamir. A Practical Attack on Broadcast RC4. FSE 2001, pages 152–164, vol. 2355, Lecture Notes in Computer Science, Springer.

[108] I. Mantin. A Practical Attack on the Fixed RC4 in the WEP Mode. ASI-ACRYPT 2005, pages 395–411, volume 3788, Lecture Notes in Computer Science, Springer.

[109] I. Mantin. Predicting and Distinguishing Attacks on RC4 Keystream Generator. EUROCRYPT 2005, pages 491–506, vol. 3494, Lecture Notes in Computer Science, Springer.

[110] I. Mantin. Analysis of the stream cipher RC4. Master's Thesis, The Weizmann Institute of Science, Israel, 2001.

[111] M. Matsui. Key Collisions of the RC4 Stream Cipher. FSE 2009, pages 38–50, vol. 5665, Lecture Notes in Computer Science, Springer.

[112] D. P. Matthews Jr. Methods and Apparatus for Accelerating ARC4 Processing. US Patent no. 7403615, Morgan Hill, CA, July 2008. Available at http://www.freepatentsonline.com/7403615.html [last accessed on April 30, 2011].

[113] A. Maximov. Two Linear Distinguishing Attacks on VMPC and RC4A and Weakness of RC4 Family of Stream Ciphers. FSE 2005, pages 342–358, vol. 3557, Lecture Notes in Computer Science, Springer.

[114] A. Maximov. Two Linear Distinguishing Attacks on VMPC and RC4A and Weakness of RC4 Family of Stream Ciphers (Corrected). IACR Eprint Server, eprint.iacr.org, number 2007/070, February 22, 2007. This is a corrected version of [113].

[115] A. Maximov. Cryptanalysis of the "Grain" Family of Stream Ciphers. ACM Symposium on Information, Computation and Communications Security (ASIACCS), pages 283–288, 2006.

[116] A. Maximov and D. Khovratovich. New State Recovery Attack on RC4. CRYPTO 2008, pages 297–316, vol. 5157, Lecture Notes in Computer Science, Springer.

[117] A. Maximov and D. Khovratovich. New State Recovery Attack on RC4. IACR Eprint Server, eprint.iacr.org, number 2008/017, January 10, 2008. This is an extended version of [116].

[118] M. E. McKague. Design and Analysis of RC4-like Stream Ciphers. Master's Thesis, University of Waterloo, Canada, 2005.

[119] A. J. Menezes, P. C. van Oorschot and S. A. Vanstone. *Handbook of Applied Cryptography*. CRC Press, 5th Printing, 2001.

[120] A. J. Menezes and N. Koblitz. A Survey of Public-key Cryptosystems. *SIAM Review*, pages 599–634, vol. 46, 2004.

[121] V. Miller. Use of Elliptic Curves in Cryptography. CRYPTO 1985, pages 417–426, vol. 218, Lecture Notes in Computer Science, Springer.

[122] I. Mironov. (Not So) Random Shuffles of RC4. CRYPTO 2002, pages 304–319, vol. 2442, Lecture Notes in Computer Science, Springer.

[123] S. Mister and S. E. Tavares. Cryptanalysis of RC4-like Ciphers. SAC 1998, pages 131–143, vol. 1999, Lecture Notes in Computer Science, Springer.

[124] V. Moen, H. Raddum and K. J. Hole. Weakness in the Temporal Key Hash of WPA. *Mobile Computing and Communications Review*, vol. 8, pages 76–83, 2004.

[125] M. Molloy and B. Reed. The Size of the Giant Component of a Random Graph with a Given Degree Sequence. *Combinatorics, Probability and Computing*, vol. 7, pages 295–305, 1998.

[126] Y. Nawaz, K. C. Gupta and G. Gong. A 32-bit RC4-like keystream generator. Technical Report CACR 2005-19, Center for Applied Cryptographic Research, University of Waterloo, 2005. Also appears in IACR Eprint Server, eprint.iacr.org, number 2005/175, June 12, 2005.

[127] New European Schemes for Signatures, Integrity, and Encryption. Available at https://www.cosic.esat.kuleuven.be/nessie [last accessed on April 30, 2011].

[128] S ONeil, B. Gittins and H. Landman. VEST Hardware-Dedicated Stream Ciphers. Available at www.ecrypt.eu.org/stream/ciphers/vest/vest.pdf.

[129] M. A. Nielsen and I. L. Chuang. *Quantum Computation and Quantum Information*, Cambridge University Press. First edition, 2000.

[130] G. Paul. Analysis and Design of RC4 and Its Variants. Ph.D. Thesis, Jadavpur University, 2008.

[131] G. Paul, S. Rathi and S. Maitra. On Non-negligible Bias of the First Output Byte of RC4 towards the First Three Bytes of the Secret Key. Proceedings of the International Workshop on Coding and Cryptography (WCC) 2007, pages 285–294.

[132] G. Paul, S. Rathi and S. Maitra. On Non-negligible Bias of the First Output Byte of RC4 towards the First Three Bytes of the Secret Key. *Designs, Codes and Cryptography*, pages 123–134, vol. 49, no. 1–3, December, 2008. This is an extended version of [131].

[133] G. Paul and S. Maitra. Permutation after RC4 Key Scheduling Reveals the Secret Key. SAC 2007, pages 360–377, vol. 4876, Lecture Notes in Computer Science, Springer.

[134] G. Paul and S. Maitra. RC4 State Information at Any Stage Reveals the Secret Key. IACR Eprint Server, eprint.iacr.org, number 2007/208, June 1, 2007. This is an extended version of [133].

[135] G. Paul, S. Maitra and R. Srivastava. On Non-Randomness of the Permutation after RC4 Key Scheduling. Applied Algebra, Algebraic Algorithms, and Error Correcting Codes (AAECC) 2007, pages 100–109, vol. 4851, Lecture Notes in Computer Science, Springer.

[136] G. Paul and S. Maitra. On Biases of Permutation and Keystream Bytes of RC4 towards the Secret Key. *Cryptography and Communications*, pages 225–268, vol. 1, no. 2., September, 2009. This is an extended version of [133], [103] with some additional materials.

[137] G. Paul, S. Maitra and S. Raizada. A Theoretical Analysis of the Structure of HC-128. 6th International Workshop on Security (IWSEC) 2011, pages 161–177, vol. 7038, Lecture Notes in Computer Science, Springer.

[138] S. Paul and B. Preneel. Analysis of Non-fortuitous Predictive States of the RC4 Keystream Generator. INDOCRYPT 2003, pages 52–67, vol. 2904, Lecture Notes in Computer Science, Springer.

[139] S. Paul and B. Preneel. A New Weakness in the RC4 Keystream Generator and an Approach to Improve the Security of the Cipher. FSE 2004, pages 245–259, vol. 3017, Lecture Notes in Computer Science, Springer.

[140] S. Paul, B. Preneel and G. Sekar. Distinguishing Attacks on the Stream Cipher Py. FSE 2006, pages 405–421, vol. 4047, Lecture Notes in Computer Science, Springer.

[141] T. P. Pedersen. Signing Contracts and Paying Electronically. *Lectures on Data Security*, pages 134–157, vol. 1561, Lecture Notes in Computer Science, Springer.

[142] B. Preneel and O.A. Logachev (Eds.). *Boolean Functions in Cryptology and Information Security*. NATO Science for Peace and Security Series: Information and Communication Security, vol. 18, IOS Press, July 2008.

[143] E. M. Reingold, J. Nievergelt and N. Deo. *Combinatorial Algorithms: Theorey and Practice*. Prentice Hall, 1977.

[144] R. L. Rivest, A. Shamir and L. Adleman. A Method for Obtaining Digital Signatures and Public Key Cryptosystems. *Communications of the ACM*, pages 120–126, vol. 21, 1978.

[145] M. Robshaw and O. Billet (Eds.). New Stream Cipher Designs: The eSTREAM Finalists. Vol. 4986, Lecture Notes in Computer Science, Springer, 2008.

[146] A. Roos. A Class of Weak Keys in the RC4 Stream Cipher. Two posts in sci.crypt, message-id `43u1eh\$1j3@hermes.is.co.za` and `44ebge\$11f@hermes.is.co.za`, 1995. Available at `http://groups.google.com/group/sci.crypt.research/msg/078aa9249d76eacc?dmode=source` [last accessed on April 30, 2011].

[147] RSA Lab Report. RSA Security Response to Weaknesses in Key Scheduling Algorithm of RC4. Available at `http://www.rsa.com/rsalabs/node.asp?id=2009` [last accessed on April 30, 2011].

[148] F. Roberts and B. Tesman Applied Combinatorics. Prentice Hall, NJ. Second Edition, April 2003.

[149] M. J. B. Robshaw. Stream Ciphers. Technical Report TR-701, version 2.0, RSA Laboratories, 1995.

[150] R. A. Rueppel. *Analysis and Design of Stream Ciphers*. Springer, New York, Inc., New York, NY, 1986.

[151] G. Sekar, S. Paul and B. Preneel. Weakness in the Pseudorandom Bit Generation Algorithms of the Stream Ciphers TPypy and TPy. 2007. Available at `http://www.ecrypt.eu.org/stream/pyp2.html` [last accessed on April 30, 2011].

[152] G. Sekar, S. Paul and B. Preneel. New Attacks on the Stream Cipher TPy6 and Design of New Ciphers the TPy6-A and the TPy6-B. Western European Workshop on Research in Cryptology (WEWoRC 2007), Bochum, Germany, July 4–6, 2007, pages 127–141, vol. 4945, Lecture Notes in Computer Science, Springer.

[153] G. Sekar, S. Paul and B. Preneel. New Weaknesses in the Keystream Generation Algorithms of the Stream Ciphers TPy and Py. ISC 2007, pages 249–262, vol. 4779, Lecture Notes in Computer Science, Springer.

[154] G. Sekar, S. Paul and B. Preneel. Related-Key Attacks on the Py-Family of Ciphers and an Approach to Repair the Weaknesses. INDOCRYPT 2007, pages 58–72, vol. 4859, Lecture Notes in Computer Science, Springer.

[155] S. Sen Gupta, K. Sinha, S. Maitra and B. P. Sinha. One Byte per Clock: A Novel RC4 Hardware. INDOCRYPT 2010, pages 347–363, vol. 6498, Lecture Notes in Computer Science, Springer.

[156] S. Sen Gupta, S. Maitra, G. Paul and S. Sarkar. Proof of Empirical RC4 Biases and New Key Correlations. Accepted in SAC 2011. To appear in Lecture Notes in Computer Science, Springer.

[157] S. Sen Gupta, S. Maitra, G. Paul and S. Sarkar. Work in progress. April 2011. [This work would be a part of the Ph.D. Thesis of Sourav Sen Gupta.]

[158] P. Sepehrdad, S. Vaudenay and M. Vuagnoux. Discovery and Exploitation of New Biases in RC4. SAC 2010, pages 74–91, vol. 6544, Lecture Notes in Computer Science, Springer.

[159] P. Sepehrdad, S. Vaudenay and M. Vuagnoux. Statistical Attack on RC4 – Distinguishing WPA. EUROCRYPT 2011, pages 343–363, vol. 6632, Lecture Notes in Computer Science, Springer.

[160] A. Shamir. How to Share a Secret. *Communications of the ACM*, pages 612–613, vol. 22, 1979.

[161] A. Shamir. Identity-based Cryptosystems and Signature Schemes. CRYPTO 1984, pages 47–53, vol. 196, Lecture Notes in Computer Science, Springer.

[162] A. Shamir. Stream Ciphers: Dead or Alive? ASIACRYPT 2004, page 78 (invited talk), vol. 3329, Lecture Notes in Computer Science, Springer.

[163] C. E. Shannon. A Mathematical Theory of Communication. *Bell Systems Technical Journal*, pages 623–656, vol. 27, 1948.

[164] C. E. Shannon. Communication Theory of Secrecy Systems. *Bell Systems Technical Journal*, pages 656–715, vol. 28, 1949.

[165] Y. Shiraishi, T. Ohigashi, and M. Morii. An Improved Internal-state Reconstruction Method of a Stream Cipher RC4. Proceedings of Communication, Network, and Information Security (Editor: M.H. Hamza), Track 440–088, December 10–12, 2003, New York, USA.

[166] P. Shor. Algorithms for Quantum Computation: Discrete Logarithms and Factoring. Foundations of Computer Science (FOCS) 1994, page 124–134, IEEE Computer Society Press.

[167] P. Shor. Polynomial-Time Algorithms for Prime Factorization and Discrete Logarithms on a Quantum Computer. *SIAM Journal on Computing*, pages 1484–1509, vol. 26, 1997. This is an extended version of the paper [166].

[168] T. Siegenthaler. Decrypting a class of stream ciphers using ciphertext only. *IEEE Transactions on Computers*, pages 81–85, vol. C-34, 1985.

[169] J. Silverman. *A Friendly Introduction to Number Theory*. Prentice Hall, NJ. Second Edition, 2001.

[170] S. Singh. *The Code Book: The Science of Secrecy from Ancient Egypt to Quantum Cryptography*. Anchor, 2000.

[171] R. J. Solomonoff. A Formal Theory of Inductive Inference. *Information and Control*, pages 1–22, vol 7(1), 1964.

[172] O. Staffelbach and W. Meier. Cryptographic Significance of the Carry for Ciphers Based on Integer Addition. CRYPTO 1990, pages 601–614, vol. 537, Lecture Notes in Computer Science, Springer.

[173] P. Stankovski, S. Ruj, M. Hell and T. Johansson. Improved Distinguishers for HC-128. *Designs, Codes and Cryptography*, Online First, 13 August, 2011. DOI: 10.1007/s10623-011-9550-9.

[174] D. R. Stinson. *Cryptography: Theory and Practice*. Chapman & Hall / CRC, Third Edition, 2005.

[175] A. Stubblefield, J. Ioannidis and A. D. Rubin. Using the Fluhrer, Mantin, and Shamir Attack to Break WEP. AT&T Labs Technical Report TD-4ZCPZZ, August 6, 2001.

[176] A. I. Saltykov. The number of components in a random bipartite graph. *Diskretnaya Matematika*, vol. 7, no. 4, 1995, pages 86–94.

[177] E. Tews. Attacks on the WEP protocol. IACR Eprint Server, eprint.iacr.org, number 2007/471, December 15, 2007. [last accessed on April 30, 2011].

[178] E. Tews and M. Beck. Practical Attacks Aganist WEP and WPA. Proceedings of the Second ACM Conference on Wireless Network Security (WISEC 2009), Zurich, Switzerland, pages 79–86, ACM, 2009.

[179] E.Tews and A. Klein. *Attacks on Wireless LANs About the security of IEEE 802.11 based wireless networks*. VDM Verlag. December 19, 2008.

[180] E. Tews, R. P. Weinmann and A. Pyshkin. Breaking 104 bit WEP in less than 60 seconds. WISA 2007, pages 188202, vol. 4867, Lecture Notes in Computer Science, Springer.

[181] V. Tomasevic, S. Bojanic and O. Nieto-Taladriz. Finding an internal state of RC4 stream cipher. *Information Sciences*, pages 1715–1727, vol. 177, 2007.

[182] Y. Tsunoo, T. Saito, H. Kubo, M. Shigeri, T. Suzaki and T. Kawabata. The Most Efficient Distinguishing Attack on VMPC and RC4A. SKEW 2005. Available at http://www.ecrypt.eu.org/stream/papers.html [last accessed on April 30, 2011].

[183] Y. Tsunoo, T. Saito, H. Kubo and T. Suzaki. A Distinguishing Attack on a Fast Software-Implemented RC4-Like Stream Cipher. *IEEE Transactions on Information Theory*, page 3250–3255, vol. 53, issue 9, September 2007.

[184] S. Vaudenay and M. Vuagnoux. Passive-Only Key Recovery Attacks on RC4. SAC 2007, pages 344–359, vol. 4876, Lecture Notes in Computer Science, Springer.

[185] D. Wagner. My RC4 weak keys. Post in sci.crypt, message-id `447o11$cbj@cnn.Princeton.EDU`, 26 September, 1995. Available at `http://www.cs.berkeley.edu/~daw/my-posts/my-rc4-weak-keys` [last accessed on April 30, 2011].

[186] H. Wu. Cryptanalysis of a 32-bit RC4-like Stream Cipher. IACR Eprint Server, eprint.iacr.org, number 2005/219, July 6, 2005. [last accessed on April 30, 2011].

[187] H. Wu. The Stream Cipher HC-128. Available at `http://www.ecrypt.eu.org/stream/hcp3.html` [last accessed on April 30, 2011].

[188] H. Wu. A New Stream Cipher HC-256. FSE 2004, pages 226–244, vol. 3017, Lecture Notes in Computer Science, Springer. The full version is available at `http://eprint.iacr.org/2004/092.pdf` [last accessed on April 30, 2011].

[189] H. Wu and B. Preneel. Attacking the IV Setup of Py and Pypy. August, 2006. Available at `http://www.ecrypt.eu.org/stream/pyp2.html` [last accessed on April 30, 2011].

[190] H. Wu and B. Preneel. Key Recovery Attack on Py and Pypy with Chosen IVs. 2006. Available at `http://www.ecrypt.eu.org/stream/pyp2.html` [last accessed on April 30, 2011].

[191] H. Wu and B. Preneel. Differential Cryptanalysis of the Stream Ciphers Py, Py6 and Pypy. EUROCRYPT 2007, pages 276–290, vol. 4515, Lecture Notes in Computer Science, Springer.

[192] B. Zoltak. VMPC One-Way Function and Stream Cipher. FSE 2004, pages 210–225, vol. 3017, Lecture Notes in Computer Science, Springer.

Index